Handling and Mapping Geographic Information

SCIENCES

*Geography and Demography*, Field Director – Denise Pumain

---

*Cartography*, Subject Head – Colette Cauvin-Reymond

# Handling and Mapping Geographic Information

*Coordinated by*
Claire Cunty
Hélène Mathian

ISTE

WILEY

First published 2024 in Great Britain and the United States by ISTE Ltd and John Wiley & Sons, Inc.



ISTE Ltd
27-37 St George's Road
London SW19 4EU
UK

www.iste.co.uk

John Wiley & Sons, Inc.
111 River Street
Hoboken, NJ 07030
USA

www.wiley.com



Library of Congress Control Number: 2023923684

British Library Cataloguing-in-Publication Data
A CIP record for this book is available from the British Library
ISBN 978-1-78945-161-0

ERC code:
SH2 Institutions, Values, Environment and Space
*SH2_11 Human, economic and social geography*
*SH2_12 GIS, spatial analysis; big data in political, geographical and legal studies*

# Contents

**Foreword** . . . . . xi
Colette CAUVIN-REYMOND

**Introduction** . . . . . xvii
Claire CUNTY and Hélène MATHIAN

**Chapter 1. Exploring Statistical Relationships with Maps and Charts** . . . . . 1
Jean-Benoît BOURON, Claire CUNTY, Hélène MATHIAN and Myriam BARON

1.1. Introduction . . . . . 1
1.2. Mapping the world: which world? what data? . . . . . 4
1.2.1. Planispheres, projections and centers . . . . . 5
1.2.2. Grid-related issues . . . . . 7
1.2.3. Data at global scale: a composite material . . . . . 9
1.3. Exploring data and relationships with maps and charts . . . . . 11
1.3.1. Exploring statistical relationships and spatial organizations . . . . . 12
1.3.2. Formalizing a relationship: from a statistical model to the map of model deviations . . . . . 16
1.4. Describing statistical relationships between several variables . . . . . 18
1.5. Conclusion . . . . . 25
1.6. References . . . . . 26

**Chapter 2. Heterogeneous Data Integration and Geoweb Cartographic Representations** . . . . . 29
Marianne GUÉROIS and Malika MADELIN

2.1. Introduction. . . . . 29
2.2. New data: from production to visualization . . . . . 32
2.2.1. New data production methods and new scientific applications . . 33
2.2.2. Data with high spatial and temporal resolution but poorer attributes and quality? . . . . . 35
2.2.3. Visualizing new data on Web platforms . . . . . 41
2.3. New data, traditional data: why and how to integrate them? . . . . . 45
2.3.1. Data heterogeneity, integration, interoperability: preamble to vocabulary development . . . . . 46
2.3.2. Visual overlay of heterogeneous data, facilitated by advances in technical interoperability . . . . . 48
2.3.3. Term-to-term matching to assess the quality and enrich the attributes of new data . . . . . 50
2.3.4. Aggregations for combining heterogeneous data within "pivotal" spatiotemporal units . . . . . 52
2.3.5. Interpolated data measuring two continuous phenomena to compare them within a common grid . . . . . 55
2.4. Conclusion . . . . . 59
2.5. References . . . . . 61

**Chapter 3. Environmental Data and Cartographic Objects** . . . . . 65
Étienne COSSART

3.1. Introduction. . . . . 65
3.1.1. Defining cartographic objects: positioning the problem . . . . . 65
3.1.2. Specific environmental data. . . . . 66
3.2. Building cartographic objects: deconstructing to reconstruct. . . . . 67
3.2.1. From geographic object to cartographic object. . . . . 67
3.2.2. A few reminders on set theory . . . . . 68
3.2.3. Defining objects and collecting data. . . . . 71
3.3. Dealing with disparate and incomplete data: examples from environmental geography . . . . . 74
3.3.1. Defining cartographic objects from incomplete data . . . . . 74
3.3.2. Defining cartographic objects at reference scales . . . . . 82
3.4. Conclusion . . . . . 86
3.5. References . . . . . 87

**Chapter 4. Mapping and Identifying Geographic Configurations: The Example of Segregation** . . . . . 89
Sylvestre DUROUDIER

4.1. Introduction. . . . . 89
4.2. Mapping: rendering spatial configurations visible or invisible. . . . . 91
4.2.1. Defining the study framework . . . . . 91
4.2.2. The scale of spatial configurations: a question of geographic mesh . . . . . 93
4.2.3. How the map is created . . . . . 99
4.3. How to measure a phenomenon so as to reveal its forms . . . . . 104
4.3.1. The dependence of measures to the definition of categories . . . . 104
4.3.2. The indices approach . . . . . 107
4.3.3. Multivariate analysis approach: qualifying segregated neighborhoods. . . . . 112
4.4. Capturing spatial forms using dynamic approaches . . . . . 116
4.4.1. Why a dynamic approach to spatial morphologies? . . . . . 116
4.4.2. Approaches using spatial indices or spatial autocorrelation . . . . 117
4.4.3. Approach based on discontinuities. . . . . 120
4.4.4. Approach based on population potential . . . . . 122
4.5. Conclusion . . . . . 125
4.6. References . . . . . 126

**Chapter 5. Map and Statistical Model to Explore Spatial Heterogeneity** . . . . . 131
Mohamed HILAL and Julie LE GALLO

5.1. Introduction. . . . . 131
5.2. From raw open-source data to statistical data. . . . . 133
5.2.1. Open-source data with a complex structure. . . . . 133
5.2.2. Object of study and scale of analysis . . . . . 136
5.3. Preliminary explorations of spatial variations. . . . . 139
5.3.1. Maps to explore the spatial structure of each variable . . . . . 139
5.3.2. Testing the hypothesis of an administrative or spatial effect in spatial organization . . . . . 143
5.4. Analyzing relationships statistically and rendering a map . . . . . 149
5.4.1. Hedonic regression principles. . . . . 150
5.4.2. Model with no spatial attributes . . . . . 151
5.4.3. Spatial model estimated by GWR . . . . . 158
5.5. Conclusion . . . . . 168
5.6. References . . . . . 169

**Chapter 6. Mapping Time** . . . . . 173
Claire CUNTY and Hélène MATHIAN

6.1. Introduction. . . . . 173
6.2. Formalization. . . . . 175
6.2.1. Spatial objects and their temporal component . . . . . 175
6.2.2. From spatiotemporal objects to spatiotemporal data . . . . . 177
6.2.3. From geographic data to cartographic data . . . . . 179
6.2.4. Visualizing time. . . . . 181
6.3. Monitoring territorial changes. . . . . 184
6.3.1. Visualizing temporal phenomena. . . . . 184
6.3.2. Representing changes . . . . . 190
6.4. Representing phenomena associated with movement . . . . . 191
6.4.1. Representing movements: from points to trajectories. . . . . 192
6.4.3. Shape changes. . . . . 196
6.5. Representing temporality. . . . . 197
6.5.1. Tracking the spatial organization of events. . . . . 198
6.5.2. Representing lifespan, duration and change . . . . . 202
6.5.3. Representing space-time. . . . . 204
6.6. Conclusion . . . . . 206
6.7. References . . . . . 209

**Chapter 7. Cartograms, Anamorphic Maps: Transformed Territories**. . . . . 217
Anne-Christine BRONNER

7.1. Introduction. . . . . 217
7.2. Cartograms to represent count data associated with areal units . . . . . 219
7.2.1. Why use cartograms?. . . . . 220
7.2.2. Links between methods and data. . . . . 223
7.2.3. Anatomy of the cartogram: methods. . . . . 223
7.2.4. The piezopleth map to represent rates associated with areal units. . 233
7.3. Anamorphic map for the representation of space-time . . . . . 234
7.3.1. Accessibility to a place (unipolar accessibility) . . . . . 237
7.3.2. Accessibility between all places (multipolar accessibility). . . . . 243
7.3.3. Azimuthal transformation . . . . . 246
7.4. Anamorphic maps, cartograms: cross-cutting reflections on common principles and reading difficulties . . . . . 247
7.4.1. Some principles underlying all methods . . . . . 247
7.4.2. Anamorphic maps layout . . . . . 249
7.4.3. Reading and understanding anamorphic maps: some difficulties . . 250

7.5. Conclusion . . . . . . . . . . . . . . . . . . . . . . . . . . . . 251
7.6. References . . . . . . . . . . . . . . . . . . . . . . . . . . . . 252

**Chapter 8. Exploration, Aggregation and Spatiotemporal Visualization of Big Data** . . . . . . . . . . . . . . . . . . . . 257
Claude GRASLAND

8.1. Introduction. . . . . . . . . . . . . . . . . . . . . . . . . . . . 257
8.2. Defining the object of study and selecting the corpus . . . . . . . . . . . 259
8.2.1. Press news: a multidimensional object . . . . . . . . . . . . . . . . 259
8.2.2. Defining international news. . . . . . . . . . . . . . . . . . . . . 261
8.2.3. Corpus definition based on RSS news feeds . . . . . . . . . . . . . 263
8.3. Crossing the "who" and "what" dimensions . . . . . . . . . . . . . . . 264
8.4. Crossing the "who", "what" and "when" dimensions . . . . . . . . . . . 266
8.4.1. Detecting trends. . . . . . . . . . . . . . . . . . . . . . . . . . 266
8.4.2. Analysis of seasonal variations . . . . . . . . . . . . . . . . . . . 268
8.4.3. Analysis of weekly variations. . . . . . . . . . . . . . . . . . . . 269
8.5. Crossing the "who", "what" and "where" dimensions. . . . . . . . . . . 270
8.5.1. Mapping the distribution of international news by country . . . . 271
8.5.2. Identifying and mapping specific national features . . . . . . . . . 273
8.6. Graphs to represent co-location relationships. . . . . . . . . . . . . . . 275
8.6.1. Measuring and visualizing association links . . . . . . . . . . . . . 276
8.6.2. From co-citations to global regionalization. . . . . . . . . . . . . . 278
8.7. Conclusion . . . . . . . . . . . . . . . . . . . . . . . . . . . . 282
8.8. References . . . . . . . . . . . . . . . . . . . . . . . . . . . . 284

**Conclusion**. . . . . . . . . . . . . . . . . . . . . . . . . . . . . 287
Claire CUNTY and Hélène MATHIAN

**List of Authors** . . . . . . . . . . . . . . . . . . . . . . . . . . 291

**Index** . . . . . . . . . . . . . . . . . . . . . . . . . . . . . . . 293

# Foreword

**Colette CAUVIN-REYMOND**
*LIVE, CNRS, Université de Strasbourg, France*

Within the framework of the ISTE SCIENCES encyclopedia and, more specifically, of the "geography and demography" field, four works are devoted to cartography. Cartography is a scientific and artistic discipline (Robinson 1952, 1953), and it is indispensable to any person, organization or institution that needs to process and represent geographic data to bring out its spatial characteristics. With current technological changes, the continuous increase in the use of Web 2.0 and the appearance of social networks, cartography is undergoing profound changes.

This discipline has, in fact, undergone many more or less fundamental changes over time, both conceptually and technically, the two being intimately linked, as de Rosnay (2008) writes: "Scientific progress and technological progress feed each other". If its junction with statistics, as early as the end of the 19th century, had introduced new processes to translate localized phenomenon, the advent of the computer, and subsequently the birth of computerized cartography, could be considered a revolution. The latter was coined in an article by Tobler in 1959, simultaneously announcing the steps for the construction of an automated map and the basic principles of what would become geographic information systems (GIS).

Between this date and the beginning of the 1980s, this "new" cartography was developed in two directions: on the one hand, computer cartography, which reproduced what had previously been done manually, and, on the other hand, computer-assisted cartography, which opened up new avenues for creating innovative representations or introducing methods of surface analysis[1]. In the early 1980s, microcomputers appeared, whose immense capacities are now available to a large number of people with minimum knowledge. “Making” maps seems to be easier, updates become simpler, and software, of varying quality, is multiplying. The availability of the Internet, and especially the Web in 1994, further simplified the diffusion of graphic documents, which multiplied.

In addition to these new features, a fundamental change involved the communication paradigm, which is essential in the field. Despite automation, the arrival of the Internet and the Web, the logic of moving from an author/producer (the one who conceives, who represents) to a reader/user, sometimes with feedback, remains predominant; the reader, the one who consults, who looks at the map and uses it, remains passive, even if, thanks to animation and especially interactivity, they can move their document around, zoom over it, fly over it, etc.; but they cannot modify it. However, with the arrival of Web 2.0, the Semantic Web and the advent of the Geoweb from 2005 to 2010, and especially 2015, a new turning point is taking place: the reader/user, whoever they may be, becomes active. They can modify the maps on the Web and can even create them. This is the beginning of a new period for cartography, which will be further developed in the works for the ISTE encyclopedia, in order to present an updated state of this science, with its transformations for the period 2010–2022.

The chosen date of 2010 as the threshold is explained by the fact that many works on cartography, both in French (Béguin and Pumain 1994) and English (Cauvin et al. 2010a, 2010b, 2010c; Slocum et al. 2009), were published in the previous years, exposing the characteristics of the field just before the “real” start of the Semantic Web, highlighting the steps necessary to produce a map, as well as the successive choices that must necessarily be made. Indeed, the construction of a map requires at least three main steps, each of which has a very specific role. The first step is to create a localized database from the geographic information provided. The second stage ensures the processing and transformation of the data in the database, insisting either on the locations or on the thematic data, or on both simultaneously, with or without taking time into account; it leads to the determination of the representation mode adopted. As for the third stage, that of communication and diffusion, it is not only based on technical solutions, on semiotic choices, on the adoption of particular actions, etc., but also on the knowledge of elements of visual perception and cognition.

1 This approach largely corresponds to the analytical cartography introduced by Tobler (1976, 2000).

These three stages[2] are interrelated and are linked to both the challenges and the relationship between the recipient/user of the map. These three stages guided the selection of the four works chosen for the ISTE encyclopedia, because they start from the state of cartographic science at a period just before important changes. It will therefore be easier to highlight, describe and characterize the specific and original features of the years 2010–2022. One volume is devoted to each of the stages, and a fourth shows cartography from a historical aspect, thus facilitating a better understanding of the changes over time. Each volume can be read independently of the others and in an order that the reader wishes, according to his or her desires and expectations. The same is true for each chapter.

The aim of the historical volume is "to outline the history of cartography as it is done, and as it continues to evolve, by proposing a synopsis of the reflections and their modifications over the last forty years, in order to advance a history of cartography that takes into account current reflections and research, and above all, that opens up new avenues to explore".

The volume on geographic information and cartography focuses on data, its characteristics and its use in cartography during a period where digital techniques are gradually, but fundamentally, recomposing contemporary societies. Data acquisition is no longer an activity reserved for specialists. The sources of geodata are diversifying, citizens are directly and often voluntarily providing geographic information (GPS readings, cell phones, connected mobile objects, X (formerly Twitter), etc.); networking is immediate. The search for information for the production of maps is therefore profoundly transformed and reveals major issues in terms of digital society.

The volume on the processing and mapping of geographic information has a double objective. The first is to present methods and techniques for processing and transforming information to produce maps that respond specifically to each problem. These procedures take into account the spatial, thematic and temporal components of this information, and focus on spatial and spatiotemporal processes using indicators and models. The second objective is to highlight the importance of the interdependence of the different stages of cartographic construction, in which the processing stage is central. On the one hand, the latter must take into account the data to be processed, selected at the previous stage. On the other hand, it specifies, for the next stage, the original cartographic representations that will result from the selected processing. These two objectives are systematically concretized with examples that provide a clear understanding of the methods and their contribution to the understanding of the phenomenon under study.

---

2 A similar pattern is found in the latest version of Robinson's reference book from 1995.

The volume on communication, however, deals with the transmission of the map, because the latter is not a document that we produce so that it can remain in our pocket. It must become visible, and eventually audible and sensible, by appealing to graphic variables or others types of variables, possible actions and presenting display device properties. This volume will certainly be the easiest for detecting the specificities of the current period, with the role of choices at all stages, the interweaving of these stages and the collaborative work.

With the same tools and the same means, specialists and non-specialists will produce maps by taking what is at their disposal but with distinct training, which induces risks. Indeed, the new cartographic processes that are constantly emerging have many positive aspects: ease of use, diversity and attractiveness. However, like all new developments, their advantages have a downside; each of these processes has specific properties that are important to know at least in order to apply them wisely and avoid mistakes. The four books presented here show that we are moving toward new logics, new ways of conceiving, translating, processing and cartographically transcribing geographic information, without necessarily having to reject what already exists. The world of maps does not escape the general evolution in which we are all participating, with our wealth and our worries. May these books help everyone to be aware of what they produce and to use all these tools with discernment.

## References

Béguin, M. and Pumain, D. (1994). *La représentation des données géographiques. Statistique et cartographie*. Cursus, Paris.

Cauvin, C., Escobar, F., Serradj, A. (2010a). *Thematic Cartography 1: Thematic Cartography and Transformations*. ISTE Ltd, London, and John Wiley & Sons, New York.

Cauvin, C., Escobar, F., Serradj, A. (2010b). *Thematic Cartography 2: Cartography and the Impact of the Quantitative Revolution*. ISTE Ltd, London, and John Wiley & Sons, New York.

Cauvin, C., Escobar, F., Serradj, A. (2010c). *Thematic Cartography 3: New Approaches in Thematic Cartography*. ISTE Ltd, London, and John Wiley & Sons, New York.

Robinson, A.H. (1952). The look of maps. An examination of cartographic design. PhD Thesis, University of Wisconsin, Madison.

Robinson, A.H. (1953). *Elements of Cartography*. Wiley, New York.

Robinson, A.H., Morrison, J., Muehrcke, P.C., Kimerling, A.J., Gutpill, S.C. (1995). *Elements of Cartography*. Wiley, New York.

Rosnay, J. (2008). *2020, les scenarios du futur : comprendre le monde qui vient*. Fayard, Paris.

Slocum, T.A., McMaster, R.B., Kessler, F.C., Howard, H.H. (2009). *Thematic Cartography and Geography Visualization*. Prentice Hall, Hoboken.

Tobler, W.R. (1959). Automation and cartography. *The Geographical Review*, 49(4), 526–534.

Tobler, W.R. (1976). Analytical cartography. *The American Cartographer*, 3, 21–31.

Tobler, W.R. (2000). The development of analytical cartography. *Cartography and Geographic Information Science*, 27(3), 189–194.

# Introduction

**Claire CUNTY[1] and Hélène MATHIAN[2]**

*[1] EVS, Université Lumière Lyon 2, France*

*[2] EVS, CNRS, Lyon, France*

> *Mapping is perhaps the most formative and creative act of any design process, first disclosing and then staging the conditions for the emergence of new realities.*
> James Corner, 1999

## I.1. Maps and mapping

The map is more than just a "simple" representation of space and must be seen as a succession of cartographic operations (Besse and Tiberghien 2017). It is both the graphic result and the set of cartographic operations that led to the resulting product.

> To operate is to transform a material by following a certain number of formal rules (rules that are not necessarily determined a priori, but which on the contrary can be invented or redefined in the course of the operation itself), in order to obtain a result, an object, a product. Every operation is a "formation" and a transformation [...] (Besse and Tiberghien 2017, pp. 14–16).

Formally, the transition from geographic information or geographic data to maps involves several stages of data "transformation", operating at different levels. Cauvin et al. (2010) define a categorization of these different "transformations": entity transformations, thematic and geometric attribute transformations,

*Handling and Mapping Geographic Information*,
coordinated by Claire CUNTY and Hélène MATHIAN. © ISTE Ltd 2024.

cartographic transformations, semiotic transformations and finally, display transformations (Figure I.1).

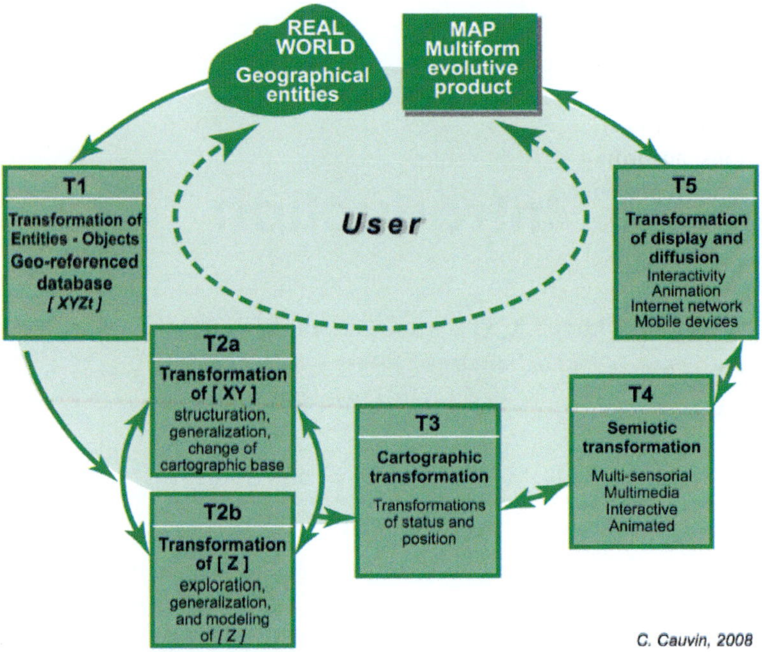


**Figure I.1.** *The map as a sequence of transformations (Cauvin et al. 2010)*

The aim of this volume is to present the various operations involved in the articulation between geographic information and the map, within the purview of *processing* (T2a, T2b, T3 in Figure I.1). We have chosen to present these within an operational framework. All the examples and developments presented in the chapters are strongly rooted in a *spatial analysis* approach, that is to say, with the objective of describing and modeling spatial structures. In all the chapters, the authors review the operations involved in handling geographical data, from a pragmatic point of view, in relation to a geographical question, focusing on a number of these contextualized choices. All of these operations are associated with choices and methods that "orient" the data in a certain direction (by giving it a meaning), shedding light on a geographical question by taking into account the constraints of the medium and the target audience. The questions and developments associated with these operations are being renewed within the technological context (widespread geo-referencing of data, Big Data, etc.) and methodological context (democratization of programming libraries, technological evolution of visualization environments, etc.), while necessarily retaining conceptual reflections.

Geographical data[1], processing and mapping are linked in different ways depending on the role of the map: it can be the result, an intermediate stage in an analysis, or an interface for data exploration. The categorization proposed by MacEachren (1994) is very useful and widely used to position cartographic productions according to their objectives (Figure I.2). He defines four uses of cartographic representations based on three dimensions: the level of knowledge of the phenomenon addressed, the level of interactivity with the map and the target audience. Most of the analysis approaches developed in the chapters of this volume fall into the exploration/confirmation phase.

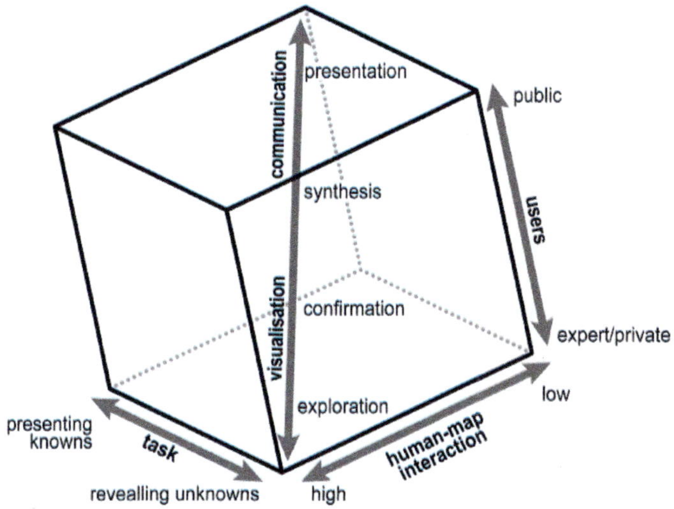


**Figure I.2.** *Different visualization objectives, according to MacEachren (1994)*

Some chapters develop a chain of processes responding to a specific geographical question and show how the map accompanies this process (Chapters 4, 5 and 8); others offer an updated or contextualized view of a specific methodological question (Chapters 1 and 3); still others present an inventory of solutions to a specific methodological question (Chapters 2, 6 and 7) (Figure I.3).

1 Throughout the volume, "geographic information" and "geographic data" will be inter-used with the same meaning. Thereafter, in the introduction, the term "data" will implicitly mean "geographic data".

Together, they cover a wide range of processing operations that link data and maps, and allow us to discuss the various challenges involved.

In this introduction, we propose two methodological readings of the contents of this volume. At the end of the introduction, a "map" of the chapters is proposed, showing the specific features of each chapter in relation to the various issues raised above and in the course of the two readings that follow (Figure I.3).

## I.2. Preliminary reading: through the prism of a few operations and methods

A preliminary reading of the chapters should be based on a few key concepts that underlie the differentiation of practices.

These entry points correspond to different operations. The sequence of these operations generally follows a pattern: (1) geographic information, (2) processing and (3) visualization (the orange outer circle in Figure I.3). However, this sequence is often not so linear, and there is frequent to-ing and fro-ing between operations, especially with the development of digital information visualization interfaces, which enable operations to be interactively linked. For example, an initial hypothesis on a question may lead to the mobilization of data, followed by a type of processing and visualization. The result sometimes leads to reconsideration of one or all of the operations.

Therefore, we propose an initial reading in light of this sequence of operations to clarify the processes discussed or used in the various chapters that concern entities and data, treatments and methods mobilized and re-presentation.

### I.2.1. *Linking between data and entities*

Moving from a phenomenon to its measurement requires the conceptual clarification of the geographical entities being produced. For example, observing the spread of Covid (a phenomenon) requires us to make explicit the objects (mobile individuals, hospitals, towns, etc.) and operations that will be carried out to define the *geographical entities* that will enable us to approach the representation of the phenomenon. This conceptualization stage enables us to give meaning to the data being mobilized to illustrate a given phenomenon. This can be omitted when the aim is simply to *show data*, without any real reflection or link to a phenomenon or its analysis, as demonstrated by the success of dashboards (business intelligence tools) which automatically visualize data: the map is the data, and the processing is essentially used to manage technical issues of visualization within interactive environments.

Focusing on data processing in the sense of data analysis, the chapters in this volume all address the question examining the transition from *entity design* and associated issues to *information construction*, contextualizing it in relation to a variety of uses.

Certain chapters are based on a formalism, today considered as classic in the field of GIS, which consists of identifying, when describing entities, what belongs to the geographic dimension and what to the attribute dimension. A geographic object is defined in terms of a geographical dimension (XY), a thematic component (Z) and, where appropriate, a temporal dimension (T). Chapters 6 and 7 discuss specific transformations of these dimensions, and Chapter 8 outlines an extension for links between places.

Other chapters adopt an operational approach to data; the formalizations used focus on the issues associated with data interoperability, making it possible to give meaning to entities. This question is particularly relevant at a time of data "deluge", encountered with Big Data, or new data, from the Web, such as those mobilized in Chapters 2 and 8. Chapter 2 offers an epistemological perspective on this new paradigm and discusses methods for integrating it into geographic entities. This point is also addressed in Chapter 3 by formalizing the processing operations (and representations) that enable the integration of environmental "objects" of diverse origins. Chapter 5 presents another complexity of data construction, with the example of land management data (DVF: demande de valeurs foncières / land values request). Other related issues are discussed, such as the harmonization of international data to ensure comparability (Chapters 1 and 8), and statistical categories and their legitimacy (Chapter 4).

Hence, the variety of examples throughout the volume enables us to cover a wide range of data types and associated issues: administrative data, sampled or continuous data, big data, qualitative data, quantitative data, linkage data, etc. The examples also offer a wide range of levels of complexity. They also offer a wide range of levels of complexity: from simple location data with a single attribute (Chapter 2) to data requiring location reconstruction (Chapter 8) through disambiguation (natural language processing [NLP][2]) and whose form is a link between geographical entities. All the chapters make this entity-data link explicit, thus ensuring the articulation between the thematic question and its empirical and technical operationalization.

2 Also known as automatic language processing (ALP).

### I.2.2. From data to map

#### I.2.2.1. Creating indicators

Perhaps the most recurrent operation is that of creating indicators and plotting them on the map. This involves operations on both the thematic and geometric components. These correspond to "traceable" operations, involving a simple calculation (a rate of change between two dates, a duration calculation, a discretization), a spatial calculation (accessibility) or the result of a statistical method involving assumptions, and a chain of more complex operations involving optimizations, for example (calculation of a principal component analysis [PCA] factor or regression residuals). These operations are dealt with in all the chapters, justifying them in relation to the issue addressed.

Before or during the creation of indicators, the question of geographical level arises (e.g. At what level should real estate transaction prices be analyzed? Which geographical level best reflects ethnic segregation?). These questions lead to aggregation operations (from neighborhood to municipality, for example) or simplification of geometric information (do we calculate accessibility at the centroid of the zone or at the edge of the polygon associated with the zone?)

More often than not, the question of the indicator and that of the geographical level are inseparable and refer back to the previous discussion on the definition of the entities observed. Chapter 4 discusses this issue in particular, and the fact that the indicators created may be "sensitive" to the size and/or shape of grids, a problem known as MAUP (modifiable areal unit problem). The question arises differently depending on whether the grids are neutral "containers" with no diversity of shape or size (which can be assimilated to homogeneous grids), and the interest is solely in content, or whether they are administrative (or power) grids that involve a certain complexity, as they are neutral neither thematically nor geometrically (shape and size).

#### I.2.2.2. Analysis methods

In the field of spatial analysis, the creation of indicators mobilizes formal, quantitative, statistical and mathematical methods. These are essentially statistical methods used to identify, describe and explain the spatial organization of a phenomenon. These processes can be used at different stages or for different purposes. For example, spatial interpolation can be a method for "constructing information" (sample data, Chapter 2), for reconstructing spatial organization (Chapters 3 and 4) or for simplifying information (point density, Chapter 2). Methods and maps are interrelated. Depending on the case, the methods inform the map, or the map as a data arrangement is integrated with the method. It seems important here to introduce a distinction between the types of methods that are mobilized throughout the chapters, and which can, depending on the context, attribute different meanings to the map:

– classical statistics versus spatial statistics: when the methods employed are derived from classical statistics, the spatial dimension is not integrated into the processing a priori; visualization (transferring thematic information onto geographical space) can then sometimes reveal spatial effects a posteriori. Chapters 1, 3, 4 and 8 are typical examples. When methods are derived from spatial statistics, the spatial dimension (location, spacing, neighborhood, etc.) is formalized directly within the method. Chapters 4 and 5 are emblematic in this respect, with methods incorporating distances, neighborhoods and accessibility. Chapters 2, 6 and 7 provide further light on this type of approach;

– descriptive versus explanatory statistics: the properties of spatial data challenge the use of inferential statistical methods: dependent data (spatial autocorrelation), existing at several geographical levels (MAUP) and which may be considered as unique cases. Several chapters revisit these issues and propose different approaches (Chapters 4 and 5);

– exploratory versus confirmatory analysis: finally, the book presents several cases where the map is integrated into the analysis process, either as part of an iterative process (Chapters 5, 7, 8) or within an interactive environment (Chapters 1 and 2). In this case, the map is both a representation and a data selection operator, positioning the analyst in a research approach based on exploration rather than confirmation of statistical hypotheses.

### I.2.3. *Several visualizations*

Maps play different roles depending on when they are used in information processing (Figure I.2), but the forms they take are equally diverse. These forms are linked to semiotic and semiological choices. All chapters take for granted the sign system defined by Bertin (1967) in graphical semiology, which associates certain signs with the level of measurement they convey (quantitative, ordered, different). Interested readers can refer to fundamental works in cartography (Cauvin et al. 2010; Béguin and Pumain 2017). The volume as a whole presents a wide variety of cartographic forms. As the object of the volume is the cartographic production process, their esthetics or perception is not discussed, except in Chapters 6 and 7.

The choice of cartographic representation is linked to the type of spatial structure under analysis:

– first of all, spatial structures can emerge because of the transfer of the thematic dimension onto the geographical space. However, the spatial distribution of weights/stocks (proportional figure maps) will not be graphically translated in the same way as that of relative data (choropleth maps). The analyses derived from these representations are different and complementary (Chapter 4). Anamorphosis (Chapter 7) illustrates one way out of the "simple" transfer process, enabling both dimensions to be visualized simultaneously. Generally speaking, transfer operations allow spatial structures to emerge, even when several thematic dimensions are taken into account simultaneously (Chapters 1, 3, 4 and 5);

– when the relationships between locations are at the heart of the analysis, these are represented cartographically by lines joining the connected locations. This is proposed in Chapter 8 (see also Mericskay (2023), Chapter 4). Chapter 4 depicts another form of relationship, emphasizing differentiation between neighboring units with "discontinuity" maps;

– finally, the analysis of processes and dynamics leads us to integrate the temporal dimension into the representation. To this end, a variety of graphical methods, including animation, are used to integrate time into cartographic visualization (Chapter 6).

These initial cartographic forms, which most often serve to reveal structures, may be succeeded by maps in which the cartographic choices seek, through graphic metaphor, to highlight the spatial and sometimes spatiotemporal structure of the phenomenon (partitioning the world in Chapter 1, delineating borders in Chapter 3, discontinuity in Chapter 4, spreading epidemics in Chapter 6, accessibility in Chapter 7). These maps feature greater graphic creativity and have the objective of conveying a message.

The development of visualization platforms enables the "raw" visualization of databases, with the aim of exploring the data itself (Chapter 2). This automation is based on cartographic know-how and associated processing. But this systematization, with no underlying question and no semiological adaptation to the question, goes hand in hand with a certain cartographic standardization that must be examined.

## I.3. Secondary reading: relationships between the chapters

As in Figure I.1, the chapters can be organized in a loop, as they have been designed with a progression in mind. They have been designed in pairs to highlight certain stages, and the result shows that a number of other dimensions come into resonance with other pairs of chapter (Figure I.3).

– Chapters 1 and 8: One opens and the other closes this volume. These address the question of *relationships*: Chapter 1 looks at statistical relationships and Chapter 8 looks at the spatial relationships between entities. Both deal with a specific cartographic scale, that of the world, which raises cartographic questions that will be illuminated in two different ways. These two chapters introduce the discussion on *data exploration*, wherein the map is a tool that enables us to question the complexity of relationships, to go a step further in the thinking and in the expression of hypotheses, while also being the medium for understanding their organization.

– Chapters 2 and 3: The focus of these two chapters is on the first stages of transformation, namely, *data*: its design, acquisition and integration. The chapters discuss these issues in an original way, approaching the link between *cartographic objects and data* from a radically opposed point of view. Chapter 2 situates itself in the new paradigm of geographic information, referred to here as *new Web data*. It raises the following question: how can these data, which are produced in contexts often far from a protocol precisely adapted to an analysis, or in a participatory context, be integrated so as to serve analysis? Chapter 3 revisits the *construction of cartographic objects issues* for the observation of continuous phenomena, mainly in the context of the environmental field.

– Chapters 4 and 5: Each of these two chapters offers an extremely pedagogical approach, mobilizing *statistical modeling* to reveal forms of organization, that of racial segregation in Chapter 4 and that of a form of social segregation in Chapter 5. The methodological approach is slightly different, with Chapter 4 being more *descriptive* and Chapter 5 being more *inferential*. Both chapters discuss information construction, spatial zonings identification, MAUP and spatial autocorrelation. They clearly illustrate the range of choices that can be made during a data processing workflow, and how the map is at once data, a graphic representation and the result. The two chapters revisit this practice at each stage.

– Chapters 6 and 7: Each of these chapters offers a *panorama* of transformations associated with a specific dimension – that of *time* for Chapter 6 and that of *space* for Chapter 7. Chapter 6 proposes a *formalization of the data* used as an input to this review of cartographic solutions for integrating time. Each visualization is discussed in relation to the choices made on the data and the desired effects. Chapter 7 puts forward a series of *methodological formalizations* underpinning cartographic transformations and discusses the effects of these transformations on the basis of a single example. Both chapters deal with the question of *change*, of state in Chapter 6 and of form in Chapter 7. The two chapters converge on questions of *movement*.

## I.4. A map of the chapters

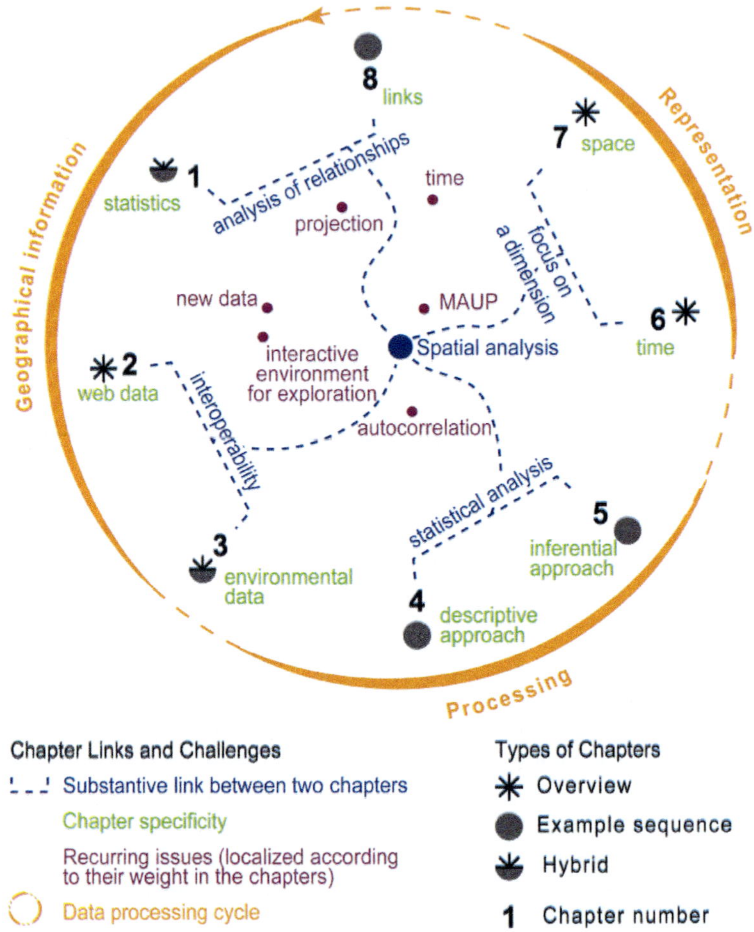


**Figure I.3.** *Chapter map: how they are linked, on which concept they rely*

Figure I.3 positions the chapters in relation to each other, and to the various points raised in this introduction.

– First, the chapters are positioned along a continuum identifying the key stages of a cycle that links the data to the map (orange circle). Of course, all the chapters cover all stages; even so, they have been positioned here to reflect as close as possible the key insights they elucidate.

– Signs remind us of the approach of the chapters: between overviews, unfolding examples and formal presentations.

– The chapters are linked by a dendrogram illustrating the proximities between chapters that the second reading has just described (blue). What unites them all is the fact that they mobilize *formal methods and models* for these transformations and are part of a *spatial analysis* approach whose aim is to reveal shapes and organizations.

– Concepts (green) highlight the specific features of each chapter, over and above the similarities mentioned above.

– A number of other concepts (purple) are covered in the chapters and are common to all. They are positioned according to their importance in the chapters.

Therefore, as this "map" illustrates, the chapters interconnect, and complement each other, refer to each other, sometimes repeat each other, and every so often respond to each other. Together, they cover a wide range of cartographic practices and productions, provide opportunities for reflection in many directions.

## I.5. Acknowledgments

The content of all the chapters was the subject of exchanges between all the authors. Each chapter of the book was finalized through an iterative process of internal cross-assessment, including external experts (Marc Bourgeois, Julie Vallée, Sébastien Oliveau, Lena Sanders, Nicolas Lambert, Clarisse Didelon), the book's two coordinators and the Cartography theme manager, Colette Cauvin-Reymond. Special thanks to the external reviewers for their willingness to be involved and their helpful suggestions, and to Colette Cauvin-Reymond for her invaluable advice and support. The whole process of writing the book gave rise to rich methodological exchanges. The assembled chapters reflect this richness and diversity of approaches to the map. We would like to thank the authors once again for their original contributions, which shed light on the different ways in which the map is used in our "in vivo" approaches.

## I.6. References

Béguin, M. and Pumain, D. (2017). *La représentation des données géographiques : statistiques et cartographie*. Armand Colin, Paris.

Bertin, J. (1967). *Sémiologie graphique : les diagrammes, les réseaux, les cartes*. Mouton et Gauthier-Villars, Paris.

Besse, J.-M. and Tiberghien, G. (2017). *Opérations cartographiques*. Actes Sud, Arles.

Cauvin, C., Escobar, F., Serradj, A. (2010). *Thematic Cartography 3. New Approaches in Thematic Cartography*. ISTE Ltd, London and John Wiley & Sons, New York.

Corner, J. (1999). The agency of mapping. In *Mappings*, Cosgrove, D. (ed.). Reaktion Books, London.

MacEachren, A.M. (1994). Visualization in modern cartography: Setting the agenda. In *Visualization in Modern Cartography*, MacEachren, A.M. and Taylor, D.R.F. (eds). Elsevier, Oxford.

# 1

# Exploring Statistical Relationships with Maps and Charts

**Jean-Benoît BOURON[1], Claire CUNTY[2], Hélène MATHIAN[3] and Myriam BARON[4,5]**

[1] *Géoconfluences, DGESCO, ENS Lyon, France*

[2] *EVS, Université Lumière Lyon 2, France*

[3] *EVS, CNRS, Lyon, France*

[4] *Lab'Urba, Université Paris-Est Créteil, Créteil, France*

[5] *IDEES, CNRS, Université de Rouen Normandie, Rouen, France*

## 1.1. Introduction

Spatial analysis involves identifying, describing and explaining the spatial organization of a phenomenon. It uses data and statistical methods to shed light on these issues. Such an approach implies a dialogue between geographical and thematic dimensions, in which maps obviously play a central role. Either methods used are derived from classical statistics[1] and do not give a specific role to space – any spatial effects are revealed a posteriori by transferring the result onto the map – or methods are derived from spatial statistics and directly integrate the geometric component of data. Here, we select first methods, and more specifically, an exploratory and descriptive one, combining data processing with graphic and cartographic representations. We have chosen to base this approach on a single

1 The term "statistics" refers both to the disciplinary field that studies phenomena by analyzing data, and to data themselves, as well as to methods used to make these data understandable.

*Handling and Mapping Geographic Information*,
coordinated by Claire CUNTY and Hélène MATHIAN. © ISTE Ltd 2024.

scale, that of the world, which raises interesting questions about the definition of entities in relation to the two points of view we are interested in here.

From a thematic point of view, that is from a geographic and mapping perspective, a planisphere is a singular form. In the hierarchy of geographical units, it is first and foremost the macro or global scale. It is the only scale that enables us to apprehend all geographical units as a closed system. From a cartographic point of view, this scale imposes *projection* choices that do not necessarily arise at other scales. While the correct conservation of distances and areas is a well-known constraint (see Chapter 7), projecting a sphere onto a plane also transforms proximities at a scale where borders have a very strong meaning. We therefore need to think specifically about this. Finally, at this scale too, a recurrent issue in geography and cartography emerges, that of the unequal size of spatial entities (known as a modifiable areal unit problem [MAUP]; see introduction and Chapter 4).

From a methodological, that is, statistical, point of view, the entities studied (countries) form a whole that cannot be considered as a sample. Like all territories, of course, they have an existence independent of any experience. However, countries of the world are undoubtedly one of the most "familiar" sets of spatial entities, generally made familiar from an early age onwards as part of national education curriculums. Despite its complexity, this collective is understood with many a priori thematics: as such, it thus constitutes a set of "cases", rather than a set of "representations" part of a larger population. All this makes it more difficult to assimilate countries into statistical entities. Such a set is therefore particularly well-suited to illustrate and discuss the descriptive and exploratory (rather than a predictive and confirmatory) framework within which statistical methods will be mobilized.

In this context, this chapter will focus on the analysis and representation of relationships between variables. The observation of a geographical phenomenon can rarely be reduced to the analysis of a single measurement. Knowledge of any observation considered from the point of view of the "spatial" dimension implies starting by mapping a variable (an indicator). The latter must enable us to account for the distribution, in both spatial and statistical sense, of the theme (e.g. infant mortality rate by country; Covid-19 incidence rate by region; transaction prices by quarter; etc.). The map makes it possible to simultaneously apprehend the thematic and spatial variability of the variable's values, that is, both the form of heterogeneity of the values (are they concentrated around the mean value, or on the contrary, widely dispersed?) and the forms of organization that this heterogeneity draws in space: concentrations, regionalization, oppositions, discontinuities, gradients, etc. As such, one map calls for another, and we quickly enter a phase of exploration that leads us to question this variable in relation to others, trying to understand the simultaneous (or not) localizations: What is the birth rate in the least affluent countries? What is the Covid-19 mortality rate in the most densely populated regions? What is the income of residents in the most expensive neighborhoods? etc. By comparing two or three (or more) maps, we begin to analyze similarities and dissimilarities, and their geographies, which leads us to question the relationships between variables represented (Figure 1.1).

Geographic data exploration involves a sequence of processing and cartographic representations, combining phases of data comprehension, sometimes phases of transformation and phases of analysis using statistical methods to reveal and model relationships between different variables. In this exploratory approach, graphic and cartographic representations are an integral part of the analysis. Visualization stimulates questioning and drives reasoning: it is a major dimension of comprehension and explanation. The question of cartography itself would require us to revisit many basic notions of semiology and statistics: for this, we refer to some previous studies (Béguin and Pumain 1994; Cauvin et al. 2010a; Lambert and Zanin 2016).

In this chapter, we propose to discuss the steps involved in this data exploration process, using an example that draws on the global scale and, as we go along, points out the specific features of this scale: how to "divide" the world, according to the major wealth and development disparities? The thematic challenge associated with this example is as much scientific as it is pedagogical, given that major historical divisions (e.g. "Third World", "developing" countries, "North and South") have been progressively abandoned without being replaced by other terms (Bouron et al. 2022).

The aim here is essentially to outline the methodological tenants of this approach in order to show how data mining can shed light on these ways of divvying up the world. After addressing the cartographic issues involved in representing the world (section 1.2), we introduce the issues involved in exploring statistical relationships based on two variables (section 1.3). We then illustrate this question, generalized to several variables and focused on the challenge regarding "world dissection", by proposing an approach based mainly on the visualizations of statistical treatments employed (section 1.4).

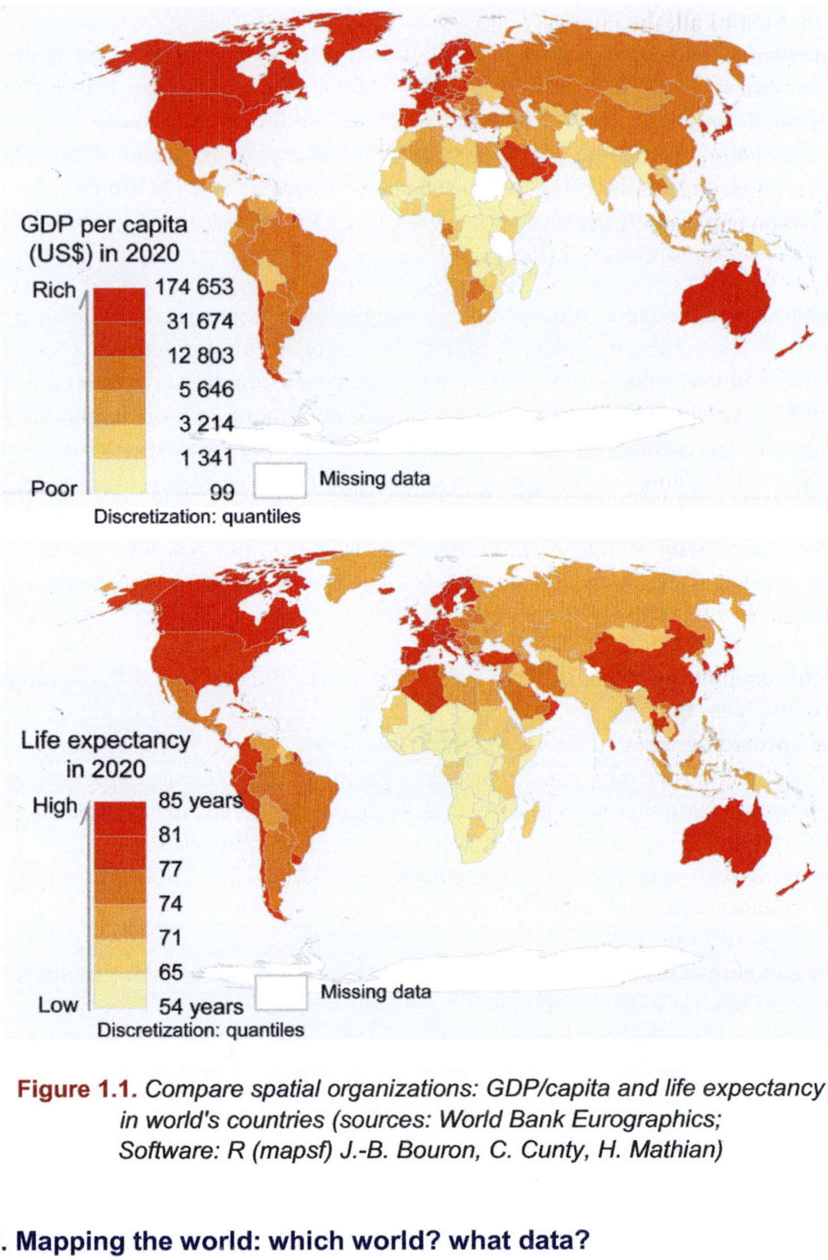


**Figure 1.1.** *Compare spatial organizations: GDP/capita and life expectancy in world's countries (sources: World Bank Eurographics; Software: R (mapsf) J.-B. Bouron, C. Cunty, H. Mathian)*

## 1.2. Mapping the world: which world? what data?

As with any cartographic approach, there are first of all the issues of the map base and its projection, the spatial grid and the associated data. We present these in the following three paragraphs and discuss their specificity considering countries of the world. First of all, the choice of projection that will allow for the transition from globe to planisphere is an important issue. As for the grids, although determined (countries, states), these can likewise be interrogated further, for instance, in terms of the different organizations of political systems (countries, federations of states). Finally, the national statistics that will be mobilized also reflect these differences and, as with any international analysis, raise the question of their thematic harmonization and synchronization.

### 1.2.1. *Planispheres, projections and centers*

The ways countries belong to bigger entities and how close or far they are from one another can be perceived differently depending on the projection chosen (Grataloup 2007; Bahoken 2022). The cartographic representation of data, depending on the choices made, orients mental images we have of the world's different countries (Lambert and Zanin 2019; Monmonier 2019). These countries can be seen, for example, as margins of the various oceans and seas, recalling that 80% of the planet is occupied by water, as shown by the Spilhaus projection. This thus seems to fit in with the continuity of the ancient world as defined by the Greeks and Romans around the Mediterranean. The question of the map's center also depends on the point of view proposed (or imposed) by the cartographer. Countries of the world can thus be viewed from the southern hemisphere, in particular Australia, providing an unexpected representation of the Indo-Pacific world for some of the inhabitants of the northern hemisphere, notably Europeans and North Americans (see Figure 1.2(a)).

When it comes to representing the countries of the world on a planisphere, a certain continuity can be demonstrated since the second half of the 19th century, with polar projections predominating. The latter were proposed as early as the 1860s, notably by Gustav Jäger, a zoologist keen to show ancient connections between great continental masses and, consequently, links that might exist between different species distributed across the world (Capdepuy 2013). The choice of the planisphere featured on the flag of the United Nations Organization (UNO) from 1945 to 1950 extends these proposals, as do the projections put forward by Jacques Bertin in the early 1950s, and even those of Christian Grataloup in the 2000s. This family of projections makes it possible to account for the interconnections between the different countries of the world, most of which are located in the northern hemisphere, without the a priori selection of a center that would orient the viewer's gaze. However, this last point has been debated, as illustrated by the successive planisphere projects for the UN flag. The American continent, or even its northern part, was first the selected focus, although it was Europe that was ultimately preferred for the final version, to better highlight the opposition between the two superpowers (Capdepuy 2011). In any case, a polar projection centered on the North Pole relegates certain countries to the periphery, in particular New Zealand, which in this case appears highly distorted. In all cases, there is no projection that can avoid focusing on one part of the world, without relegating another to the periphery.

The choice of a polar projection, to emphasize not only links between different parts of the world but also major geopolitical oppositions, almost makes us forget the deformations inherent in any projection, that is, the deformation of surfaces and/or directions when it comes to representing the Earth's geoid in two dimensions. This is particularly striking with the Gerardus Mercator projection (Figure 1.2(b)). Dating from 1569, it is widely (if not excessively) used, and not just for the navigational charts for which it was intended. It places great emphasis on spaces from mid-latitudes upwards, especially at high latitudes, showing them to be much larger than they are, like any conformal projection that preserves angles and therefore shapes to the detriment of surfaces. Russia thus appears as large as the entire African continent, whereas the latter, with its 30.4 million square kilometers, is in fact almost twice as big. Alternatives exist, such as Arno Peters' projection, presented in the early 1970s, which has many similarities with the one formalized by James Gall in the mid-19th century (Figure 1.2(c)). It corrects for differences in size between continents and countries, but at the cost of distorting the shape of the latter, like any equivalent projection. For some authors, this projection seems to favor an alter-globalist vision of the world, by considerably reducing the influence of the so-called northern countries, particularly those of Europe and North America. To remedy this, a cartographic representation of the countries of the world can be based on the projection of Richard Buckminster Fuller, created as early as 1946 (Figure 1.2(d)), which respects the surfaces and shapes of these countries. But on the other hand, we are faced with a "splintering of the world" according to Roger Brunet and Olivier Dollfus in their volume devoted to New Worlds in the latest *Géographie universelle* (1990), while Nicolas Lambert and Christine Zanin highlight the promotion of a "universal world without north and south" (2019). These examples of projection and their interpretations show that we are a long way from "cartographic production" assimilated to "a 'neutral' tool, independent of those who [make it]" and thus from an "objectivation of space" (Lévy and Lussault 2003). Mapping relationships between geographical data in an exploratory way thus begins with a comparison of the different projections likely to be selected and the identification of the ways in which they will modify the visualization of information. In this chapter, we shall be using the Equal Earth projection, which preserves surfaces and gives an idea of the Earth's rotundity. It is currently considered a good compromise for representing global-scale themes (Genevois 2018; Šavrič et al. 2019).

### 1.2.2. *Grid-related issues*

Observing the world at the level of countries[2] raises a number of questions relating to borders and the grid effects they have on cartographic images. The aim here is not to discuss territorial conflicts and/or disputed borders, even though these inevitably arise on a global scale, but to reflect in a general way on the various grids and their impact on cartographic representations. Grids are defined as a partition of space "into units whose shape and size may be regular or irregular" (Pumain 2008). In the case of national borders, and more generally in the case of administrative borders, boundaries separate irregular geographical units to which statistical variables can be associated, most often produced by state statistical agencies. These statistical variables are in fact derived from the aggregation of individual data, the aggregation criterion being the location within a geographical unit. However, for certain themes, the relevance of this aggregation can be questioned. On the one hand, these grids may differ from those of territorial governance. For example, how can we consider the United States as a whole, in the same way as France, when it is first and foremost a federation of states? On the other hand, observation at country level automatically leads to a comparison of characteristics from one country to another, considering each national characteristic as a homogeneous whole. However, in some cases, the differences observed within a country for a given variable can be much greater than the differences between countries. For example, differences in literacy rates among adult women in a country like Niger will be more marked between urban and rural areas within the country than with neighboring countries such as Chad or Mali. Geographers were quick to identify the problems associated with using and interpreting data aggregated within variable-sized grid cells under the name of modifiable areal unit problem, or MAUP (Openshaw 1983).

The grid is used to transfer thematic information to produce a map, providing information on the distribution and spatial organization of the phenomenon under study. In the case of choropleth maps, the variability of spatial units' sizes will lead to a bias in the perception of small countries, to the benefit of those with a larger surface area. Admittedly, the smallest countries in terms of surface area are often those with the fewest inhabitants: seven of the world's 15 least populous countries are also among the smallest; but these small countries are also frequently the most densely populated and the richest: the countries at the top of the various rankings by GDP per capita are all "small countries", in terms of surface area (e.g. Liechtenstein, Luxembourg and Monaco).

2 We use the term country here, even if we often implicitly refer to states. We will consider here the 193 countries currently recognized by the UN.

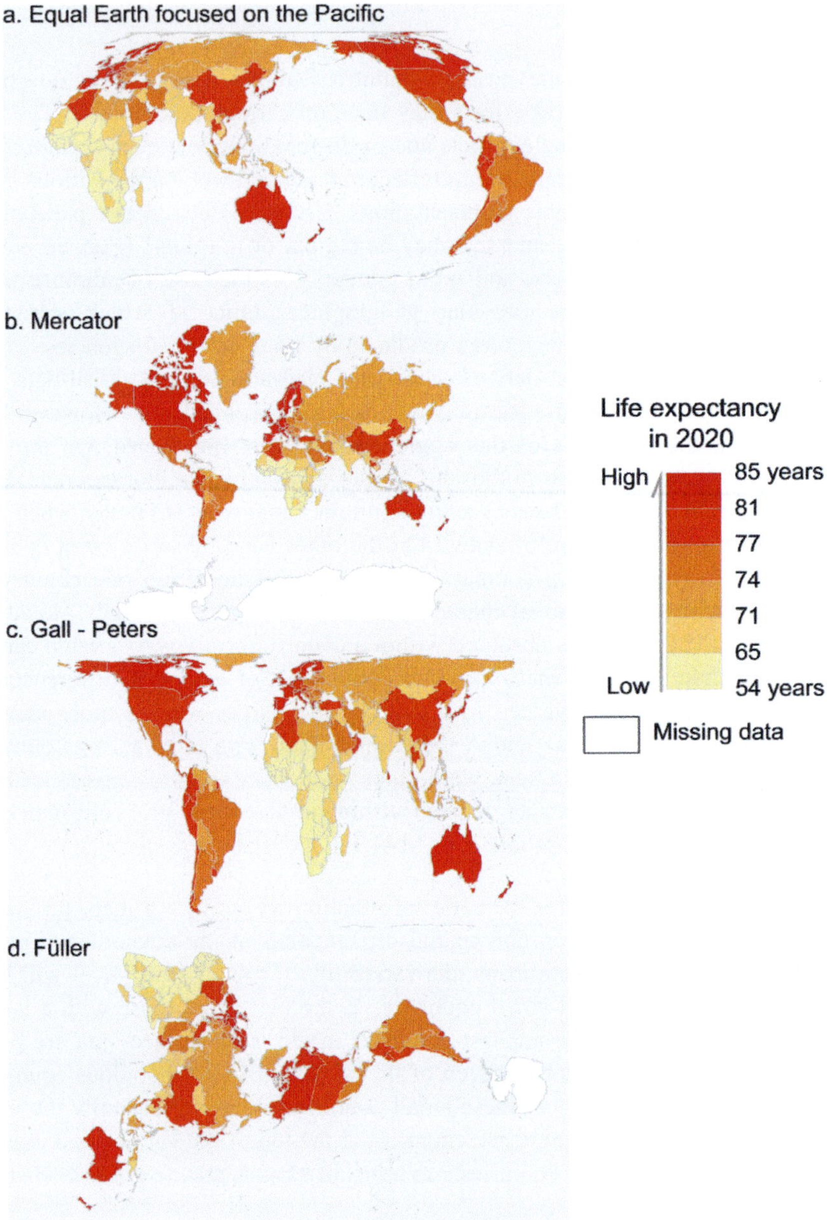


**Figure 1.2.** *The world in all its forms and projections. Variable represented: life expectancy (source: World Bank Eurographics; Software: R (mapsf) J.-B. Bouron, C. Cunty, H. Mathian)*

In this type of cartographic representation, space is seen as a support on which the cartographer simply transfers thematic information. The map thus plays the role of a revealer (Cauvin et al. 2010b, pp. 66–67) of spatial structures. When observing these structures, we need to keep in mind all the underlying elements mentioned above to avoid misinterpretation.

### 1.2.3. *Data at global scale: a composite material*

Thematic maps at global scale are well known in geography and cartography, and have been since the very first geography lessons at school or college, so much so that this type of representation can be taken as a given and not as a construct. Yet that is precisely what it is, and it is just as important to question the way in which this construct is produced as much as it is questioned at other scales. Collecting data on a global scale is far from self-evident and involves a number of pitfalls.

One of these difficulties lies in conditions under which data represented are produced. Most country-level data are collected by international organizations, the two main ones being the UN and the World Bank, plus the CIA, the US federal agency. These organizations have their own statistics departments, whose role is to collect data from the world's 193 States and numerous non-State territorial entities, and harmonize them in order to produce databases.

Let us take the most basic element of any national statistic: population. This is the basis for all data expressed on a per capita basis. None of the three agencies mentioned above has a world-wide census service. They therefore depend on national census services. At this stage, it is already clear that not all countries are in a position to carry out an exhaustive and reliable census. What are the working conditions of census-takers in failed States where the State no longer controls the administration? In informal settlements where the population does not necessarily wish to answer questions from public officials? In inaccessible mountainous areas? In war-torn regions? In regions where nomadism dominates? It is easy to see why statistics departments receive data of very uneven reliability. They are therefore obliged to correct them, based on previous censuses, compilation with other sources and their own knowledge of the local situation in each country. The question arises as to the circumstances in which these corrections are made, all the more so as estimates are added to these corrections: indeed, since each country has its own census time step, the data collected are at different dates. The United States has had a decennial census for two centuries. Some countries take their census when they can. In France, before the adoption of a rotating annual census system in 2004, intercensal periods varied between 5 and 9 years. The new French system, the only one of its kind in the world, is not without its critics in terms of reliability (Dumont

2018). In India, the census is decennial, but Covid-19 has delayed the 2021 census: in 2023 or 2024, data from the latest census will be available, more than 10 years after the previous census, carried out in 2011.

Usually, a thematic planisphere announces a variable at a specific date: for example, the number of cars per capita in 2020, for a map produced in 2022. This simple fact is already a bit of an arrangement with reality, since the 2020 population, on a global scale, combines results from recent censuses, corrected censuses, estimates from older censuses and estimates from corrected older censuses. This is not to claim that the world population figures by country are fanciful: these figures are not reality, but a reconstruction of reality.

And even then, we are only talking about the most common data, the total population. For other indicators, the quality of data collection is even more problematic. For example, the World Bank provides data on the ratio of girls to boys enrolled in elementary school at the latest available date. This date varies widely: 2020 for Algeria, 2015 for Angola, 1998 for Haiti and 1985 for Lebanon. Out of 220 territorial entities, only 104 have a figure for 2019 or 2020.

The situation is even more complicated for composite indicators, which are the result of a second statistical construction from raw data, which we have just seen to be of variable reliability. Therefore, the HDI (Human Development Index) depends on several factors, including the level of education, measured by the average number of years of schooling for adults aged 25 and over, and by the expected (and therefore projected) number of years of schooling for school-age children. We could similarly deconstruct GDP, another "totem" of international comparisons.

In our example, we have sought to use the crudest possible indicators, even if we have had to resort to composite indicators such as GDP per capita. By using these data, despite their inherent weaknesses developed above, we also want to affirm that these indicators remain, for the time being, the most acceptable way of describing the countries of the world according to a series of harmonized variables. They are the "least bad" tools for comparing countries. Our condition is that the use of these global indicators should be systematically combined with other indicators, measuring environmental impacts for example. Measuring a population's happiness, its footprint on the environment, does not yet give rise to the production of satisfactory global statistics: we still see maps of $CO_2$ emissions per capita based on domestic emissions, without taking into account "imported" emissions; China seems to have higher emissions per capita than Germany, without taking into account the fact that China's emissions largely fuel the consumption of countries richer than itself. Under these conditions, the choice of variables to classify the world's countries according to their level of wealth and development becomes fraught with challenges.

In our example, variables were selected by iteration, testing different combinations of variables. Eight variables were selected according to a number of criteria: in addition to their relevance to the question posed (the level of wealth and human development), these criteria mainly concerned the (relative) reliability of the producing organization, the availability of recent series and the low number of missing data (Table 1.1). The first phase of this work hence involved exploring the relationships between the variables.

| Variable name | Variable label |
|---|---|
| **Fertilr19** | Fertility rate in 2019* – Source: World Bank |
| **infmortr20** | Infant mortality rate in 2020* – Source: World Bank |
| ***gdpHab2020*** | GDP per capita in 2020 – Source: UN (national accounts) |
| **TcrGDP0020** | GDP per capita growth rate between 2000 and 2020** – Source: UN (national accounts) |
| **ConsoHholds** | Household consumption per capita in US$ in 2020* – Source: World Bank |
| **fcfHabave10yrs** | Gross fixed capital formation in US $ per capita (10-year average 2010–2020) – Source: World Bank |
| **T10B50** | Income gap between richest 10% and poorest 50% – Source: Chancel Lucas et al. (2021) |
| ***Lifexp20*** | Life expectancy at birth 2020* – Source: World Bank |
| **LogGDP2020** | Logarithm of GDP per capita in 2020 calculated from GDP per capita (gdpHab2020) |
| *Data date. For each country, the most recent date from 2014 is used. Data prior to 2014 are considered missing. **Not to be confused with annual GDP growth. This is GDP per capita growth over 20 years. | |

**Table 1.1.** *Data mobilized (blue: the data analyzed in section 1.4; italics: the other data mobilized in the exploratory phase)*

## 1.3. Exploring data and relationships with maps and charts

Analyzing a phenomenon generally requires the mobilization of several variables. This may involve analyzing the spatial distribution of an indicator constructed from several variables, as in the case of the HDI, widely used to assess the economic and human development of the world's countries. However, if we wish to examine the differentiation between countries on the basis of different perspectives, we need to call upon several variables and study them simultaneously. While the first step is obviously to map each variable separately, in order to integrate all similarities and any dissemblance between geographical distributions, this type of approach must be accompanied by statistical methods to help understand data.

Figure 1.1 illustrates the temptation to see spatial coincidences when comparing the two maps. However, this analysis is only visual and needs to be systematically confirmed. This is what the statistical visualization of this relation offers, in the form of a scatter plot (Figure 1.3(a)). By associating statistical graphs with map analysis, we can validate and confirm a spatial resemblance that is visually assessed in a more qualitative way ("countries in the south tend to be the least wealthy, with low life expectancy, while further north this is variable"). More realistically, associating maps with a statistical relationship makes it possible to specify the geography of this relationship, to which several spatial configurations may (or may not) correspond.

The aim of this section is not to go into specialized statistical treatments, which are the subject of other chapters in this book (see Chapters 3–5), but rather to introduce how statistical and cartographic representations form the basis of spatial analysis reasoning, enabling us to account for spatial forms such as gradients and regions. The multiplication of viewpoints, through various maps and numerous graphs, encourages an exploratory outlook.

This exploratory position is well known in geography and statistics (Tukey 1977), combining abduction, induction and deduction (Banos 2005). Observation enables us to identify a phenomenon through empirical exploration (induction), then to formulate hypotheses relating to the explanation of this phenomenon (abduction), and finally to formalize them at a theoretical level (deduction). In this context, we present a simple example of data exploration in an interactive environment (section 1.3.1). The identification of a relationship between two variables raises the issue of its generalization. This is the second stage of the approach (section 1.3.2) known as hypothetico-deductive. It leads to the formalization of the phenomenon (the model) and the calculation of new dimensions to be explored (i.e. any deviations from the model).

### 1.3.1. *Exploring statistical relationships and spatial organizations*

The fact that two variables show spatial coincidences visible on two maps (Figure 1.1) is interpreted as the fact that the two phenomena are connected, either directly (causal link) or indirectly (one or more common factors explain the two organizations). It may also be fortuitous (Charre 1995). The statistical graph used to represent a relationship is a Cartesian graph, called a "scatterplot", which cross-references the values of each of the variables, and the intensity of the relationship is assessed by the significance of the correlation coefficient (R, Pearson's coefficient). Figure 1.3 shows three emblematic cases of the existence or absence of a relationship: at country level, GDP per capita (in log) and life expectancy vary together and in the same direction (Figure 1.3(a)). In this case, the relationship is one of direct, multi-fold causality. Figure 1.3(b) shows a strong covariation between a mortality rate and life expectancy, one that is inverse (or negative). As the mortality rate is one of the components of life expectancy, this relationship reflects a form of redundancy. Finally, Figure 1.3(c) shows that there is no relationship between the level of GDP in 2020 and its growth rate between 2000 and 2020, illustrating the fact that the level of GDP has no influence on its growth, as economic dynamics are highly variable depending on the country's economy, irrespective of the level.

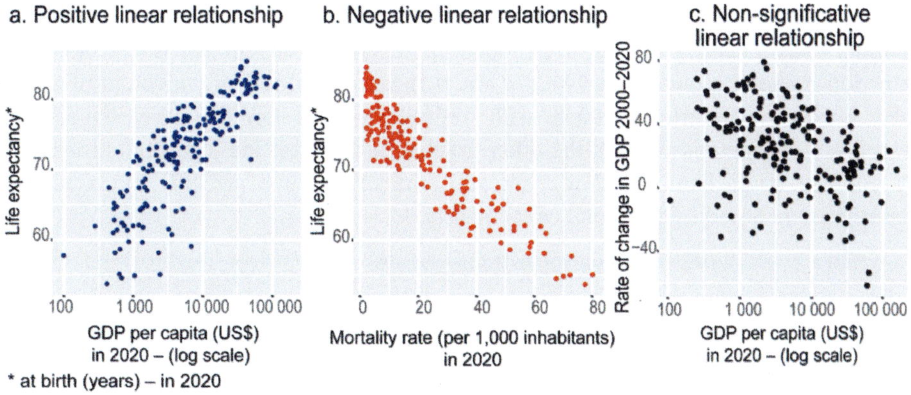


**Figure 1.3.** *Statistical linear correlations: three emblematic cases (sources: UN, World Bank, J.-B. Bouron, C. Cunty, H. Mathian)*

While these statistical graphs can be used to confirm covariation hypotheses, they cannot be dissociated from the maps. The latter reveals the spatial relationships, that is the configurations associated with this statistical relationship: a single relationship can correspond to several spatial organizations, and any spatial organization is the result of a process in which space is never neutral. Geographical proximity, contiguity, etc., explain the spatial interactions that shape the spatial configurations of the themes observed. Combining graphs and maps offers a wealth of possibilities for questioning data, stimulating questions at various levels: which countries, with equivalent and rather low GDPs, have different life expectancies?

What diversity of GDP growth rates does this group of countries of similar economic and geographical levels present?

In the trend initiated by Tukey (exploratory data analysis [EDA]), exploratory spatial data analysis (ESDA), adopted by a large number of quantitative geographers, has led to the development of geographical data exploration environments combining maps and graphs, visualizing geographical shapes and statistical relationships. By offering the possibility of interacting with graphics through selections that are visible on different cartographic representations, these environments allow for a differentiated understanding of relationships concerning space, the interrogation of data to test hypotheses, or the pushing of investigations to render unexpected observations. These environments incorporate a large number of general statistical and spatial methods for analyzing, simplifying and modeling data, thus helping to identify statistical relationships and explore their various spatialities, at different levels (Anselin 1995; Unwin 1998). Figure 1.4 shows an example of this multi-window approach, integrating maps and interacting graphics. A set of countries has been selected on the inequality map, in this case the most unequal countries (by clicking on the corresponding legend box). On the one hand, this selection makes it possible to visually assert the geographical organization of these inequalities by underlining them with yellow shading. On the other hand, this selection has repercussions on the other graphs, in this case the distribution of GDP by country (right-hand window), showing the diversity of wealth levels in the countries concerned by this level of inequality.

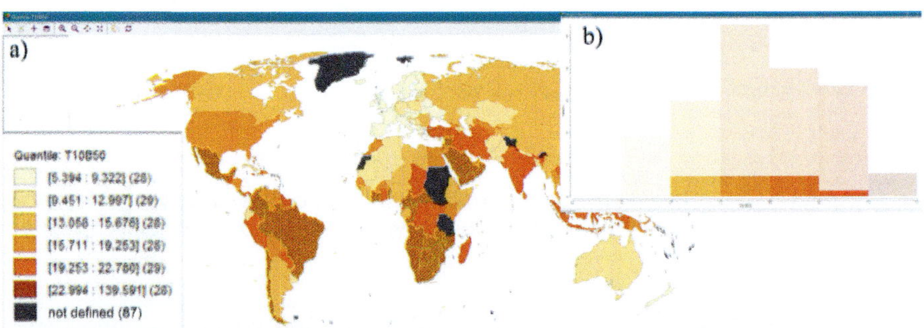


*The selection of the most unequal countries on map (a) is reflected in the GDP distribution (the part of histogram (b) representing the selected countries is highlighted).*

**Figure 1.4.** *Example of two interactive windows, dynamically linked in Geoda software (Anselin 2003)*

The selection of a set of geographically close countries could also have been selected and displayed on the map in order to visualize the specificity of this subset within the statistical distribution. These environments allow multiple windows to be opened, revealing "partial" or "regionalized" relationships. Figure 1.5 generalizes this type of approach to several windows and more than two variables. We can look at the variability of one variable, or a covariability of two variables, with the values of another fixed. Here, we select from the scatterplot the intermediate GDP values that present a wide diversity of life expectancy values (Figures 1.5(a) and (b)). Figure 1.5(c) shows the position of this subset within the distribution of a third variable (in this case, inequality), and the map shows the location of the selection (Figure 1.5(d)). In the same way, we can reason the other way round by selecting a geographical region and visualizing the positions of this subset in the statistical graphs. These statistical exploration environments make it possible to visually interrogate the associations between variables and their spatial translations, putting the map back at the heart of the analysis process, and with it, a reflection on space.

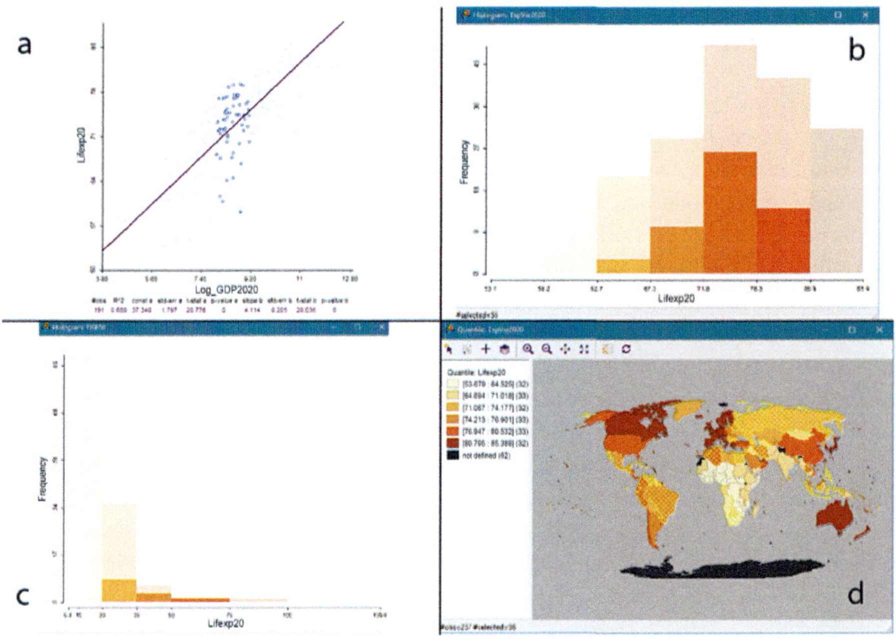


**Figure 1.5.** *Exploring relationships between several variables by selecting and visualizing positions from both a statistical and a geographical point of view*

REMARKS ON FIGURE 1.5.– *(a) Scatterplot showing GDP and life expectancy in 2020, (b) distribution graph showing life expectancy in 2020, (c) distribution graph showing the inequality indicator 2020 and (d) map showing life expectancy in 2020.*

### 1.3.2. Formalizing a relationship: from a statistical model to the map of model deviations

Moving from the stage of identifying a similarity or coincidence of spatial forms between two maps to that of identifying the relationship between the two associated variables amounts to generalizing a relationship observed across all the units studied. In our case, this translates, for example, from an empirical observation such as "the least wealthy countries have low life expectancies, while the richest have high life expectancies" to an assertion such as "knowing a country's GDP per capita, it is possible to deduce its life expectancy" (Figure 1.5(a)).

Given the *strong identity* of the entities we are working on, and the exploratory nature of our approach, a descriptive approach may suffice to account for the relationship. A *confirmatory* approach, associated with the formalization of a model (regression line), would be of interest for the purpose of evolution analysis (model parameters could be compared over time). Another interest in formalizing the relationship lies in the analysis of deviations from this model, and in particular their cartographic representation, which may reveal other explanatory factors. Here is an example of how a statistically processed graph can be used in conjunction with a map to illustrate both a relationship and that which lies outside it.

Among variables we used in this work on the division of the world, some were highly correlated, such as household consumption and GDP per capita ($R^2 = 0.86$)[3]. Identifying a strong linear correlation makes it possible to formalize a relationship (ConsoHholds = 0.47 × GDP + 1204), associate it with a quality of representativeness ($R^2$) and, depending on the case, infer this relationship on a larger population. In our case, the population (all countries in the world) is exhaustive, and the aim is not to predict consumption on the basis of GDP. However, the formalization of the model allows us to calculate the slope of the relationship (0.47), which can be interpreted as the average share of GDP in this consumption[4].

3 This means that 86% of differences in household consumption can be "statistically explained" by differences in GDP.

4 On average, household consumption (over US$ 1,204) accounts for 47% of GDP.

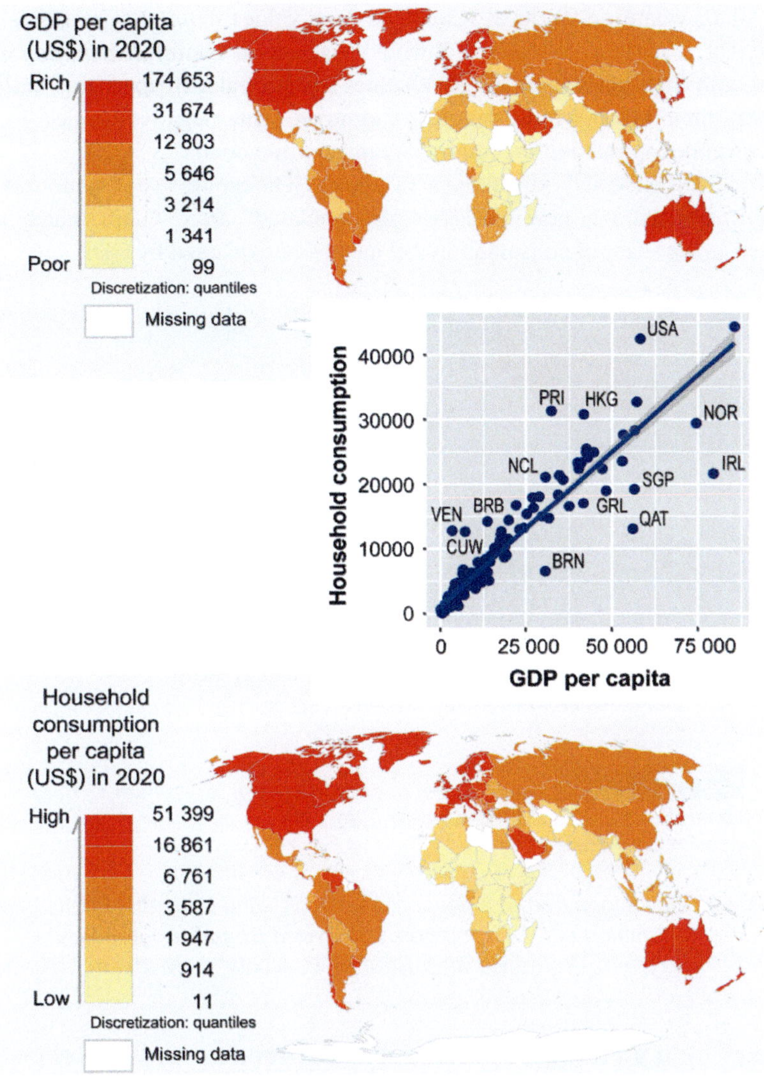


**Figure 1.6.** *GDP in 2020 and relationship with household consumption (sources: UN, World Bank Eurographics; Software: R (mapsf) J.-B. Bouron, C. Cunty, H. Mathian)*

Identifying deviations from this model is also a source of knowledge. For example, the two maps in Figure 1.6 are very similar. However, if we represent the deviations from this statistical relationship, we show a different geography (Figure 1.7): countries in the same class show the same consumption behavior relative to their GDP, whatever the level. Countries with values close to zero (light colors) are those that "consume" as their GDP would suggest, on average. More interestingly, countries with high positive or negative values (darker values) "consume" differently from what their GDP would suggest. Countries with negative values (in blue) "under-consume" in relation to their GDP/capita, while countries with positive values (in red) "over-consume" in relation to their GDP/capita. The values of these differences constitute a new variable, "de-correlated" from GDP, which can be interpreted as the share of consumption that cannot be explained by GDP.

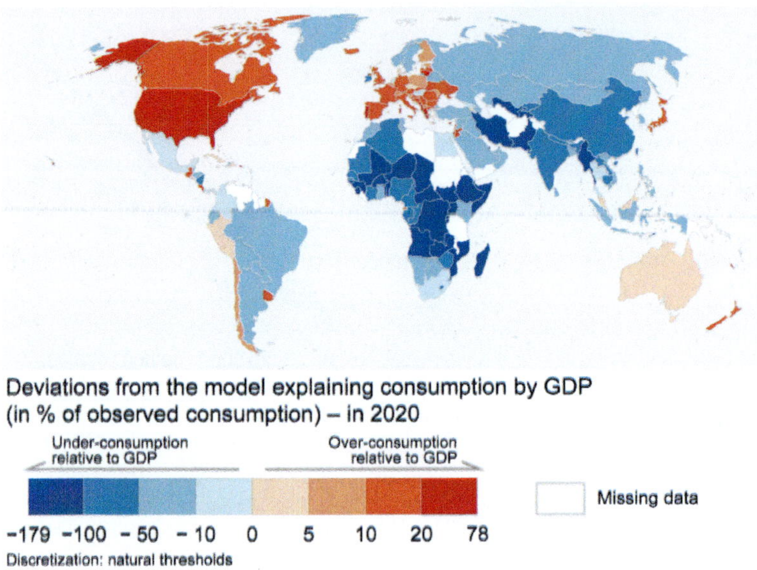


**Figure 1.7.** *Deviations from the linear model estimating household consumption as a function of GDP (sources: UN, World Bank Eurographics; software: R (mapsf) J.-B. Bouron, C. Cunty, H. Mathian)*

## 1.4. Describing statistical relationships between several variables

The dialogue between cartographic and graphical visualizations derived from statistical processing presented above enables us, through the multiplicity of viewpoints, to increase our knowledge of the geographical structuring of the phenomenon observed. We propose here to give another illustration based on a question requiring multivariate analysis.

Multivariate data processing is marked by graphics. These include Bertin's matrix experiments (1967), in which the analyst graphically reorganizes the data table, successively and iteratively permuting variables and individuals, to visually highlight the structure of the information. These are also data analysis methods, widely used by geographers since the 1970s–1980s, whose success is due to numerous graphic representations they produce: correlation graphs, factor maps, classification trees, etc. In this way, data analysis methods enable us to synthesize multivariate data, without any a priori assumptions, to identify structures of differentiation on the basis of relationships, to rank them and to categorize data. Applied to geographical data, they enable us to understand the structures of relationships between variables associated with the spatial organizations observed (Sanders 1990).

They are particularly well suited to the very general question presented here (section 1.2). Dividing up the world according to major wealth and development divides means classifying the world's countries according to a number of variables that reflect their level of wealth and development. The countries described by seven variables[5] (see Table 1.1) were then subjected to a hierarchical ascending classification (HAC). HAC is one of the so-called unsupervised classification methods, that is, structures emerge without any prior expert intervention, with the exception of upstream variable selection. The advantage of this method over others (e.g. k-means) is that it produces a classification tree (dendrogram) offering several levels of partitioning. The choice of a partition, based on this dendrogram, will be made according to various considerations: the number of classes, the proportion of information it represents, as well as our ability to interpret the differences or similarities it reveals and to distinguish them on a map. A partition into a large number of classes will bring greater precision, but the differences between classes will be more difficult to interpret.

A six-class partition was our first choice, as it offered a good compromise in terms of knowledge on differentiation factors (71%) and class structuring: too few classes reveal simple shapes; the more classes there are, the more difficult it becomes to communicate their differences and understand their distribution on a map. Once again, the result is interpreted by combining maps and graphs, that is, the statistical result (the classification tree, Figure 1.8) on which the choice of partition is based, and the associated cartographic result (Figure 1.9). By studying statistical profiles of the associated classes at each stage of the tree, we can identify major factors explaining branch separation (Figure 1.8).

---

5 The selection of the seven variables is the result of an analysis exploring two-to-two relationships and consistency choices.

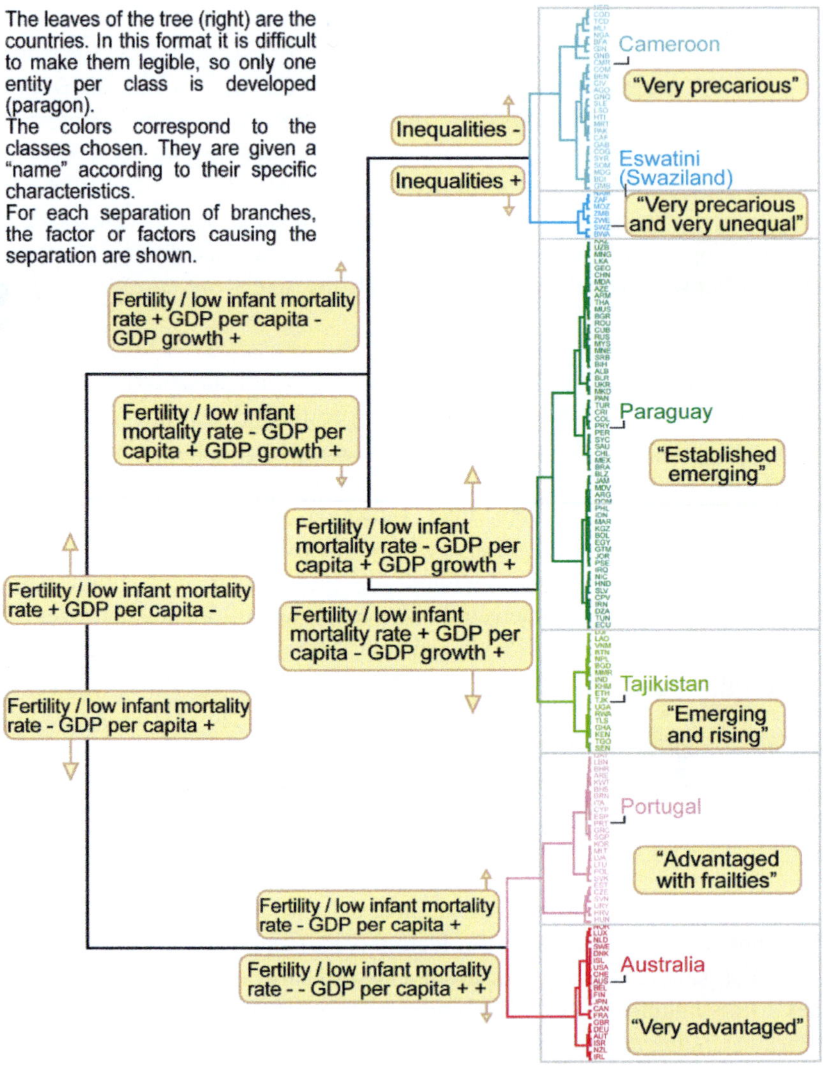


**Figure 1.8.** *Result of hierarchical classification: the dendrogram traces the structuring of classes from near to near*

REMARKS ON FIGURE 1.8.– *The classification tree is "cut" along a "cut line" running from left to right. The number of associated classes is given by the number of branches the line intersects (sources: UN, World Bank; production: H. Mathian, J.-B. Bouron, Géoconfluences, 2022).*

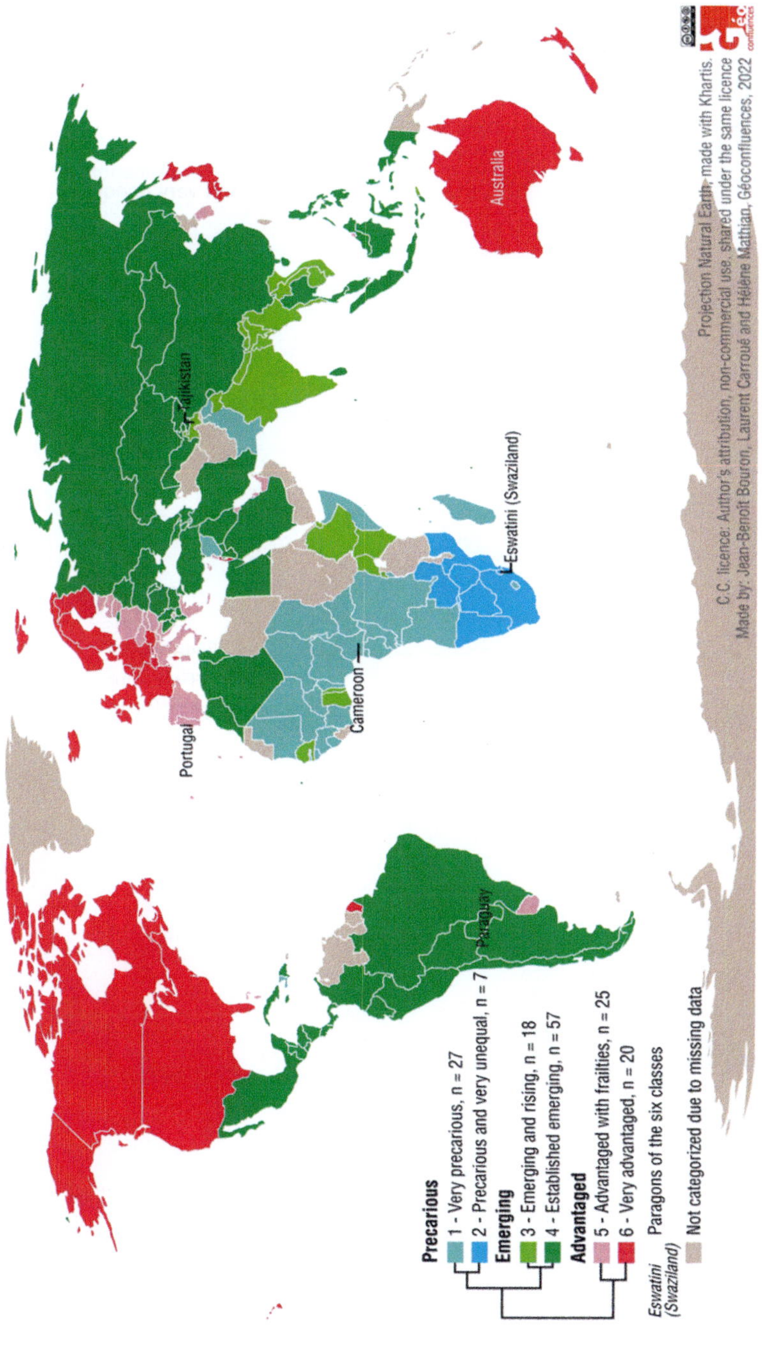


**Figure 1.9.** *World partitioning into six classes (illustrating 71% of the differences between countries)*

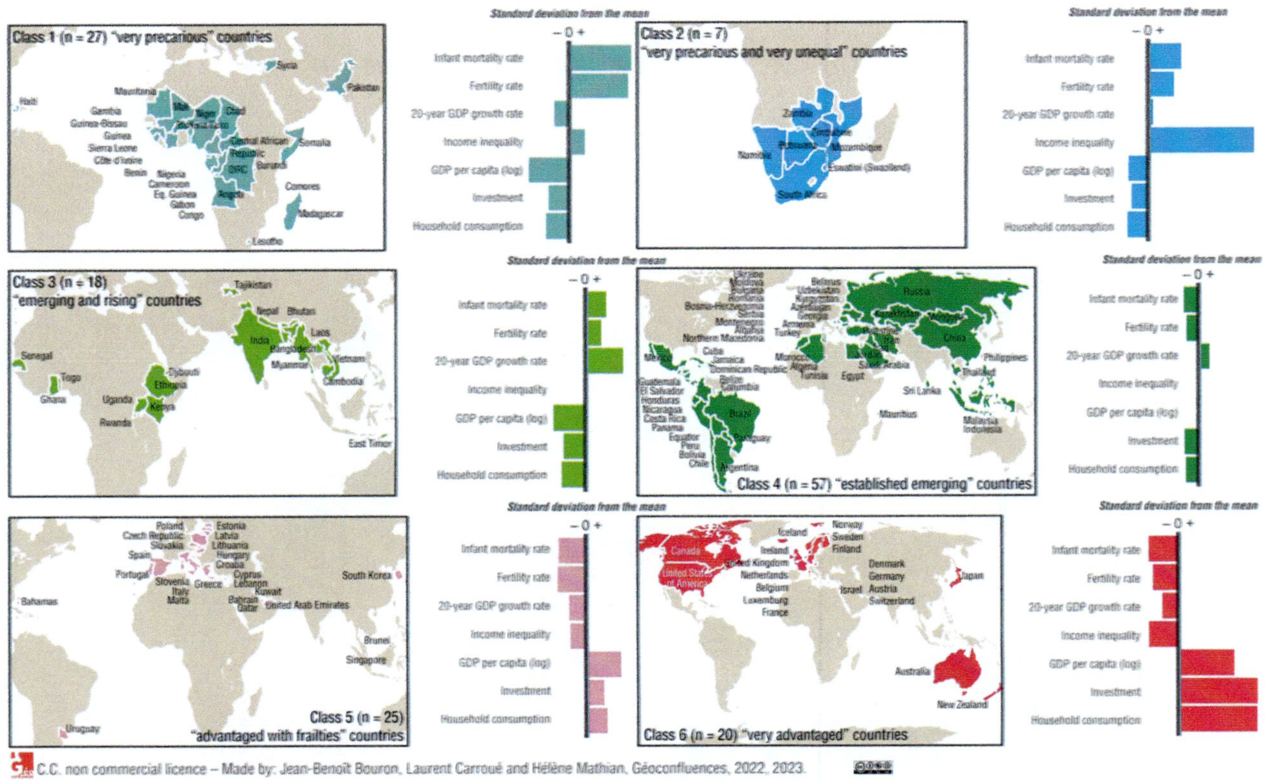


**Figure 1.10.** *Small-multiple maps of the six country classes (sources: J.-B. Bouron, L. Carroué, H. Mathian, Géoconfluences)*

Following the principle of multiple MacEachren maps, the map in Figure 1.9 can be broken down into six maps, as shown in Figure 1.10. Each map is associated with its average statistical profile relative to the seven variables employed. It is these profiles that allow us to interpret the "texture" of the class (Box 1.2). Isolating each class undoubtedly enables us to better grasp its geographical extent, and the multiplicity or otherwise of its locations. However, it is no longer possible to read the geographical proximities between classes, continuities or discontinuities, and this is what we are trying to understand by understanding social issues that may be linked to these spatial discontinuities. When reading a planisphere, we always come up against the issue of the great variability in the surface area of countries. In our case, several Class 4 countries are among the largest in the world (Russia, China, Brazil, Argentina, Mexico, Indonesia), which accentuates the visual importance of this class, reinforced by the fact that it is the one with the most countries. Conversely, the relatively small size of most of the countries in Class 5 makes it less visible than Class 6, which has five fewer countries, but some of which are very large (Canada, the United States and Australia). At first glance, you might think that all the countries in this class are in Europe, but Israel, Uruguay, Lebanon, Qatar and the United Arab Emirates are also in this class.

**Description of classes**

Analysis of bar charts (Figure 1.10), which show deviations from the mean for each variable, provides a detailed description of the classes. Based on the six average profiles, three main groups emerge. We present here only broad generalities, and refer to the article by Bouron et al. (2022) for further details:

– the first group of so-called "precarious countries": this group is made up of Classes 1 and 2, corresponding to two groups of poor countries with "very precarious countries" (27 countries in Class 1) and another class that is also precarious and characterized by high inequality (7 "very precarious countries and very unequal" in Class 2);

– a second group called "emerging" countries due to their characteristic as being in the process identified as emergent: rising per capita wealth, declining demographic markers of poverty, but continuing high levels of internal inequalities. It is made up of Classes 3 and 4, which represent 46% of the countries referenced in our study. Several economic and demographic giants are included (China, India, Russia, Brazil, Mexico, Indonesia, Malaysia, Thailand, Vietnam, Egypt, Turkey, etc.), accounting for two-thirds of the world's population and the bulk of per capita GDP growth. Class 3 is made up of 18 "emerging and rising countries". Class 4 is by far the most numerous: 57 of the 154 countries for which data are available. These are the "established emerging countries";

– a final group of "advantaged" countries is comprised of Classes 5 and 6. Class 5 comprises 25 countries that we have qualified as "advantaged with frailties" because their indicators are intermediate, between the very advantaged and the "established emerging" countries. Class 6 corresponds to the "very advantaged". These 20 countries have the lowest infant mortality rates, and the highest consumption, investment and GDP per capita values of the six groups.

**Box 1.1.** *Interpretation of classes obtained*

However, it is not uncommon to try to classify a set of entities whose differentiations tend to be organized along a continuum: in particular, when working at aggregated levels such as territories, it is rare to observe phenomena that are structured a priori into classes, which are easy to detect. The classification tree finely reveals the way in which characteristics are hierarchized to differentiate this continuum. In an exploratory approach, these differentiations can be apprehended by declining the sequence of partitions, in keeping with the idea of multiple maps. Figure 1.11 illustrates four partitions derived from the upper part of the tree. These can be read kinetically, as if a ruler was running from left to right across the tree, showing as it goes along the spatial organization revealed by the partition. At each stage, a class splits into two, linked to a new differentiation feature that refines this division, both thematically and spatially. Figure 1.11(a) shows a world split in two. What is interesting about this first partition is that it can be compared to the "North-South" boundary that still appeared on many planispheres up until the end of the 2000s. If we are to believe the seven indicators, this boundary has shifted, with Russia in the "South" and Uruguay in the "North". Admittedly, in detail, the opposition between these two groups identified by Willy Brandt in 1980 has not completely disappeared. Nevertheless, this division into two classes only reflects 39% of differentiations between countries.

In three classes (Figure 1.11(b)), one group stands out (in blue), almost all of whose members are African states, reflecting its strong regionalization. This distribution into three groups can be seen as a map of emergence: when we subtract the advantaged (in red) and the precarious (in blue), we are left with countries (in green) characterized by a rise in GDP per capita coupled with a rise in inequality, as well as lower fertility and infant mortality rate than those of poorer countries. A breakdown into three classes illustrates 56% of the differentiations. In the next step (Figure 1.11(c)), the group of advantaged countries is subdivided, isolating a sub-group (light pink) on the basis of lower economic indicators but more evenly distributed incomes. This group includes ten former socialist countries that joined EU in 2004, as well as four southern European countries weakened by recent economic crises (Portugal, Spain, Italy and Greece). Finally, the last stage (Figure 1.11(d)) reveals an additional class: a small group of Southern African countries (dark blue) stand out very clearly from the group of "poor" countries (light blue), mainly due to a very accentuated indicator: among the highest income inequalities in the world.

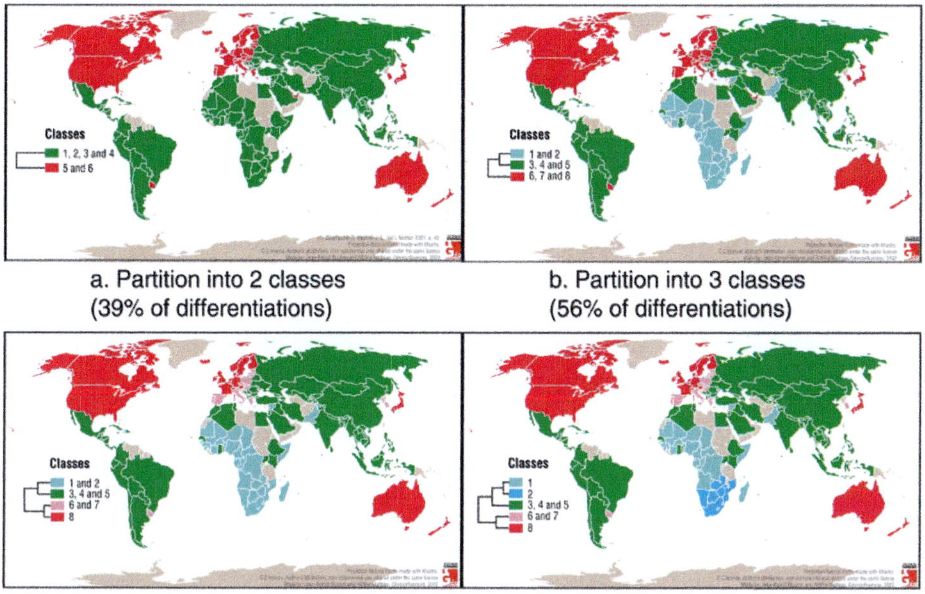


a. Partition into 2 classes (39% of differentiations)

b. Partition into 3 classes (56% of differentiations)

c. Partition into 4 classes (62% of differentiations)

d. Partition into 5 classes (67% of differentiations)

**Figure 1.11.** *Differentiation dynamics of successive splits in the hierarchical tree before the chosen partition into six classes*

## 1.5. Conclusion

This chapter could be considered as a kind of preamble to all other chapters of this book. It lays the foundations for an approach in which cartographic representations are both the product of statistical processing applied to geographic data, and key elements in a process combining exploration, confirmation and description phases. This chapter provides an opportunity to examine some of these stages through applied concrete case studies: the choice of representations and reflection on the association of geometries, the mobilization of statistical methods in an exploratory and descriptive framework and the production of indicators based on modeling. Introducing these insights by working on countries of the world places the approach in a special position, both statistically and geographically, and cartographically. Working on countries encourages us to return to the *identity* of entities, which is not the case when working on a set of sensors, or a set of pixels, or even a set of statistical census grids, such as community subdivisions.

In this approach, a map is therefore the result of numerous iterations alternating statistical processing and visualization, during which multiple choices and transformations are made, which are of course facilitated in interactive environments. However, this approach must be adapted to the geographic question at hand. In all cases, it allows us to enter into a virtuous cycle in which visualizations serve data and the understanding of the phenomenon.

## 1.6. References

Anselin, L. (1995). Local indicators of spatial association: LISA. *Geographical Analysis*, 27(2), 93–115.

Anselin L. (2003). GeoDa [Online]. Available at: https://geodacenter.github.io/.

Bahoken, F. (2022). Représenter la mondialisation par des flux, le rôle de la distance cartographique perçue. *Mappemonde*, 133 [Online]. Available at: http://journals.openedition.org/mappemonde/7558.

Banos, A. (2005). La voie de l'étonnement : favoriser l'abduction dans les Systèmes d'Information Géographique. In *Apport des SIG à la recherche*, Fotsing, J.-M. (ed.). Presses Universitaires d'Orléans, Orléans.

Béguin, M. and Pumain, D. (1994). *La représentation des données géographiques. Statistique et cartographie*. Armand Colin, Paris.

Bertin, J. (1967). *Sémiologie graphique*. Gauthier-Villars, Paris.

Bouron, J.-B., Carroué, L., Mathian, H. (2022). Représenter et découper le monde : dépasser la limite Nord-Sud pour penser les inégalités de richesse et le développement. *Géoconfluences* [Online]. Available at: https://geoconfluences.ens-lyon.fr/informations-scientifiques/dossiers-thematiques/inegalites/articles/rdecoupage-economique-mondial.

Brandt, W. (1980). *Nord-Sud : un programme de survie. Rapport de la Commission indépendante sur les problèmes de développement international*. Gallimard, Paris.

Brunet, R. and Dollfus, O. (1990). *Mondes nouveaux. Géographie universelle*. Hachette/Reclus, Paris/Montpellier.

Capdepuy, V. (2011). Le monde de l'ONU. Réflexions sur une carte-drapeau. *M@ppemonde*, 102 [Online]. Available at: http://mappemonde-archive.mgm.fr/num30/mois/moi11201.html.

Capdepuy, V. (2013). Le monde en étoile : genèse d'une projection cartographique. *Ma@ppemonde*, 111 [Online]. Available at: http://mappemonde-archive.mgm.fr/num39/lieux/lieux13301.html.

Cauvin, C., Escobar, F., Serradj, A. (2010a). *Thematic Cartography 1. Thematic Cartography and Transformations.* ISTE Ltd, London and John Wiley & Sons, New York.

Cauvin, C., Escobar, F., Serradj, A. (2010b). *Thematic Cartography 2. Cartography and the Impact of the Quantitative Revolution.* ISTE Ltd, London and John Wiley & Sons, New York.

Chancel, L., Piketty, T., Saez, E., Zucman, G. (eds) (2021). Rapport sur les inégalités mondiales 2022. Report, Word Inequality Lab.

Charre, J. (1995). *Statistique et territoire*. RECLUS, Montpellier.

Dumont, G.-F. (2018). Une exception française : son recensement de la population. Quelle méthode ? Quelles insuffisances ? Comment l'améliorer ? *Les Analyses de Populations & Avenir*, 3, 1–226.

Genevois, S. (2018). La projection Equal Earth, un bon compromis ? [Online]. Available at: https://cartonumerique.blogspot.com/2018/11/la-projection-equal-earth.html.

Grataloup, C. (2007). *Géohistoire de la mondialisation. Le temps long du monde.* Armand Colin, Paris.

Lambert, N. and Zanin, C. (2016). *Manuel de cartographie. Principes, méthodes, applications.* Armand Colin, Paris.

Lévy, J. and Lussault, M. (2003). *Dictionnaire de la géographie et de l'espace des sociétés.* Belin, Paris.

Monmonnier, M. (2019). *Comment faire mentir les cartes*. Éditions Autrement, Paris.

Morel, J. (2021). *Les cartes en question. Petit guide pour apprendre à lire et interpréter les cartes*. Éditions Autrement, Paris.

Openshaw, S. (1983). *The Modifiable Areal Unit Problem*. Geo Books, Norwick.

Pumain, D. (2008). Maillage [Online]. Available at: https://hypergeo.eu/maillage/.

Sanders, L. (1990). *L'analyse des données appliquée à la géographie*. RECLUS, Montpellier.

Šavrič, B., Patterson, T., Jenny, B. (2019). The Equal Earth map projection. *International Journal of Geographical Information Science*, 33, 454–465.

Tukey, J. (1977). *Exploratory Data Analysis*. Addison-Wesley, Reading.

Unwin, A. (1998). Exploratory spatial data analysis with local statistics. *The Statistician*, 47(3), 415–421.

# 2

# Heterogeneous Data Integration and Geoweb Cartographic Representations

**Marianne GUÉROIS[1] and Malika MADELIN[2]**

[1] *Géographie-cités, RIATE, Université Paris Cité, France*

[2] *PRODIG, Université Paris Cité, France*

## 2.1. Introduction

With the digital revolution, geographic data abound in many fields, contributing to a "data deluge" (Kitchin 2014a; Singleton and Arribas-Bel 2021). The proliferation of these "new" localized data, whether from sensors and technical transmitters, web services platforms or participatory approaches, has led to the production of original cartographic representations, often disseminated via visualization interfaces, which play an essential role in the exploration of online information. In addition to the many studies that question the access to these data or the nature of the cartographic innovation triggered by their emergence, a major issue, less often addressed, lies in their interoperability with so-called traditional or conventional data, such as census data. Such integration is essential to giving meaning to the new data, whether by assessing their quality, enriching their thematic attributes or refining the analysis of their spatial distribution by linking them to contextual variables. These are a prerequisite for the production of maps that are

*Handling and Mapping Geographic Information*,
coordinated by Claire CUNTY and Hélène MATHIAN. 

relevant to the issues at hand. However, although these combinations are unprecedented in terms of the type of the natively digital information it brings into play, it is an extension of earlier lines of enquiries.

Indeed, the notion of heterogeneous data integration, broadly defined as the process of making data compatible from different sources, was developed as early as the 1990s. It paved the way for a succession of technical and regulatory innovations that contributed to the creation of geographic information infrastructures and encouraged the overlay of heterogeneous datasets (Mustière 2014). It makes it possible, for example, to view cadastral parcels on an aerial photograph, the Corine Land Cover land-use database, the old Cassini map, etc., on the *Géoportail* website (portal of France's Institut géographique national). The digital maps by Geoweb[1] generation thus complete the progress made throughout this process of technical integration, allowing the aggregation of multiple information and web content, even if users often have little freedom on the parametrization of geovisualizations (Mericskay 2021).

In this chapter, we take as our starting point the multiplication and diversification of data resulting from the digital revolution, and ask how these transformations contribute to renewing the challenges of heterogeneous data integration: Which aspects of this integration have progressed the most with the advent of the Geoweb? In what way does the fine spatiotemporal resolution of the new data offer unprecedented opportunities for integrating data from different sources?

To answer these questions, we first need to look at what characterizes new data, among the many definitions and categorizations that have been proposed in recent years. In particular, to what extent are access, direct visualization and exploration of spatial distribution facilitated by the distribution of reconfigured cartographic representations on a multitude of web platforms?

Second, we will illustrate the need for interoperability between new and conventional data, using several thematic examples and diverse data from the commercial service platform *Airbnb*, the *OpenStreetMap* collaborative mapping project and *Netatmo* weather stations (see Table 2.1). In particular, we shall examine how older works on the coupling of heterogeneous data can resolve the main obstacles to such integration, and what their cartographic representation brings to the geographic analysis of new data.

---

1 Geoweb: according to Haklay et al. (2008), geographic applications accessible via the Web.

| | How geographic information is produced | Objectives according to audience | Distribution and access | Spatial, attribute, and temporal components |
|---|---|---|---|---|
| ***Airbnb* rentals**<br><br>Short-term real estate rentals | Entry by hosts and by renters and landlords | *Airbnb:* Facilitating the host-renter relationship<br>General public: individual digital service<br>Scientists: analysis of the transformation of the urban real estate market, etc. | Direct access to visualization (website)<br>Paid access to data: *AirDNA* for any territory<br>Free access to webscrapped data by *InsideAirbnb* (sample of central cities) | Point data (spatial resolution to within 100 m)<br>Attributes: description, price, availability, etc.<br>*Airbnb*: near-real-time updates; *AirDNA:* monthly aggregations; *InsideAirbnb*: monthly repeatability |
| ***OpenStreetMap* equipment**<br><br>Commercial service facilities, etc. | Entry by contributors | *OpenStreetMap:* building a cartographic commons<br>General public: source for various online applications<br>Scientists: access to equipment, etc. | Direct access to visualization (website)<br>Free access to contributory database. facilitated by API[(2)], software plug-ins, automatic archiving by intermediaries (e.g. Geofabrik) | Point and/or polygon data<br>Numerous attributes. *key:amenity,* whose value describes the type of equipment.<br>Almost real-time or daily updating (archiving) |
| ***Netatmo* stations**<br><br>Fixed measurements of weather data | Automatic transmission to a server via Wi-Fi | *Netatmo:* selling sensors<br>General public: visualization, comparison of weather data<br>Scientists: analysis of urban heat island, precipitation, etc. | Cost of one station (~170 €)<br>Direct access to visualization (website)<br>Free access to recent data, facilitated by an API[(2)] | Point data with geographic localization via Wi-Fi, smartphones or declarative<br>Attributes: timestamp, temperature, humidity, pressure, etc.<br>Update ~ 5 min. |
| ***AirCasting* sensors**<br><br>Mobile measurement of air pollution in particles | Automatic sending to a server via smartphone | *Habitatmap* behind *AirCasting:* NOG[(1)] in environmental justice<br>General public: measuring air pollution on a route<br>Scientists: analysis of pollution according to city morphology, etc. | Sensor cost (~250 €)<br>Direct access to visualization (website)<br>Free access to database via website (archiving) or API[(2)] | Point data with geographic localization via smartphone<br>Attributes: timestamp, particle concentration, temperature, humidity<br>Recordings every second, during a session |
| ***Sensor. Community* sensors**<br><br>Fixed measurements of particulate air pollution | Automatic transmission to a server via Wi-Fi | *Sensor.Community:* open source participatory science project<br>General public: measuring air pollution at home<br>Scientists: analysis of spatio-temporal variability, etc. | Sensor cost (~50 €)<br>Direct access to visualization (website)<br>Free access to database via website (archiving) or API[(2)] | Point data with declarative geographic location<br>Attributes: timestamp, particle concentration, temperature, humidity<br>Update ~ 2 min. |

(1) Non-governmental organization; (2) Application Programming Interface

**Table 2.1.** *Main characteristics of the datasets used in this chapter*

## 2.2. New data: from production to visualization

Search on your smartphone for restaurants around your location, measure your exposure to air pollution on your daily commute, visualize your itinerary on a dedicated map app, etc. These new applications of geographic information are made possible by the fairly recent and rapid development of digital data production, management and processing capacities. They have contributed to the proliferation of data, referred to here as "new" (see Box 2.1), particularly spatially referenced data. First, we shall present these new data, their diversity and the plurality of their producers. Then we shall revisit their innovative character for territorial analysis, before questioning the ways in which web applications for cartographic visualization orient their uses.

> There are many ways to describe the new, natively digital, data that have emerged in the information age. The first and most frequently used term is *big data*, about which there is an abundance of literature, even if the limits remain unclear. Many authors take up Laney's 2001 definition of the "3 Vs" (cited, for example, by Kitchin 2014a): beyond their *Volume*, big data are *Varied* in their formats and structuring, and characterized by a *Velocity* of capture. Others add *Value* for the ability to generate profit, and *Veracity* to express their inherent unreliability and the level of accuracy that varies according to sources (Bastin and Tubaro 2018). Kitchin (2014a) proposes to complete this definition by adding other characteristics: the completeness of the data, their fine resolution and their flexibility, enabling them to be easily modified and extended. These massive data are often defined in opposition to *small data*, still referred to as scarce data (Miller and Goodchild 2015), which are low volume and produced from samples, mainly by institutional bodies and researchers (Kitchin 2014a). For Duroudier et al. (2020), one of the "main methodological challenges today is to hybridize traditional small data methods providing rich semantic information, with big data methods enabling more massive and longer collections" (p. 128, authors' translation). For scientific purposes, the new data are also referred to as *unconventional* to emphasize their opposition to so-called traditional, referenced data, which come mainly from statistical offices and official agencies. These were often constructed for commercial purposes and were neither designed nor documented with scientific measurement and analysis in mind. This opportunistic and not originally intended for such use is echoed by Arribas-Bel (2014) in his definition of *emerging data*: (1) they are accidental for scientists; (2) moreover, they offer very broad thematic coverage, due to the growing ubiquity of the services or technologies from which they derive; (3) finally, they would be distinguished by their degree of openness (legal and financial access), which represents a debatable point, due to a sometimes very high technical and financial entry cost.

**Box 2.1.** *A variety of terms to describe new data*

### 2.2.1. *New data production methods and new scientific applications*

The explosion of new data is often presented in terms of major thematic areas and in relation to the technical objects that support them (Li et al. 2016): mobile data, linked to global positioning systems (GPS), smart card ticketing systems and social networks, service platforms and volunteered geographic information (VGI). To this, we can add a category corresponding to images, whether high-resolution satellite images, photos/videos from social networks or video surveillance. The development of these technological innovations and the Internet, as well as the miniaturization of acquisition devices, have facilitated the accessibility and automation of data collection and production, often with fine spatial and even temporal resolution. Linked to these new modes of acquisition, data are increasingly centered on the individual, who leaves digital footprints of their movements and activities on networks and becomes a producer, voluntarily or unwittingly, of geographical information: they will report an object abandoned in the street with a smartphone, quantify their exposure to air pollution with a personal sensor, give their opinion of a restaurant, or a robot will share their location on a social network. Kitchin (2014a) emphasizes the differences generated by the modes of production of new data, whether this production is directed, that is, operated by a human being, automated (automated surveillance, sensors) or volunteered (citizen science, social media, etc.).

The fields of applications, while varied, are often illustrated by examples from the “smart city” and its management (vehicle fleets, lighting, etc.), self-quantification and individual digital services (physical activities, home automation, connected vehicles, real estate rentals, etc.), environmental monitoring (animals equipped with transmitters, plant species location, etc.), disaster and emergency management (tweets, drone images, etc.) or even the construction of commons such as the *OpenStreetMap* database (www.openstreetmap.org) or various citizen science projects (Madelin 2020). This panorama testifies to the diversity of uses and actors involved, even if the processes of collecting, storing and accessing new data are very often managed by private providers and even by digital giants and oriented toward private and commercial interests (Kitchin 2014a; Thatcher et al. 2016), whereas the producers of conventional databases are traditionally from the public sector.

To illustrate the diversity of possible uses for these data, let us take the example of *Netatmo*’s connected personal weather stations (see Table 2.1). These stations, purchased by private individuals, measure air temperature and humidity and automatically send the data to a server via Wi-Fi every 5 min. The company then provides a mapping platform to visualize recent data (see Figure 2.1; www.weathermap.netatmo.com/). The company’s aim is not to sell data, but to sell sensors, among other things, via this cartographic display. It also enables weather station owners to view their own data and that of neighboring stations. Finally, timestamped and localized meteorological measurements can also be used, in a scientific approach, to study the urban heat island[2], its spatial delimitation and temporal variability (Madelin and Dupuis 2020).

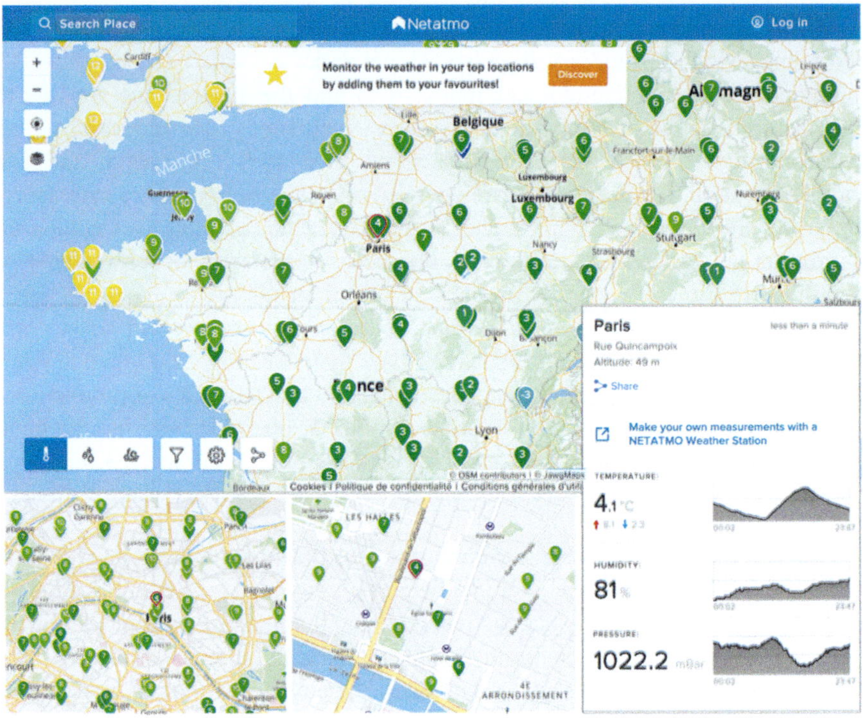


**Figure 2.1.** *Netatmo visualization platform for individual weather station data (www.weathermap.netatmo.com), with multiple zooms and details for one station (circled in fuchsia on the maps)*

This access to data is not always so easy. Short-term rental company *Airbnb* offers a web service platform (www.Airbnb.com), whose information (attributes describing the rental, price, availability, etc.) is entered by individuals (hosts, renters) and updated in near real time (see Table 2.1). Motivated by the search for a rental in a given location and on a given date, a person will visualize the result of

2 The urban heat island translates into night-time air temperatures that are generally higher in urban city centers than in the surrounding countryside areas, especially on nights with a clear sky and calm wind.

their search on a map of rentals represented by points, with a resolution to within 100 m, and their associated information. However, it is more difficult for local authorities wishing to quantify the phenomenon of short-term rentals on their territory, or for researchers wishing to analyze the repercussions on the evolution of the real estate market, to obtain the data. These stakeholders can turn to the company AirDNA (www.airdna.co), which sells data on a monthly aggregated basis for any given territory, or to the InsideAirbnb website (www.insideAirbnb.com, see Figure 2.8), which provides free access to web scraped data for several cities on an almost monthly basis. Generally speaking, while public data are increasingly Open Data, disseminated and freely reusable, new data from the private sector are difficult to share, or are shared via application programming interfaces (APIs), which often require a significant technical input cost, and whose terms and conditions are defined by the disseminator, sometimes in a non-transparent way. One example is X's (formerly Twitter's) API, which only gives access to a selection of tweets.

Unlike traditional modes of production, where data collection is primarily for scientific purposes and aimed at producing knowledge and statistics, the acquisition of new data is often oriented toward meeting operational demands (transportation management, targeted advertising, etc.). Scientists then use these data opportunistically, without any precise knowledge of the production and collection process, or of the selections and modifications made to the data. In this way, data can be "diverted" for scientific purposes.

### 2.2.2. Data with high spatial and temporal resolution but poorer attributes and quality?

Beyond the abundance of ideas on these new data, a certain consensus is emerging on their very high spatial and temporal resolution, which constitutes a richness for most authors (Li et al. 2016; Ferreira and Vale 2020). This creation of geographic information very often relies on geolocation and timestamping systems (GPS, telephone relays, Wi-Fi terminals, etc.): the vast majority of these new data are therefore point-based spatiotemporal data, whose precision and accuracy depend upon the sensor used and the reception of the signal (known problems for GPS in relation to the masks of relief, buildings, etc., and with atmospheric conditions). As an illustration, let us take the example of *AirCasting*'s air pollution sensors (see Table 2.1), which can be used to measure fine particulate matter concentrations while on the move. As the sensor is connected to a smartphone, a session associates a timestamp and a geographical position with each air pollution measurement, at a time resolution of 1 s. The data are then shared and made available on a mapping platform (www.aircasting.habitatmap.org; see Figure 2.2). They can be used to locate potential sources of pollution and to analyze spatiotemporal variability as a function of street width, orientation and so on. In some cases, however, this high spatial resolution is called into question due to the greater semantic poverty and lower quality of the information.

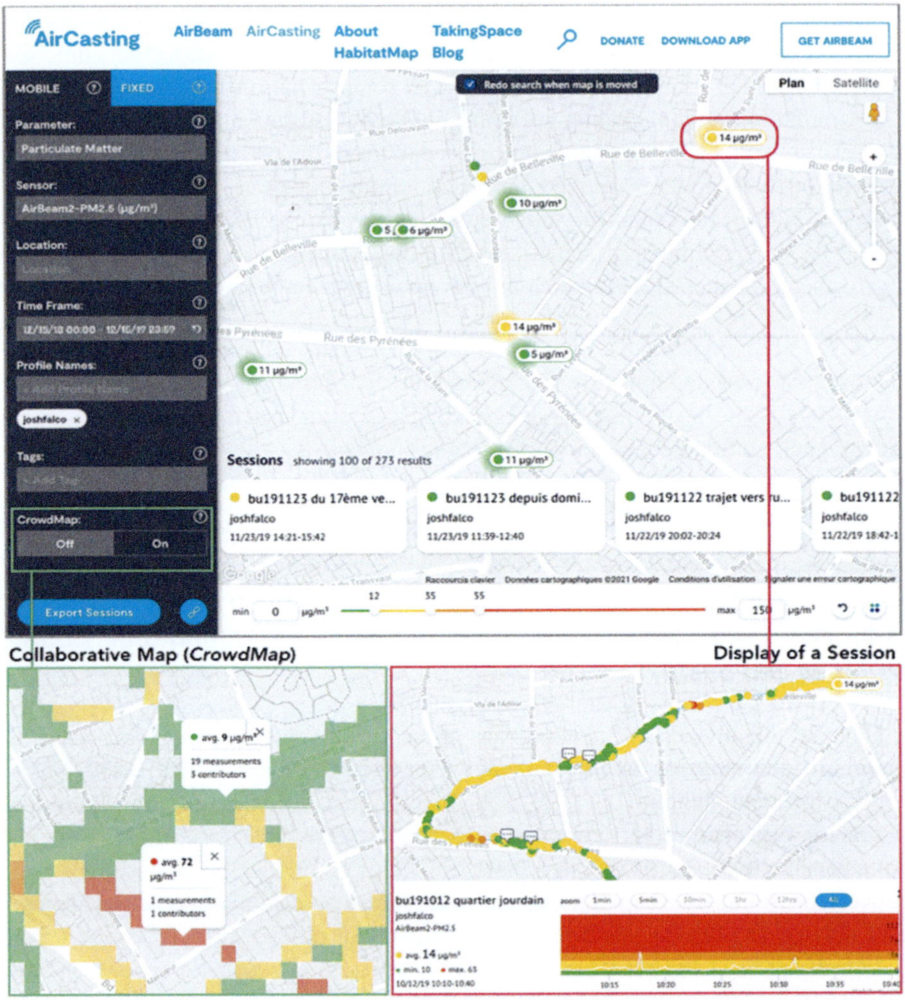


**Figure 2.2.** *AirCasting data visualization platform (www.aircasting.habitatmap.org) for mobile air pollution sensors, with below an extract of the collaborative map and details for one session*

#### 2.2.2.1. *High spatial and temporal resolution*

From a spatial point of view, access to increasingly disaggregated, often individual[3], information is frequently enhanced. For example, is the air temperature in Paris better understood from a reference weather station located in an urban park (in this case, Montsouris), or from 1,500 individual *Netatmo* stations, positioned in a wide variety of environments, but not standardized, thus requiring a solid critical analysis of the data? Is not a person's individual exposure to air pollution best estimated from individual sensors measuring fine particle concentrations in the various microenvironments they travel through, once the data have been validated?

It is equally important to mention the richness of this information from a temporal point of view. To take the example of mobile pollution sensors (*AirCasting*), the sessions record several pieces of data in a single minute, whereas the data from reference air quality monitoring stations or the resulting models go down to an hourly basis at most. Rental availability data on the *Airbnb* platform or contributions on *OpenStreetMap* are updated in near real time. However, this temporal dimension is often tricky to manipulate. On the one hand, it is rarely directly delivered as an attribute in itself, within these new data (which would be processed by aggregating detailed time-series from frequent recordings). It often appears in the form of metadata, which is not always easy to interpret: in the case of *OpenStreetMap*, there is a field (tag) providing information on the creation, modification or suppression of registered elements, but this information is not always well informed in practice. On web service platforms, it can also be difficult to distinguish between the different temporalities of services: for example, in the case of information about *Airbnb* rentals, it is difficult to distinguish between the calendar of availability of accommodation, that of actual rentals, and that of comments by visitors a posteriori. On the other hand, access to temporal depth, essential for studying the evolution of a phenomenon and/or territory over time, requires organizing regular recording of versions, which evolve continuously on web service platforms or in contributory databases, managed, in some cases, by intermediary actors. For example, the *InsideAirbnb* website archives files on "scraped" *Airbnb* rentals for a growing number of metropolitan municipalities, with

---

3 Individual data are considered here in the narrow sense of an elementary entity, referring not only to a person, but also to any material or immaterial observation considered at an elementary, indivisible level of information (a household, a piece of equipment, a measurement of pollutant concentration, etc.).

a frequency initially annual and now monthly. No doubt due to the volume of data generated, this archiving is now limited in time. *AirDNA* sells retrospective data, enabling us to observe the spatiotemporal spread of *Airbnb* rentals, for example from central Paris to municipalities in the inner suburbs (*Institut Paris Région 2021*). The *Geofabrik* website (www.geofabrik.de) facilitates access to *OpenStreetMap* data, not only by making them freely available via files broken down according to geographic sub-areas, with weekly updates, but also by archiving older extractions (up to 2014 for French regions). However, there is no guarantee that this archiving will last.

#### 2.2.2.2. *Quality: semantic variability and spatial representativeness*

As a counterbalance to this richness in spatial and temporal resolution, a second characteristic is that of quality, which in many respects is inferior to that of conventional data, and calls into question the use of such data for scientific purposes. Behind this frequent criticism, at least two aspects are often mentioned and can be distinguished.

First and foremost, it is the lack of standardization of information and the poverty of metadata that characterizes them, due to their frequently "accidental" nature (Arribas-Bel 2014), or at least not oriented a priori toward a scientific purpose.

In the example of the connected weather station, a person may choose to monitor the humidity and temperature of the cellar, or place the station on a windowsill, on the fifth floor, in full sunlight. The conditions under which these measurements are acquired are far removed from the normative conventions of meteorological measurements (2 m away, in a weather shelter, far from heat sources and buildings, etc.), which then makes it difficult to compare these data in a scientific approach and requires significant data pre-processing work (Madelin and Dupuis 2020). Over and above this acquisition of non-standardized data, the difficulty of using new data is mainly due to the lack of information on their production, and more generally of metadata, which then requires recourse to methods for verifying and enriching the data (see section 2.3). The position of a weather station in the sun can be inferred from data analysis (Madelin and Dupuis 2020), the mode of transport for GPS traces from speed, or the motivation for a trip from coupling with the characteristics of activity locations (Chen et al. 2016).

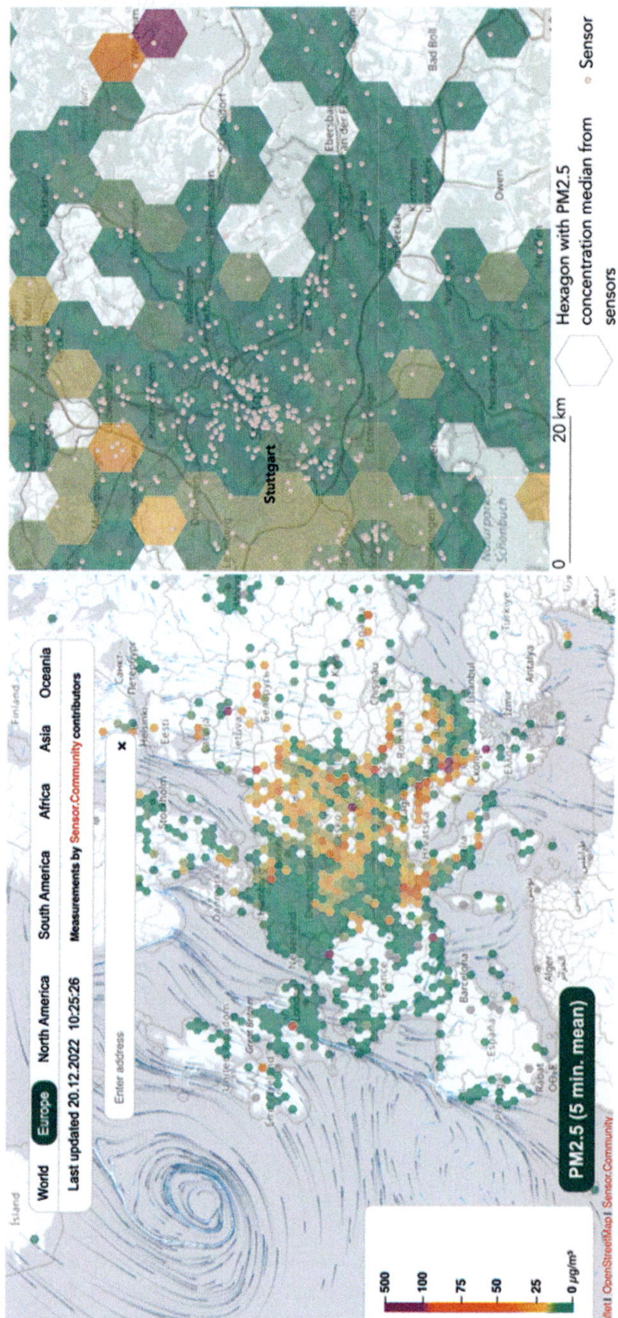


**Figure 2.3.** *Sensor.Community visualization platform for air pollution sensor data (www.sensor.community) and zoom in on the Stuttgart region, adding geolocated sensor points*

Second, quality is affected by problems of data completeness, linked either to incomplete information on attributes[4], or to the sampling of observed elements: "gaps" in the contributions of *OpenStreetMap* points of interest (POIs) due to the urban location of the main contributors, a lack of equipment of the Seine-Saint-Denis territory with individual weather stations (unlike Hauts-de-Seine), linked to the price of this connected object. More broadly, volunteered information depends on people's level of equipment, age, income and level of education. In time and space, these new data are not necessarily exhaustive or representative (Boyd and Crawford 2012), despite their bigness and variety. Small data are produced on the basis of a representative sample of the underlying survey and their purpose, whereas the automation of opportunistic data collection sometimes produces an overabundance of data, which then proves to be of little relevance. For example, *Sensor.Community*'s 14,000 fixed connected sensors (see Table 2.1) transmit a pollutant concentration every 2–3 minutes, generating a significant amount of data for daily and monthly archiving (in December 2021, over 15 billion pieces of data have been accumulated since the start of the project). In particularly well-equipped areas (such as Germany, see Figure 2.3), these fixed measurements are often redundant over time, given the low variation of pollution sources. That said, this abundance of data makes it possible to identify outliers and rule out unreliable sensors.

Beyond the appropriateness of using these new spatial data, epistemological questions are driving the geography community, like other humanities and social sciences: Is science data-driven, the "datafication" of a society where knowledge production comes more from data than theory (Kitchin 2014b; Miller and Goodchild 2015)? Is there a position, a territory for geographers to defend around Geographic Data Science (Singleton and Arribas-Bel 2021)? Are they echoes of old debates on quantitative geography and the place of the digital, on the *S* for *Science* in GIS and not just *System* (Ferreira and Vale 2020), or questioning the neutrality of data, with the development of the critical approach to data (Boyd and Crawford 2012; Thatcher et al. 2016)? Whatever the case, the data are out there, and scientists have plenty of reasons to seize them, even if they are not intended for them, even if it is only to point out their shortcomings.

---

4 These criticisms apply less to data from certain service platforms: for example, *Airbnb* hosts will often try to fill in the fields describing their rental property.

### 2.2.3. *Visualizing new data on Web platforms*

The profusion of new geographic data has led to the proliferation of web-based data visualization interfaces in the form of dashboards, interactive maps and charts. The map reconfigured in the Geoweb environment then plays an essential role in data enhancement, sometimes in participatory data collection devices, in data access and understanding, generating new challenges for cartographers (Robinson et al. 2017), in particular due to the volume of data. Cartographic interfaces enable users to "navigate through a universe of data" (Joliveau 2011) and often occupy a central place on websites, as evidenced by the display of *Sensor.Community*'s map of air pollution sensors right at the top of the site's home page, the existence of a page dedicated to the cartographic representation of *Netatmo*'s weather data, or the visual result of an *Airbnb* rental query. In addition to their role as advertising showcases, territorial marketing tools, trends, or even as a federator of a community of contributors, cartographic web platforms enable direct visualization and basic exploration of a dataset, sometimes downloading and, more rarely, observing the spatial distribution of a phenomenon. For example, the City of Paris Open Data portal (www.opendata.paris.fr) offers a "Create a Map" tool to describe a spatial dataset, with the option of previewing the data before downloading, or looking at the spatial distribution using a number of functions (see Figures 2.4A and B). Let us take the example of the *Dans ma rue* database, which contains over 875,000 spot reports of anomalies in public spaces (January 2022), entered by contributors via the *Dans ma rue* application that ran for 13 months.

The individual nature of data, combined with new geolocation tools, results in visual representations dominated by the point layout mode, well suited to the spatialization of inventories. If the aim is not only to locate a phenomenon, but also to read the values, classify them, group them and compare them, these point pattern maps are not without problems of readability, due to the overload of information[5]. However, compared with paper maps, Geoweb visualizations can provide added value in most cases. On the one hand, they allow access to information on a case-by-case basis, by selecting each entity and reading the associated tooltips or information windows, making use of a medium that is close to inventory maps. On the other hand, they can be used to increase navigation possibilities at different spatial and temporal levels of observation by aggregating or selecting information.

5 The multitude of points in small spaces can, conversely, be used to propose original maps, such as those of the language of geolocated tweets in New York City: www.ny.spatial.ly/.

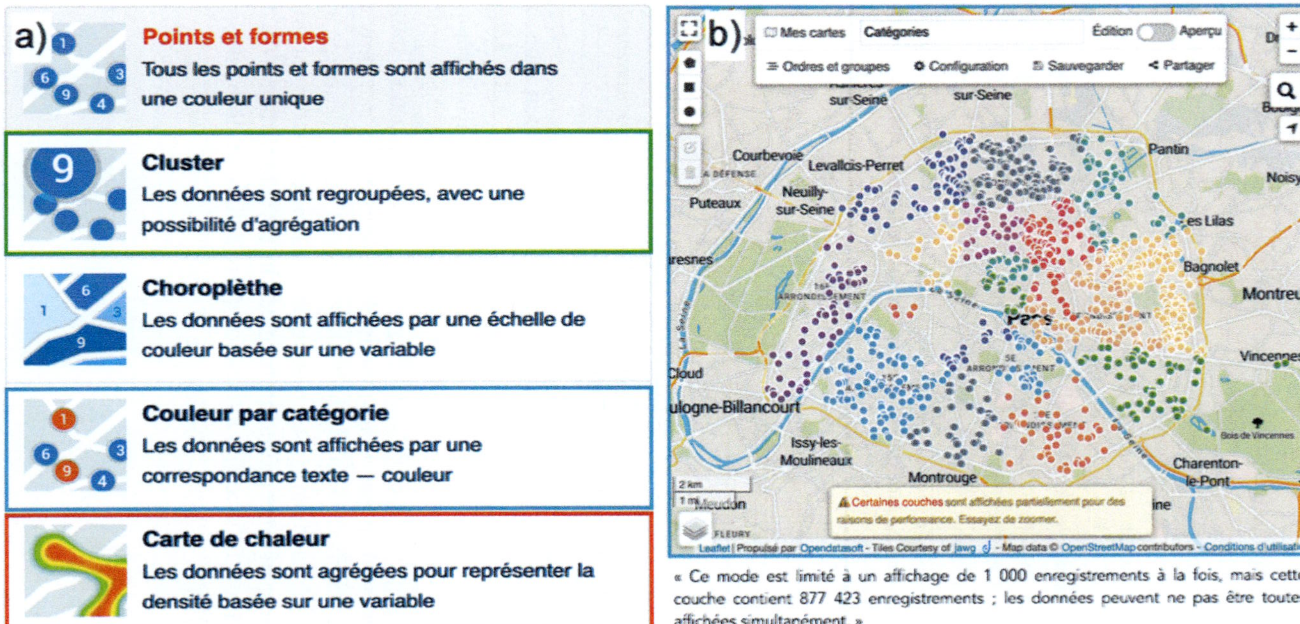


**Figure 2.4A.** *Several ways of displaying the Dans ma rue database, proposed by the tool "Create a map" on the City of Paris Open Data portal (www.opendata.paris.fr)*

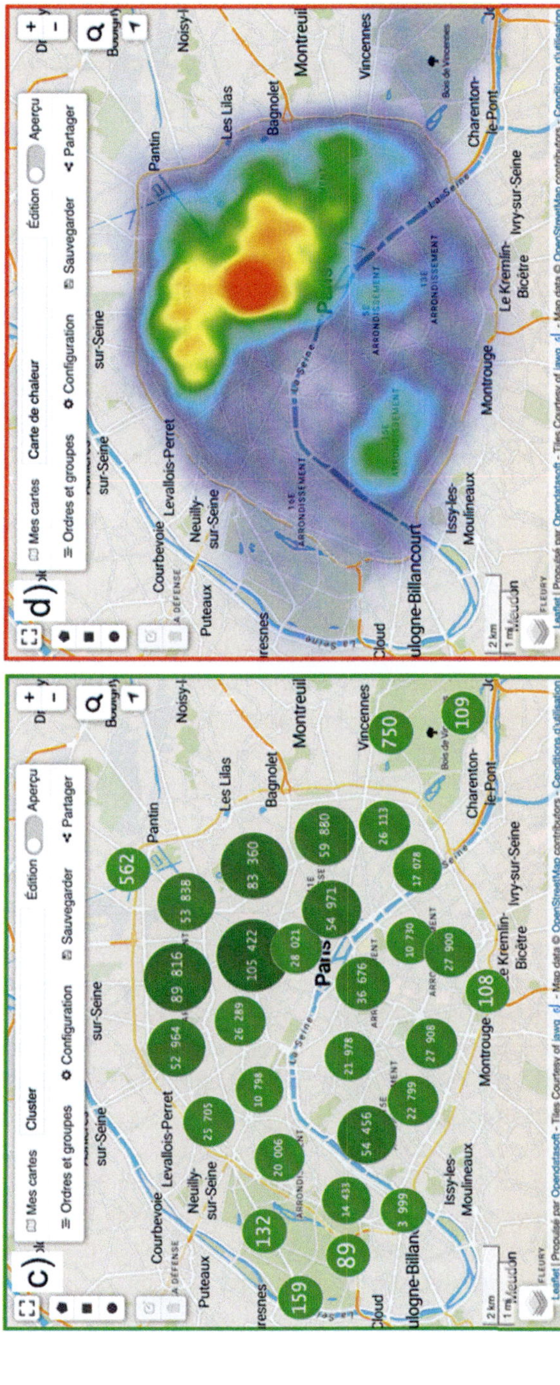


**Figure 2.4B.** *Several ways of displaying the Dans ma rue database, proposed by the tool "Create a map" of the City of Paris Open Data portal (www.opendata.paris.fr) (continued)*

REMARK ON FIGURE 2.4.– *(a) Selection of display modes, (b) "Color by category" mode (here according to the "rounding" field), with an a priori random sampling, for the sake of readability, (c) "Cluster" mode, with a representation of entity aggregates by proportional circles according to their number, complemented by a slight variation in intensity and a textual indication, and (d) "Color map" mode, with an entity density surface.*

Interactive visualization can be used to manage readability (and therefore also to give meaning, to identify trends) using different methodological solutions (Meier 2016): most often through visual aggregation of information, with great freedom given to spatial and temporal aggregation scales thanks to zoom functions, but sometimes at the cost of difficult control of organization parameters. This aggregation is based on marker groupings (clusters, as in the *Leaflet* or *GoogleMaps* libraries; see Figure 2.4B(c)) or on pre-existing zonings (grids, hexagonal tiles, etc.). In the *Sensor.Community* example, the map displayed (see Figure 2.3) represents aggregated data (in this case, medians of fine particle concentrations), in near real-time, in hexagons, the size of which varies according to the zoom, but without any visual precision on the number of sensors per hexagon; this can pose a problem in terms of the representativeness of the data displayed (especially at large zooms). The same concern applies to the collaborative air pollution map proposed on the *AirCasting* website (see Figure 2.2, bottom right), where aggregation in this case focuses on the temporal dimension, with no warning about the site. Pollution levels are variable over time, making it difficult to interpret the map produced. Aggregation, which is a fairly conventional and technically straightforward method, sometimes distances the spatial or spatiotemporal objects constructed from the phenomena under study.

In other cases, improving readability is based on the selection of the information represented. It is not always clear what the criteria are: selection may be based on a priori random sampling of the data (e.g. the preview of certain *Open Data Paris* data, or their representation by category or "choropleth"; see Figure 2.4A a and b) or be a function of a commercially oriented algorithm (the rentals offered on the *Airbnb* platform meet less-than-transparent criteria, whereas on *InsideAirbnb*, all points can be viewed according to the type of rental; see Figure 2.8). The selection of *Netatmo* weather stations on their online map partially reveals a reference grid, variable as per the level of zoom, and one of the selection criteria is the proximity of the stations to the points on the grid (see Figure 2.1). In another example, the *AirCasting* platform displays pollutant measurements taken with a portable sensor, over sequences ranging from a few seconds to a few hours, by the first measurement point alone. By clicking on the point, the user can then view the entire session, on a map with measurements, alongside a time chart (see Figure 2.2).

Finally, some applications offer smoothing functions, most often using heatmaps and moving averages. Mericskay (2016), in a state-of-the-art review on new ways of representing digital data, lists numerous plug-ins available to create heatmaps; however, these are not widely used in visualization platforms linked to data dissemination. We give here the example of the City of Paris' open data portal, which offers this mode of data visualization (see Figure 2.4B(d)), where information on the size of the neighborhood and the type of weighting is not clearly discernable.

In this way, the proliferation and diversification of native, digital, data have given rise to a wealth of information, which cartographic platforms help to disseminate and visualize, either in the raw form of point data or in a more aggregated form. Beyond this monothematic, single-source display of new, relatively stovepiped data, one of the major challenges in its exploitation lies in integrating it with complementary data from other sources: Are the highest concentrations of pollutants found near road intersections? Are the highest temperatures in cities associated with particular types of urban fabric? Do *Airbnb* rentals tend to cluster in neighborhoods with conventional tourist accommodations or in areas lacking them? What types of spaces are *OpenStreetMap* data the most complete, compared with a reference dataset, such as the BPE (*Base permanente des équipements*) from the INSEE (*Institut national de la statistique et des études économiques*)? Yet this integration of a monothematic dataset derived from web platforms and complementary data is not instantaneous and requires more or less complex transformations (Li et al. 2016).

## 2.3. New data, traditional data: why and how to integrate them?

The issue of interoperability between new and traditional data brings to the fore the notion of integrating heterogeneous spatial data, which has an abundant literature from the late 1980s onwards (Ruiz et al. 2011), in geographic information sciences in particular. In this section, we present some of the issues involved in integrating new data with traditional data, in the light of this older work. First, we return to the polysemy of the notions of heterogeneity and database integration (section 2.3.1). We then illustrate in greater detail the diversity of integration methods, depending on whether they involve the visual overlay of data or the combination of spatial, semantic and temporal information content (sections 2.3.2 and 2.3.3). Even if these methods are often situated upstream of visualization, the aim is to also question what the cartographic representation of these combinations brings to the spatial analysis of new data, how it gives meaning and how it feeds the discussion of the possibilities opened up by their scientific exploitation.

### 2.3.1. *Data heterogeneity, integration, interoperability: preamble to vocabulary development*

One of the difficulties inherent in the integration of heterogeneous datasets lies in the polysemy of the term "integration" and in the existence of an abundance of associated lexical terms. This complexity is linked both to the diversity of the disciplines involved (from the computer sciences to the humanities) and to the multiplicity of the methods employed. In particular, we need to clarify what we mean by data heterogeneity, integration and interoperability, before illustrating the related methods and their cartographic implications.

In a broad sense, data integration (or conflation) can be defined as the process of making heterogeneous datasets compatible, with the aim of making them interoperable. In a narrower sense, many studies associate the notion of spatial data integration with the bringing together of two datasets describing the same geographical fact, but with different data modeling, collection temporalities, qualities or scales (Walter and Fritsch 1999). This operation, which consists of "matching geographical objects that refer to the same real-world phenomena but which are derived from different datasets" (Shereen 2002, p. 153, authors' translation), can be more precisely described as "matching". "Term-to-term matching" is used to describe situations wherein the aim is to match each of the elementary geographic objects in the two datasets (e.g. two points corresponding to the same piece of equipment, or two lines corresponding to the same stretch of road in two databases from different sources).

While the term "interoperability" generally refers to the result of the integration process, there are several dimensions of interoperability that do not necessarily overlap: for instance, the visual matching (in the sense of cartographic overlay) of point data on shops from the *OpenStreetMap* database and shops from *INSEE*'s permanent equipment database (*BPE*) is enabled by a minimal integration operation based on the alignment of the two databases in the same spatial coordinate system (see Figure 2.5 and an example in Figure 2.9). However, it is not possible from this initial transformation to clearly identify the best documented areas in *OpenStreetMap* in relation to the *INSEE* database: the answer to this question would require a higher degree of interoperability, made possible by the term-to-term matching of commercial facilities present in both databases.

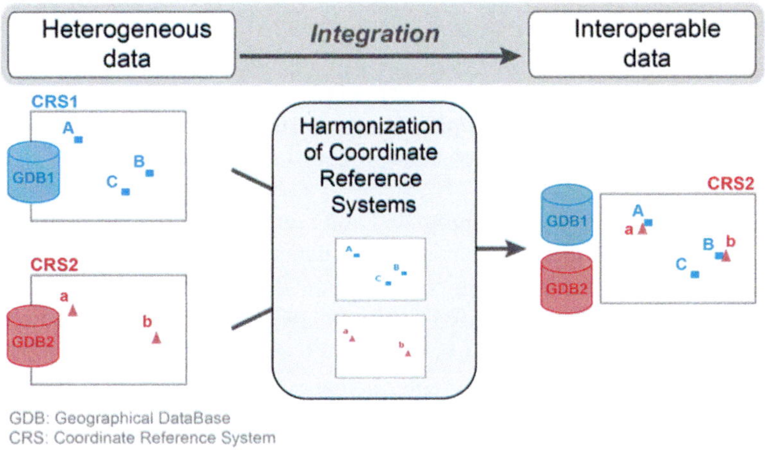


**Figure 2.5.** *Minimal integration for visual overlay: harmonization of spatial coordinate systems*

The preceding example highlights the great diversity of heterogenous sources, which have been the subject of several classification proposals in the literature, with varying degrees of detail (see Sheth and Larson 1990; Bishr 1998; Ruiz et al. 2011; Wiemann and Bernard 2016). For simplicity's sake, it is useful to distinguish two main levels of heterogeneity, relating, on the one hand, to the data's technical environment and, on the other hand, to its content. From a technical point of view, heterogeneity resides in the data environment and in the difficulties that several IT systems have in exchanging data, which for a long time was a major obstacle to data sharing and the risk of overlaps within the same information system (see Box 2.2). At a second level of heterogeneity, described by different authors as "semantic", "cognitive" or "methodological" (Sheth and Larson 1990), the obstacles to bringing databases together are due to the diversity of ways in which information is collected, or to the choices made in modeling "reality" in each of the two databases (Walter and Fritsch 1999). Some distinguish in this category conceptual heterogeneity, which refers to divergent definitions of the same phenomenon (Mathian and Sanders 2014), well illustrated, for example, by the uses of the notion of land artificialization (Ruas 2020). Another form of semantic heterogeneity refers to differences in data models, whether in terms of spatial resolution and geometric implementation, collection temporalities or attribute richness and categories: in the *OpenStreetMap* database, commercial facilities are represented using points, multipoints and/or polygons, whereas in the *INSEE BPE* they are represented solely by points. Furthermore, the classifications of commercial activities within the two databases are not always based on the same thematic categories (see section 2.3.4). Finally, there may be heterogeneity in terms of data quality (Guptill and Morrisson 1995, cited by Duroudier et al. 2020), with imprecision problems sometimes skewing the reconciliation of databases. This is the case when comparing *Airbnb* rentals, whose precise location is known to within plus or minus 100 m, with the characteristics of the neighborhoods where they are located (imprecision problem).

These crossovers are far from automated and require careful consideration of the methods best suited to their implementation, depending on data models and the analysis objectives. In the following sections, we illustrate the variety of possible solutions, based on methods that have become classic in the field of heterogeneous data integration, but which are particularly well suited to new data in that they respond to the desire to give them greater semantic depth to offer them contextual perspective and to assess their quality.

### 2.3.2. *Visual overlay of heterogeneous data, facilitated by advances in technical interoperability*

The visual overlay of datasets from different sources within the same cartographic interface represents the first degree of data interoperability, the prerequisite for the dialogue between databases. Although this is no longer a problem and is fairly straightforward today, this pre-processing stage has not always been so easy to implement (see Box 2.2).

In the 1970s and 1980s, most geographic information systems (GIS) were still considered "islands of information" (Bishr 1998), independent and not very interoperable. Differences in data exchange protocols, operating systems and file formats stood in the way of data sharing between computer systems and a visual overlay within the same information system (Mustière 2014).

In the 1990s, several standardization initiatives paved the way for standards proposed by the International Organization for Standardization (ISO) and the Open Geospatial Consortium (OGC). The concept of Spatial Data Infrastructure, well-illustrated by the INSPIRE Directive (Directive 2007/2/EC), has gradually taken hold, with the establishment of a common framework on the technical (data and metadata formats), economic (harmonization of user licenses and pricing policies) and organizational (identification of stakeholders) fronts (Noucher 2013; Mustière 2014). Among the innovations introduced, metadata standards have been of fundamental importance, both for improving the accessibility of databases and for the publication of specifications essential to the preparation of their interoperability. This is also one of the mainsprings of the greater ease with which different geographic positioning systems can be managed, essential for data overlay, in the same way as the invention of EPSG (European Petroleum Survey Group) codes and the growing facilities for "on-the-fly" projection in GIS software.

More recently, the development of the Geoweb has been accompanied by significant format improvements, which sometimes lead some authors to refer to this framework as a "major interface for data aggregation specific to the Web" (Mericskay 2021, authors' translation), completing the process initiated in the 1990s. With its "new properties of integration, sharing and mixing", the Geoweb behaves like "an aggregator of web information and content" (Mericskay 2016, p. 2, in reference to Plantin 2014, authors' translation). Drawing on Haklay et al.'s (2008) observation that interoperability between systems and databases is an essential element of the Geoweb, Mericskay (2016, p. 4) thus highlights the enrichment of the standards established by the Open Geospatial Consortium and "which have become genericized for web services (WMS, WFS, WCS) as well as for flat or geographic file formats (CSV, GeoJson, KML, GML)" (authors' translation).

**Box 2.2.** *A brief history on the technical integration of heterogeneous data*

Despite the progress made in overlaying several heterogeneous datasets, cross-referencing of new data with each other or with conventional data remains rare in visualization interfaces, whether for distribution platforms, observatories or spatial data infrastructures. In most cases, complementary data are overlaid on the data of interest to offer a cartographic basemap (often on a *GoogleMap* or *OSM* background; see Figures 2.1 to 2.4) that is overlaid on the data of interest, without combining contents, producing "autonomous", independent maps. Apart from these now-standard backgrounds, there are few data overlays that add thematic value: such is the case with *Sensor.Community*'s sensor map, where wind speed and direction, which help assess pollutant dispersion, are represented by dynamic blue lines. These "comet-like trails", proposed in 2012 by Viégas and Wattenberg (www.hint.fm/projects/wind/), are constructed from US weather forecasts (see Figure 2.3(a)).

As well as overlaying data, the integration process can be enhanced by facilitating the alignment of information, making it visually compatible. This involves bringing together information content and refers to the way in which content is represented and modeled. Behind this common objective, several integration methods can be distinguished, depending on the degree of spatial precision of the combination and the nature of the data[6].

6 This section does not deal with integration methods adapted to geometries that are not compatible with each other (geoprocessing by intersection/union/difference, disaggregation/aggregation of attribute information), which are covered in greater detail in Chapter 3 of this book.

### 2.3.3. *Term-to-term matching to assess the quality and enrich the attributes of new data*

First of all, there are numerous methods, more or less automated, for implementing reconciliations of datasets describing the same phenomenon. In their contemporary classification of integration methods (named conflation), Ruiz et al. (2011) devote a long development to the term-to-term matching of entities and summarize matching methods around three main stages: pre-processing aimed at making data comparable (map projection, file formats, degrees of generalization, etc.), identification of correspondences between entities in the two databases and finally implementation of the alignment between corresponding entities (and evaluation of differences).

One of the most frequent objectives of integration via matching is to assess the quality of a dataset, by comparing it with another so-called reference database, which describes the same phenomenon, but which is "of better quality and ideally of precisely known quality" (Mustière 2014, p. 9, authors' translation). Most of the time, it is the quality of the new data that is discussed and assessed following this matching. The case of *OpenStreetMap*'s contributory data is a good illustration of this purpose. Therefore, the comparison of features present in this database and in "external" databases describing the same phenomenon (from commercial databases or national reference databases, such as *IGN*'s *BD Topo*®) constitutes one of the classic aspects of the analysis of the quality of POIs in the *OpenStreetMap* database. Jonietz and Zipf (2016) distinguish several types of quality indicators, depending on whether they are based on a comparison of the number of objects represented in each database (completeness indices), on the accuracy of object location (via distance measurements between objects) or on the accuracy of semantic information.

The latter can be examined using various measurements on attributes (percentage of correct classification of objects, Levenshtein distance measuring the degree of similarity between two character strings). Touya et al. (2017) emphasize that none of these methods is sufficient on its own and rely on a combination of approaches, which they test on metro station data from the *OpenStreetMap* database in Paris, by comparison with *BD TOPO* data. At a very fine level of observation, they built an automated algorithm for the term-to-term matching of objects in each database based on several criteria: for each *BD TOPO* component, the "candidates" for matching are selected within a 350-m buffer zone around the reference points; then their quality is assessed on the basis of resemblance criteria based on position – Euclidean distance to the reference station – and on the name associated with the object. This alignment has shown that changes to *OpenStreetMap* features are less a result of disagreements between contributors than of improved quality and reduced heterogeneity of thematic attributes (see Figure 2.6 and an example in Figure 2.9).

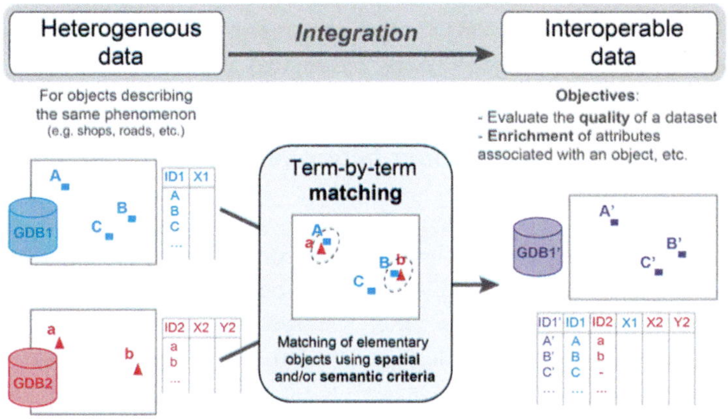


**Figure 2.6.** *Integration of heterogeneous datasets by term-to-term matching*

Conversely, it is sometimes contributory data that are used to update reference data. For example, national mapping agencies are considering whether to use *OpenStreetMap* data to update their information; this mainly concerns road network infrastructure data, which is more mature and exhaustive than data on amenities (Olteanu-Raimond et al. 2017).

Another aim of integrating datasets describing the same phenomenon is to enrich the thematic attributes of either database. This is a particular type of matching, based on the transfer of semantic information from the richer database to the more detailed geometric elements of the other. This process, sometimes referred to as "fusion" in GIS, can also lead, when the data models of the two bases are very close, to the "creation of a composite product resulting from the unification of several geometries" (Mustière 2014, authors' translation). With the emergence of new data, including GPS traces (see section 2.2.2), this goal is all the more important as the latter often lose in thematic richness or clarity of structuring what they gain in precision of spatial and/or temporal resolution. For example, as part of a study on children's daily mobility in Rennes (*Mobi'Kids* project, Duroudier et al. 2020), the itineraries of a hundred or so individuals surveyed were first deduced from GPS "footprints", recorded over several days using mobile devices, and then thematically enriched on the basis of individual interviews; the latter made it possible to associate information on the journeys of the individuals surveyed (mode, reason, accompaniment, etc.) and on the characteristics of the individuals (social profiles, perceptions of journeys). In this case, matching GPS traces with questionnaire information is relatively straightforward, with cross-referencing carried out during a face-to-face interview, via a tablet, based on a commented examination of the tracks collected. As with many other datasets that are candidates for matching, the most complex part of the process takes place upstream, when the GPS traces are pre-processed to create new objects (stops and routes).

### 2.3.4. Aggregations for combining heterogeneous data within "pivotal" spatiotemporal units

A second classic category of solutions for reconciling heterogeneous content is based on the aggregation of point entities (one of the dominant representation modes for natively digital data). In this case, the spatial (and sometimes temporal) resolution of the point dataset(s) is degraded within "pivotal" spatial units of various kinds (statistical or administrative units, geometric zonings within regular squares or hexagons, irregular geometric zones resulting from tessellation operations, etc.). These spatial aggregation operations are accompanied by various operations on the attributes depending on their nature: a sum for counts (e.g. quantity of *Airbnb* rentals or *OSM* facilities within a mesh), an average for ratios (e.g. price per person of *Airbnb* rentals within the same grid), etc. (see Figure 2.7). Such combinations at the level of local "pivotal" entities often respond to the desire to identify trends in spatial organization that would not be readable at the level of point entities.

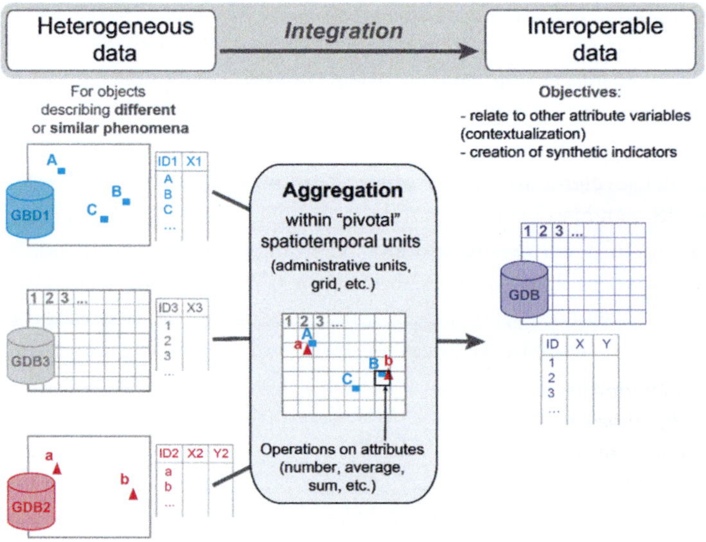


**Figure 2.7.** *Integration of heterogeneous datasets by aggregation within "pivotal" spatial units*

When transposed to new data, these integration operations are often favored to put into perspective data usually valued from a monothematic angle. Some studies of the Airbnb phenomenon illustrate this contextualization objective. Take, for example, a recent study of long-term Airbnb rentals "diverted" from the conventional rental stock in Greater London: Shabrina et al. (2022) combine point data from the *InsideAirbnb* site with different spatial units (Lower Layer Super Output Areas [LSOA], and Boroughs) corresponding to the main dissemination levels of local census statistics. The proportion of homes diverted from the market is thus estimated by dividing the total number of homes (census data) by the number of *Airbnb* homes with more than 180 days of bookings in a year (a threshold contrary to local regulations), and whose hosts have multiple ownerships (see Figure 2.8).

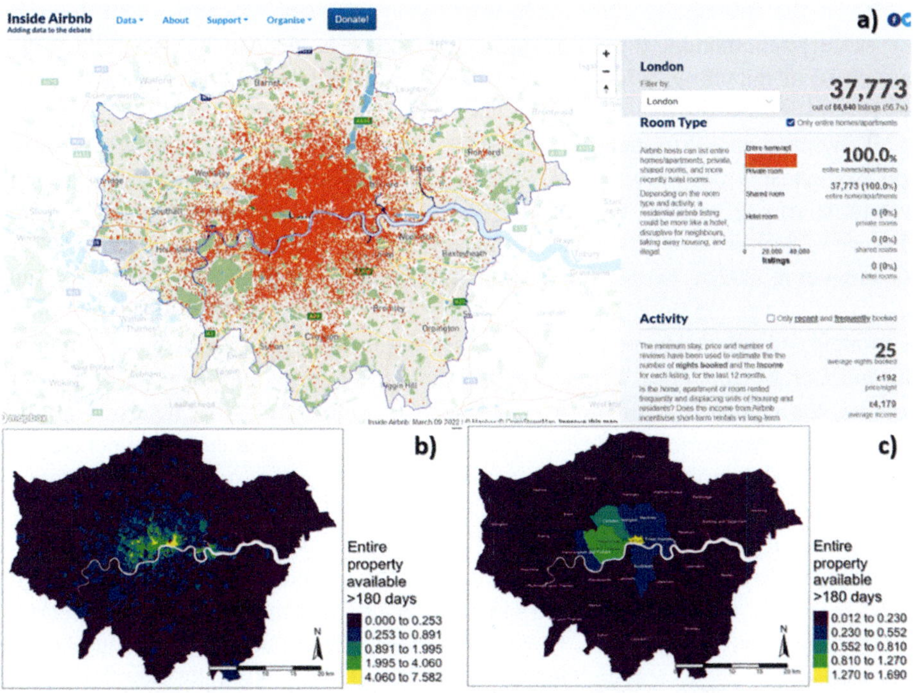


**Figure 2.8.** *(a)* InsideAirbnb*'s visualization interface for London (www.insideairbnb.com/london), (b) share in total Airbnb accommodation rented out in full more than 180 days a year, at Lower Layer Super Output Areas and (c) Borough level (Shabrina et al. 2022)*

Aggregating the number of "diverted" rentals at infra-urban level first enables us to estimate their relative weight in relation to the stock of main residences: on average, they represent a very small proportion (0.15%) of the total housing stock in Greater London, but this rate sometimes reaches 7% in certain LSOA in the city. The spatial variability of this share of diverted locations is then explored by studying the correlation with more general characteristics of the housing stock: for example, at Borough level, the share of diverted Airbnb locations in 2015 is positively correlated with the average increase in rents in the private rental stock between 2015 and 2017. One of the limitations of these aggregation operations, relatively simple in appearance but largely dependent on the choice of aggregation level, lies in the 100-m location imprecision, which can pose a problem at fine scales of analysis, when a rental "falls" on an uninhabited unit. This freedom of appreciation raises the broader question of the most appropriate level(s) of spatial aggregation.

When the two datasets are made up of elementary point entities and describe the same phenomenon, the choice of spatial aggregation can be a substitute for term-to-term matching, which is sometimes too cumbersome to carry out and not really justified from a thematic point of view. Figure 2.9(c) is taken from a study of retail facilities in a European cross-border region (Grande Région, a cross-border cooperation area between France, Germany, Luxembourg and Belgium). The aim was to determine whether the use of *OpenStreetMap* data could compensate for the absence of national amenity databases in some of the countries covered by the study (Guérois et al. 2020). On the French side, the existence of the *BPE* offered a unique opportunity to assess the quality (in the sense of completeness) of *OpenStreetMap* at a municipal level of observation. However, while it is straightforward to compare the number of amenities aggregated at community level (aggregation with shop counts), the comparison of the two databases is hampered by the heterogeneity of equipment classifications: In *OpenStreetMap*, there is no nomenclature per se, but rather objects described using tags composed of "key-value" pairs (e.g. amenity = bank), the list of which, documented in a *Wiki*, is open and non-binding. A correspondence table with the *BPE* nomenclature thus had to be created to reconstitute semantic consistency between the two databases (see Figure 2.9(b)). This semantic integration stage has enabled us to map the *OpenStreetMap* completeness rate for certain types of business (ratio between the number of *OpenStreetMap* equipment and that of the *BPE*) and to highlight the variation in this completeness according to spatial contexts (see Figure 2.9(c)): higher within inner-city communities, whether within urban areas or isolated agglomerations, lower in rural communities outside urban areas, and especially in suburban and peri-urban areas, making it difficult, for the time being, to use *OpenStreetMap* amenity data outside city centers.

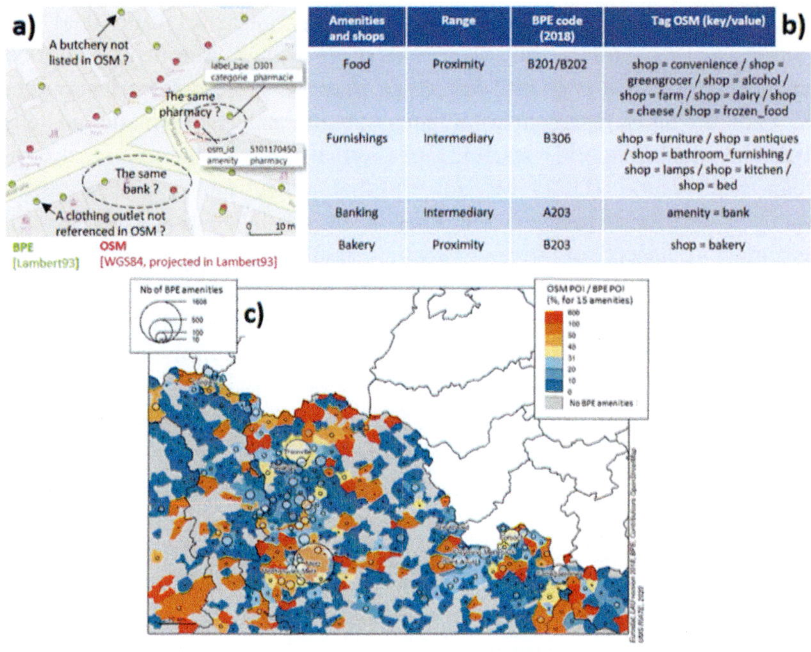


| Amenities and shops | Range | BPE code (2018) | Tag OSM (key/value) |
|---|---|---|---|
| Food | Proximity | B201/B202 | shop = convenience / shop = greengrocer / shop = alcohol / shop = farm / shop = dairy / shop = cheese / shop = frozen_food |
| Furnishings | Intermediary | B306 | shop = furniture / shop = antiques / shop = bathroom_furnishing / shop = lamps / shop = kitchen / shop = bed |
| Banking | Intermediary | A203 | amenity = bank |
| Bakery | Proximity | B203 | shop = bakery |

**Figure 2.9.** *Extract from OSM (OpenStreetMap) and BPE (INSEE)*

REMARK ON FIGURE 2.9.– *(a) Forbach city center. (b) Extract from the correspondence table between BPE commercial equipment categories and OSM tags. (c) Completeness rate of the OSM database at the France-Luxembourg-Germany border: in orange, municipalities where information is fairly complete, in densely populated town centers (OSM points of interest represent 50–100% of BPE facilities) (Guérois* et al. *2020).*

### 2.3.5. *Interpolated data measuring two continuous phenomena to compare them within a common grid*

A third category of methods for the alignment of semantic content is spatial interpolation, which applies to point data representing continuous phenomena, often biophysical in nature. Interpolation aims at estimating the values taken by the observed parameter at other non-valued points in space based on the values at the measurement points and the distance between these points (spatial estimation) (Arnaud and Emery 2000).

The comparison between Netatmo data (see section 2.2.1) and land surface temperature *MODIS* satellite data from NASA (National Aeronautics and Space Administration) provides a good example of data integration from heterogeneous sources, when the latter represent continuous phenomena. The thematic challenge is to analyze the spatiotemporal variability of thermal contrasts in cities and to study the urban heat island (UHI) by not limiting ourselves to a few reference meteorological stations. Instead, the aim is to spatialize the UHI using two complementary datasets of greater spatial resolution, one recording surface temperatures and the other air temperatures. Land surface temperature products from MODIS's AQUA and TERRA satellites provide surface temperature estimates with a spatial resolution of about 1 km and twice-daily coverage in the absence of cloud cover. *Netatmo*'s individual stations measure air temperature with a much greater temporal resolution, but for a smaller sample of measurement sites, whose representativeness may be weak. The challenge of comparing the two databases resides both in assessing the potential of the *Netatmo* collaborative database to describe the UHI, and in verifying the existence of a close positive correlation between surface and air temperatures.

As long as the observed phenomenon is captured using heterogeneous data collection methods (point station measurements for Netatmo, raster images in the case of MODIS), the two datasets are reconciled via a spatial interpolation of the Netatmo station measurements (see Figure 2.10). Based on a selection of clear-sky nights, a variogram analysis of air temperatures from 6,180 Netatmo stations was carried out, followed by kriging interpolation[7], on a grid at the same resolution as the MODIS images (~1 km) and at a given time (in this case 2 h in Coordinated Universal Time). This transition from point data to continuous fields of values means that the estimates of surface and air temperatures can now be compared using algorithms for satellite data or geostatistics for Netatmo.

Figure 2.11 shows the two median thermal anomaly images and their comparison on the 1 km grid. While differences in distribution are visible, the shape and extension of the two islands, air and surface, are fairly similar, with a shift toward lower values for the *Netatmo* median image compared with the *MODIS* image, which is explained by an over-representation of *Netatmo* stations in the center of the agglomeration (Madelin and Dupuis 2020).

7 Kriging is a family of geostatistical interpolation methods (Arnaud and Emery 2000) based on two steps: a variogram analysis to verify the existence of spatial autocorrelation, its scope, etc., followed by a model fitting to interpolate the values of the phenomenon at any point in space.

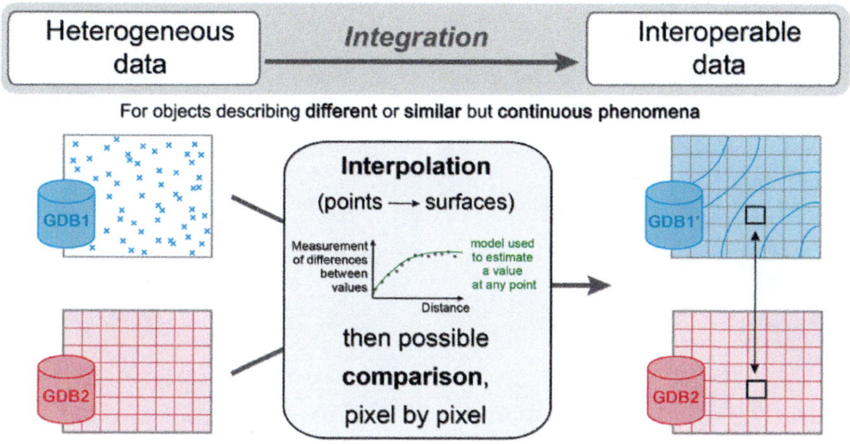


**Figure 2.10.** *Integration of heterogeneous datasets describing continuous phenomena, based on the transformation by spatial interpolation of a point dataset*

In some cases, the matching of this type of data also relies on the matching of the temporal dimension of the geographic information; made possible by the very fine temporal resolution of the databases, this matching is important due to the high temporal variability of the temperatures (at different time intervals: hourly, daily, seasonal, etc.). From a methodological point of view, this can be achieved simply by filtering the data: the Netatmo data are selected according to the time of day when the satellite passes overhead. Another, more complex approach, but that enables continuous information on hourly temperature variability to be maintained and data to be temporally aligned, is based on temporal interpolation of one of the two datasets: Figure 2.12 illustrates the case of comparing air temperatures from Netatmo sensors with those from Météo France's Montsouris station. The aim is to verify the quality of Netatmo's measurements and exclude outlier stations. Here, the two datasets are aligned by temporally interpolating the Netatmo data and then by creating a new time-discretization to match Météo France (Figure 2.12(a)). This alignment makes it possible to exclude stations whose data are a priori outside the norm: for example, stations in the sun whose daily maximum temperatures are much higher than the reference data, or a station located inside a household, with low variations (Figure 2.12(b)). Just under half of all sensors were thus excluded from the analysis (Madelin and Dupuis 2020).

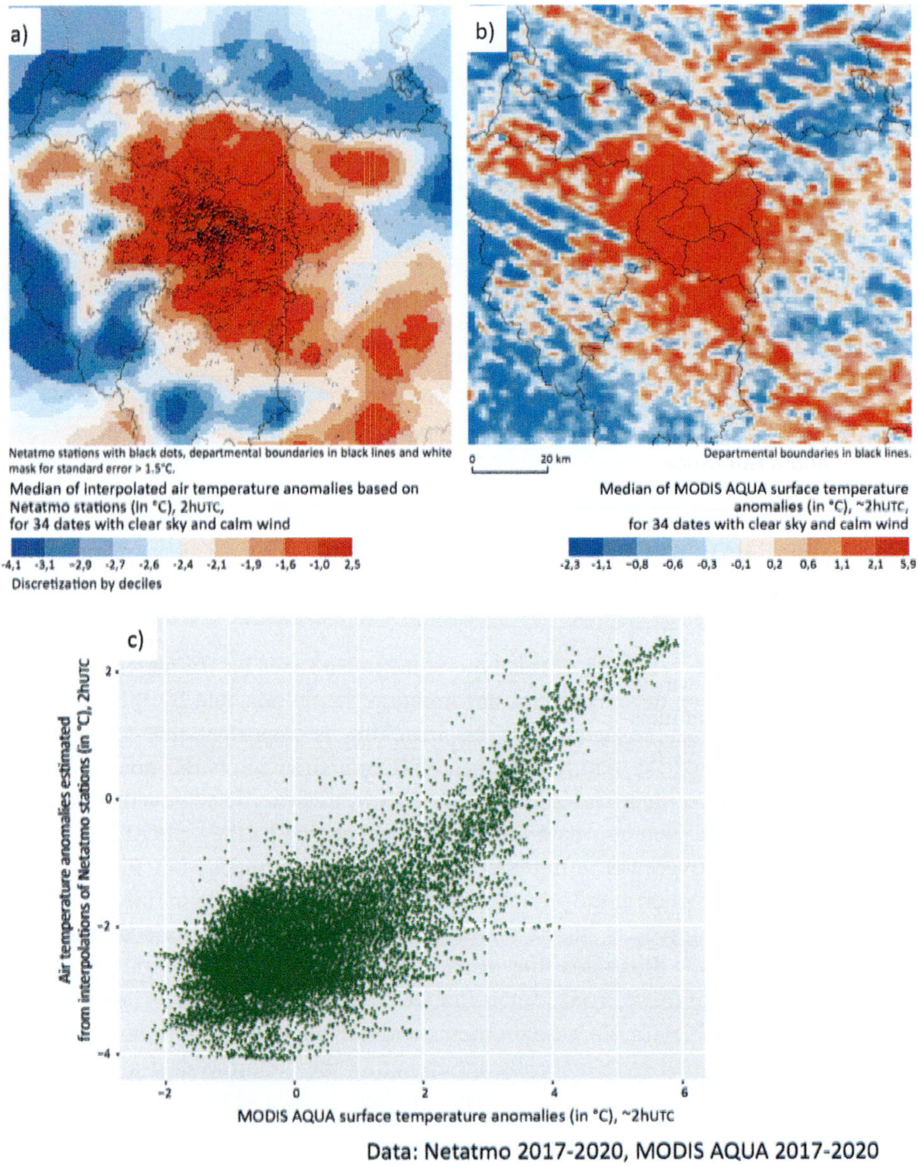


**Figure 2.11.** *Median images of (a) spatial interpolations of air temperature anomalies from Netatmo stations and (b) surface temperature anomalies from MODIS, during nights with clear sky and calm wind. (c) Cartesian plot of daily temperature anomalies from both sources*

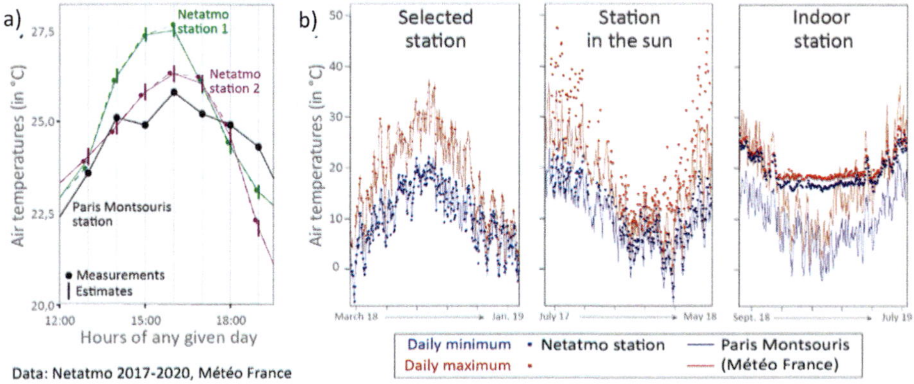


**Figure 2.12.** *(a) Temporal alignment on an hourly time interval of data from Netatmo and a reference station (b) to qualify the stations and their environment*

## 2.4. Conclusion

The challenges of integrating heterogeneous spatial data have taken on considerable importance with the proliferation of natively digital local information supplied by a wide variety of producers. These challenges are further compounded by the mass and granularity of the data. In this chapter, we have shown, using a few examples from a wide range of applications and methods, that the major advances in the technical interoperability of these data into the Geoweb, essential for their visual overlay, do not exhaust the issues linked to integration with other new data or with data of a conventional nature. Once the databases have been overlaid, matching is rarely spontaneous, and the question of how to dialogue the contents of the databases remains unresolved, as the diversity of data models, types of spatial features, degrees of spatial and temporal resolution and thematic richness necessitates recourse to various transformations and spatial analysis operations (see Figure 2.13).

This observation certainly needs to be qualified. On the one hand, the high spatial and temporal resolution of the new data greatly facilitates this integration, due to the predominance of point features, which makes it possible to multiply the choice of observation levels, from the individual to the aggregated levels relevant for analysis. In other respects, however, the lesser harmonization and structuring of new data, the poverty or weakness of associated metadata and the sometimes more incomplete nature of data oriented toward specific technical uses are all obstacles to their combination with traditional data. These characteristics accentuate the importance of pre-processing, critical data analysis and filtering of usable data. They sometimes even shift the definition and identification of the object of study to the analysis process, as these steps are less often carried out upstream of data acquisition (Li et al. 2016).

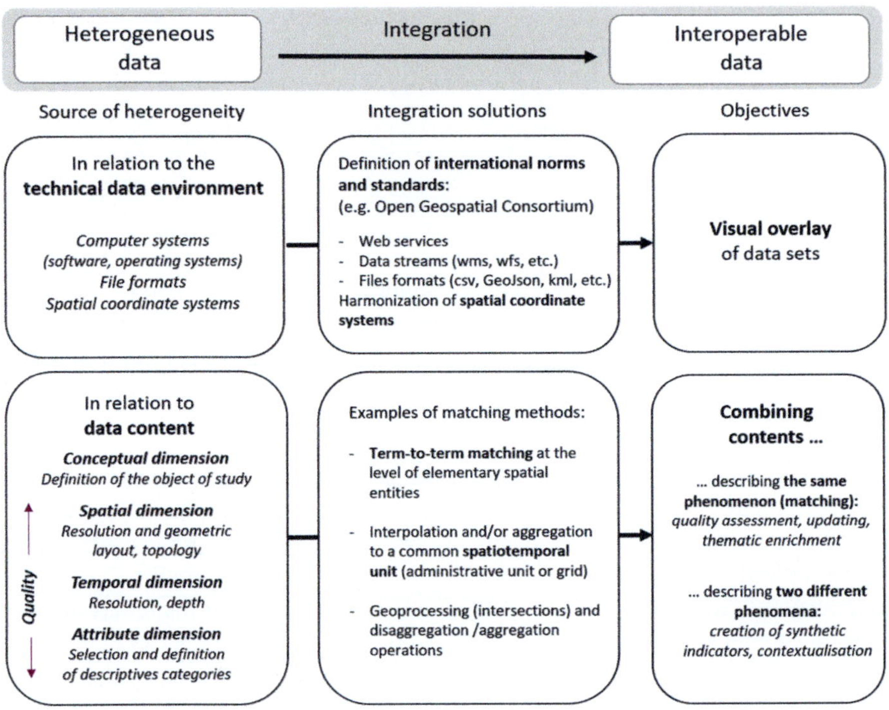


**Figure 2.13.** *From heterogeneous data to interoperable data: sources of heterogeneity, integration solutions and objectives*

The challenges of integrating new data with conventional data lie in part in improving access to the most appropriate methods and tools for making this integration, within the Geoweb itself. Some authors see the automation of integration processes as the key to increasing the number of cross-referencing possibilities available to users (Wiemann and Bernard 2016). To a certain extent, these perspectives echo the conclusions drawn by Bahoken et al. (2020) from their typological study of a Geoweb of flows and networks, when they detect the recent emergence of applications that are more flexible and open to data combination, compared to the dominant monothematic visualization interfaces. They remind us of the importance of drawing up, upstream, an accessible synthesis of the vast array of methods that have long been used to carry out such integration. This effort is also important for making progress in the international comparison of territories (e.g. large metropolitan areas) for which reference data are disseminated at spatiotemporal resolutions and in thematic categories that are often very heterogeneous from one country to another (Pecout et al. 2019). Beyond the improvement of tools built to facilitate these cartographic integrations, the place of the user in the world of big data and the automation of fine data collection remains essential: not only does it help guide the transformation and construction of data adapted to the formulated problem, and thus ensure that processing and visualization can lead to solid conclusions, but it also maintains a critical eye on the combination of personal data at the individual level, which is not without raising major privacy issues.

## 2.5. References

Arnaud, M. and Emery, X. (2000). *Estimation et interpolation spatiale : méthodes déterministes et méthodes géostatiques*. Hermès, Paris.

Arribas-Bel, D. (2014). Accidental, open and everywhere: Emerging data sources for the understanding of cities. *Applied Geography*, 49, 45–53.

Bahoken, F., Le Campion, G., Maisonobe, M., Jégou, L., Come, E. (2020). Typologie d'un Geoweb des flux et réseaux. *Geomatica*, 74(3), 154–174.

Bastin, G. and Tubaro, P. (2018). Le moment big data des sciences sociales. *Revue française de sociologie*, 59(3), 375–394.

Bishr, Y. (1998). Overcoming the semantic and other barriers to GIS interoperability. *International Journal of Geographical Information Science*, 12(4), 299–314.

Boyd, D. and Crawford, K. (2012). Critical questions for big data: Provocations for a cultural, technological, and scholarly phenomenon. *Information, Communication & Society*, 15(5), 662–679.

Chen, C., Ma, J., Susilo, Y., Liu, Y., Wang, M. (2016). The promises of big data and small data for travel behavior (aka human mobility) analysis. *Transportation Research Part C: Emerging Technologies*, 68, 285–299.

Duroudier, S., Chardonnel, S., Mericskay, B., Andre-Poyaud, I., Bede, O., Depeau, S., Devogele, T., Etienne, L., Lepetit, A., Moreau, C. et al. (2020). Diagnostic qualité et apurement des données de mobilité quotidienne issues de l'enquête mixte et longitudinale Mobi'Kids. *Revue internationale de géomatique*, 30(1/2), 127–148.

Ferreira, D. and Vale, M. (2020). Geography in the big data age: An overview of the historical resonance of current debates. *Geographical Review*, 112(2), 1–17.

Guérois, M., Ysebaert, Y., Giraud, T., Maranget, B., Hamez, G., Boquet, N., Dorkel, N. (2020). Apports des données OpenStreetMap pour l'analyse des centralités commerciales dans les espaces transfrontaliers. Rapport pour l'Agence Nationale de la Cohésion des Territoires, UMS RIATE.

Guptill, S.C. and Morrison, J.L. (1995). *Elements of Spatial Data Quality*. Elsevier, Oxford.

Haklay, M., Singleton, A., Parker, C. (2008). Web mapping 2.0: The neogeography of the Geoweb. *Geography Compass*, 2(6), 2011–2039.

Joliveau, T. (2011). Le Geoweb, un nouveau défi pour les bases de données géographiques. *L'Espace géographique*, 40(2), 154–163.

Jonietz, D. and Zipf, A. (2016). Defining fitness-for-use for crowdsourced points of interest (POI). *ISPRS International Journal of Geo-Information*, 5(9), 149.

Kitchin, R. (2014a). *The Data Revolution: Big Data, Open Data, Data Infrastructures and Their Consequences*. Sage Publications, London.

Kitchin, R. (2014b). Big data, new epistemologies and paradigm shifts. *Big Data & Society*, 1(1), 1–12.

Li, S., Dragicevic, S., Castro, F.A., Sester, M., Winter, S., Coltekin, A., Pettit, C., Jiang, B., Haworth, J., Stein, A. et al. (2016). Geospatial big data handling theory and methods: A review and research challenges. *ISPRS Journal of Photogrammetry and Remote Sensing*, 115, 119–133.

Madelin, M. (2020). La géographie biophysique participative. In *Géographie de l'environnement. La nature au temps de l'anthropocène*, Dufour, S. and Lespez, L. (eds). Armand Colin, Paris.

Madelin, M. and Dupuis, V. (2020). Intensité et spatialisation de l'îlot de chaleur urbain parisien à partir de données participatives. *Climatologie*, 17(9), 12.

Mathian, H. and Sanders, L. (2014). *Spatio-temporal Approaches: Geographic Objects and Change Process*. ISTE Ltd, London, and John Wiley & Sons, New York.

Meier, S. (2016). The marker cluster: A critical analysis and a new approach to a common web-based cartographic interface pattern. *International Journal of Agricultural and Environmental Information Systems*, 7(1), 28–43.

Mericskay, B. (2016). La cartographie à l'heure du Geoweb : retour sur les nouveaux modes de représentation spatiale des données numériques. *Cartes & géomatique*, 229, 37–50.

Mericskay, B. (2021). La géovisualisation de données massives sur le Web : entre avancées technologiques et évolutions cartographiques. *M@ppemonde*, 131, 1–27.

Miller, H.J. and Goodchild, M.F. (2015). Data-driven geography. *GeoJournal*, 80(4), 449–461.

Mustière, S. (2014). Intégration de données géographiques. Research summary, Université Paris-Est.

Noucher, M. (2013). Infrastructures de données géographiques et flux d'information environnementale : de l'outil à l'objet de recherche. *Networks and Communication Studies*, *NETCOM*, 27(1/2), 120–147.

Olteanu-Raimond, A.-M., Hart, G., Foody, G., Touya, G., Kellenberger, T., Demetriou, D. (2017). The scale of VGI in map production: A perspective on European national mapping agencies. *Transactions in GIS*, 21(1), 74–90.

Pecout, H., Grasland, C., Guérois, M., Le Goix, R., Madelin, M., Nussbaum, F., Valette, J.-F. (2019). Information territoriale locale et analyse comparative des dynamiques métropolitaines : le projet Grandes métropoles. In *Demande(s) territoriale(s)*, Lajarge, R., Cailly, L., Ruas, A., Saez, G. (eds). Karthala, Paris.

Plantin, J.C. (2014). *Participatory Mapping: New Data, New Cartography*. ISTE Ltd, London, and John Wiley & Sons, New York.

Robinson, A.C., Demšar, U., Moore, A.B., Buckley, A., Jiang, B., Field, K., Kraak, M.-J., Camboim, S.P., Sluter, C.R. (2017). Geospatial big data and cartography: Research challenges and opportunities for making maps that matter. *International Journal of Cartography*, 3(1), 32–60.

Ruas, A. (2020). Que penser du sens donné à l'artificialisation et à la désartificialisation des sols ? *Cybergeo Conversation* [Online]. Available at: https://cybergeo.hypotheses.org/688.

Ruiz, J., Javier Ariza, F., Ureña, M., Blázquez, E. (2011). Digital map conflation: A review of the process and a proposal for classification. *International Journal of Geographical Information Science*, 25(9), 1439–1466.

Shabrina, Z., Arcaute, E., Batty, M. (2022). Airbnb and its potential impact on the London housing market. *Urban Studies*, 59(1), 197–221.

Sheeren, D. (2002). L'appariement pour la constitution de bases de données géographiques multi-résolutions. Vers une interprétation des différences de représentations. *Revue internationale de géomatique*, 12(2), 151–168.

Sheth, A.P. and Larson, J.A. (1990). Federated database systems for managing distributed, heterogeneous, and autonomous databases. *ACM Computing Surveys*, 22(3), 183–236.

Singleton, A. and Arribas-Bel, D. (2021). Geographic data science. *Geographical Analysis*, 53(1), 61–75.

Thatcher, J., O'Sullivan, D., Mahmoudi, D. (2016). Data colonialism through accumulation by dispossession: New metaphors for daily data. *Environment and Planning D: Society and Space*, 34(6), 990–1006.

Touya, G., Antoniou, V., Olteanu-Raimond, A.M., Van Damme, M.D. (2017). Assessing crowdsourced POI quality: Combining methods based on reference data, history, and spatial relations. *ISPRS International Journal of Geo-Information*, 6(3), 80.

Trouillard, E. (ed.). (2021). *Les locations saisonnières en Île-de-France. État des lieux d'avant crise*. L'Institut Paris Région, Paris.

Walter, V. and Fritsch, D. (1999). Matching spatial data sets: A statistical approach. *International Journal of Geographical Information Science*, 13(5), 445–473.

Wiemann, S. and Bernard, L. (2016). Spatial data fusion in Spatial Data Infrastructures using Linked Data. *International Journal of Geographical Information Science*, 30(4), 613–636.

# 3

# Environmental Data and Cartographic Objects

**Étienne COSSART**
*EVS, Université Jean-Moulin Lyon 3, France*

## 3.1. Introduction

The painter Paul Klee used to say that "art doesn't represent the visible, it makes it visible" (author's translation). Cartography is undoubtedly an art insofar as it can correspond to this objective: it makes visible patterns of spatial organization, thus contributing to geographical reasoning.

Such a combination of geographical reasoning and cartographic representation of data appeared as early as 1972, in the first issue of the journal *L'Espace géographique*. Cartography was seen not only as a method of sketching, making it possible to grasp the structure of spatial systems (Brunet 1972), but more broadly as a method of carrying out research in geography, alternating heuristic phases of revelation (in the photographic sense of the term) of patterns of organization, from which hypotheses to be tested emerge, and phases of confirmation, during which the map represents associations, concomitances, interactions, most often in close association with the mathematical tool (Claval 1972).

### 3.1.1. *Defining cartographic objects: positioning the problem*

In the first, exploratory stage of geographic reasoning, the map is used for heuristic purposes, that is, to raise questions and working hypotheses – what we might call a temporary problematization. During this exploration phase, cartographic visualization is closely linked to the nature of the initial raw data, whether it comes from a previously assembled database (e.g. an open-source database) or from data produced by field investigation. In other words, the geometry and topology of the objects are represented as they were captured in the field or digitized by another operator, without transformation. The attribute data, unmodified, has not yet been processed, and it is only a question of reading "the organization of a phenomenon and its inscription in space" (Lambert and Zanin 2016, author's translation).

**Geometry versus topology of geographic objects**

Geometry describes the shape and position of geographic objects. It is represented by points, lines or surfaces in vector mode, or sets of pixels in vector (or raster) mode. For example, a forest area can be represented by a polygon, defined by its coordinates.

Topology defines the relationships between a geographic object and its neighbors. For example, the polygon of forest A is a neighbor of that of forest B if and only if the polygons share an edge or node.

**Box 3.1.** *Geometry and topology: review*

Following this reading phase, problematizations (possibly temporal) emerge. Choices need to be made about how to collect and store geographic data. In this second stage, cartographic objects are constructed or deconstructed in their geometry, bearing in mind that in geographic information systems (GIS) these objects are associated with attribute datasets. Data can be aggregated or, conversely, disaggregated according to the choices made; join systems need to be explained in order to associate the necessary data with the selected cartographic objects. Cartographic objects are both the representation medium and the storage medium for attribute data.

Geographers are faced with a recurring problem: these alternations in the use of cartography are based on empiricism, and therefore on a form of intuition in the choices to be made in order to represent and then explain the facts under study (Brunet 1972). Choices are inherent to the notion of model (Durand-Dastès 1993) to any modeling and simplification process. Here, we illustrate the implications of these choices, focusing on environmental data.

### 3.1.2. *Specific environmental data*

The aim of environmental geography is to survey changes in the state of the biophysical settings, to understand the extent to which human settlement contributes to these changes and, finally, to help formalize remediation strategies as part of a decision-making process. The combination of these three objectives requires a combination of heterogeneous geographic information layers, including both geographic objects that make up the physical environment and aim to characterize it (e.g. plant species, lakes, rivers and glaciers), and territorial divisions. Indeed, the characteristics (economic, demographic, social) of these divisions should enable us to grasp the ways in which human settlements affect the environment. Territorial divisions, where environmental remediation action takes place, must include environmental data that are useful to managers, that is, those which inform their decisions and possibly enable planning.

This geographer's work poses a first problem: characterizing objects in the physical environment in order to document its condition. This requires spatialization, which makes the geographical approach indispensable in environmental research. In most cases, this involves reconstructing the distribution of phenomena in the form of spatial grids, based on station measurements and spot observations, which depend to a large extent on local contexts. Are these measurements meaningful? How can we spatially represent a continuous phenomenon on the basis of these necessarily incomplete point observations? How can these be mapped?

Second, these environmental data need to be compared with territorial divisions, at different levels, to provide decision-support elements for environmental managers and practitioners. This confrontation requires spatial analysis work, in which the right balance must be struck when constructing cartographic objects. These must make sense, both in terms of reflecting the spatial variability of biophysical phenomena, and in terms of the stakeholders acting at specific scales.

In this chapter, we show how cartographic formalization imposes the choice of geometry, topology and semantics to be represented in close relation to the data processing carried out during a geo-physical approach. After recalling some general formal elements derived from set theory, we illustrate the problems involved in constructing environmental cartographic objects using two complementary examples.

## 3.2. Building cartographic objects: deconstructing to reconstruct

### 3.2.1. *From geographic object to cartographic object*

In a cartographic approach, or even more broadly in spatial analysis, the objects studied are defined as "entities that are conceptualized in order to produce knowledge" (Mathian and Sanders 2014, author's translation). Based on this definition, geographical objects have the property of being localized, localizable, knowing that this localization carries a meaning to be integrated into reasoning in order to produce knowledge with a spatial dimension.

When mapping, choices are made in relays, affecting the meaning given to the geographic object. These choices are necessary in the construction of cartographic objects, that is, geographic objects formalized for mapping. It is a question of making explicit the choices often made implicitly in what Brunet (1972) called intuition to select and represent one facet of reality rather than another.

The problems inherent in these choices are numerous in environmental geography, mobilizing incomplete and disparate data. Continuous biophysical phenomena (e.g. temperature fields and topography) have to be represented, usually on the basis of point observations (e.g. station surveys), while managing the gaps between these observation points. These fields, usually expressed in raster grids, must then be associated with discontinuous objects, represented in vector format (e.g. land-use patterns and administrative divisions).

The construction of cartographic objects from incomplete and disparate data must be based first and foremost on the desired level of generalization. Stepping back from the temptation, which often exists, to work at the smallest possible level of detail, we need to reflect on the levels of observation that make sense, for example, when they enable optimum representation of the object under study, or when they are levels at which specific stakeholders operate.

All too often, these fundamental choices remain implicit in the trial-and-error process of using software that mobilizes, processes and represents geographic information. These are based on the mathematical principles of set theory, which underlies GIS geoprocessing operations in particular. While the latter enables cartographic objects to be constructed in concrete terms, set theory enables us to return to the ways in which these objects are conceptualized.

### 3.2.2. *A few reminders on set theory*

Applications of set-theoretic vocabulary and operations now extend far beyond the realm of mathematics, as illustrated by the frequent use of the Venn diagram, a simple formalization of an intersection operation between sets (Figure 3.1).

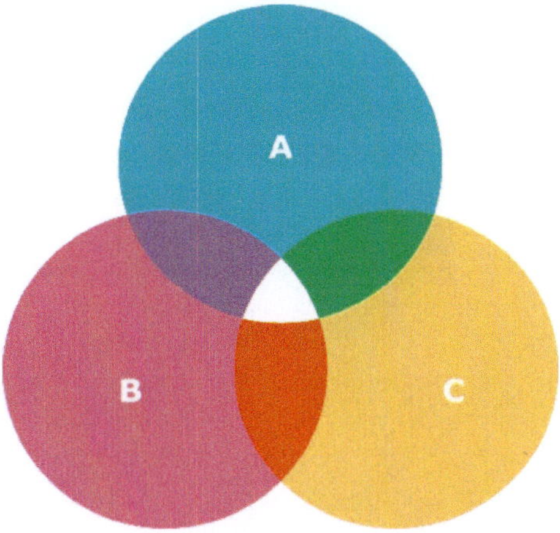


**Figure 3.1.** *Venn diagram. In geography, it is possible to construct and deconstruct sets and subsets based on the modalities of belonging to one of the three zones and subsets derived from operators based on set theory*

COMMENT.– Set theory is a theory of the modalities of belonging, founded by Georg Cantor at the end of the 19th century. In mathematics, a set is a collection of objects, called elements of the set. The principle is to start with basic objects, which can be aggregated to form sets, to which they belong according to shared properties. Sets can give rise to the creation of new sets, subsets, based on operations of union (∪), intersection (∩), difference, for example. The possibilities are endless, and make it possible to construct ad hoc sets, with specific properties.

In geography, a set is a collection of geographic objects that share a common characteristic. While this characteristic is often related to location, it can also be semantic. A set can be combined with another, or partitioned, according to elementary operations that have implications for both the geometric and attribute components of the geographic data. We illustrate these operations through the following examples:

1) a forest, seen as a collection of objects (coniferous zones, deciduous zones, clearings);

2) a glacier, seen as a portion of the topographic surface covered by a body of ice. In the latter case, the glacier's position on two watersheds (A and B) and on different altitudinal ranges enables us to define and reconstruct a posteriori a collection of objects (Figure 3.2).

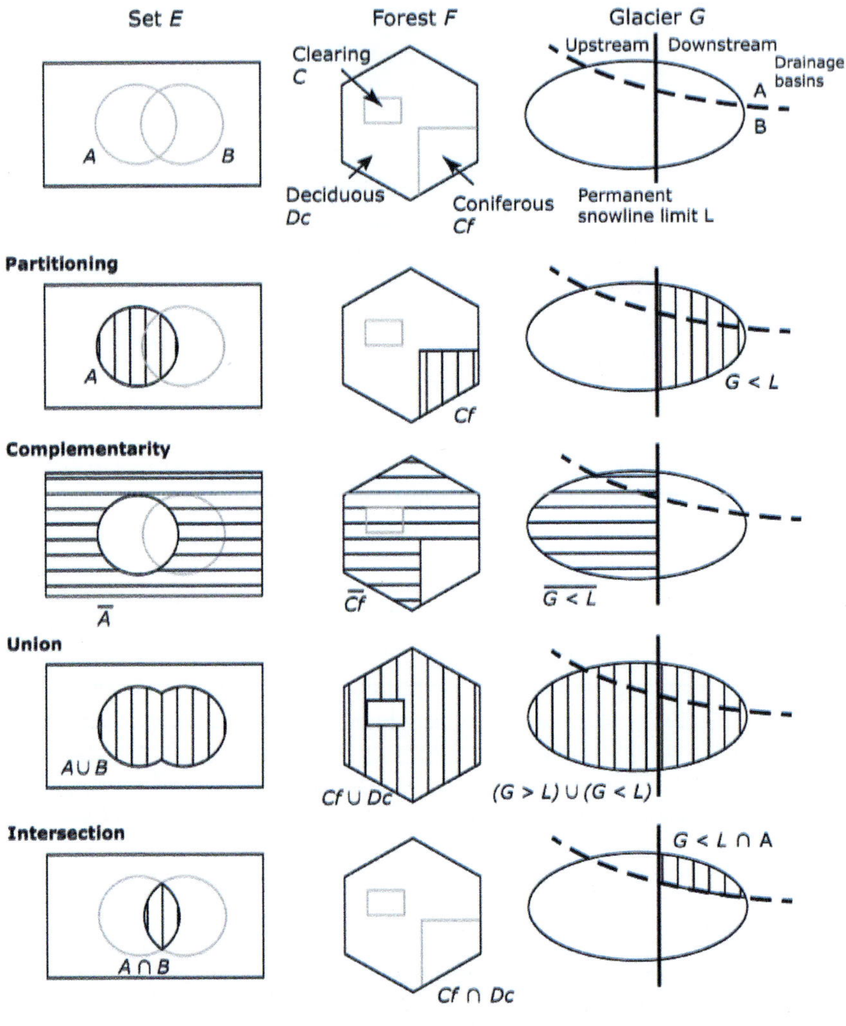


**Figure 3.2.** *Main operations that can be performed using set theory. Left: generic Venn diagram, center: fictitious forest, right: glacier. Modified and completed after Cole and King (1968)*

We present here four main operations.

– *Partitioning into subsets*: the definition of a subset is that all its elements must also belong to the set from which the subset has been defined. Beyond this general definition, we can construct the subset "conifers" from the set "forest", or the subset "portion of glacier below permanent snow altitude", that is, the lower portion of the glacier, where the snow mantle disappears during the spring and summer, and as a result the ice sheet is exposed and prone to melting.

– *Complementarity*: the complement of the subset U of the set E is the set of elements of E that do not belong to U. It is denoted $\bar{U}$. In other words, the complement of the subset "conifers" of the set "forest" corresponds to the elements of the forest that are not conifers. In this case, it will be deciduous trees and clearings. The complementary of the "portion of glacier below permanent snow altitude" subset is the set of points on the glacier above permanent snow altitude, where the ice remains protected from melting, covered by a blanket of snow.

– *Union*: the operation of joining two or more sets groups together all the elements of these sets. The key operator for this action is "or", denoted $\cup$. For example, if the "coniferous" or "deciduous" elements of the forest are sought, then the union will result in the creation of a grouping of tree elements, excluding "clearing" elements. In the case of a glacier, selecting the subset "glacier portion below permanent snow altitude" or the subset "glacier portion above permanent snow altitude" will unite a set corresponding to the initial set. In fact, by joining complementary subsets, we can recreate the initial set to which they belong.

– *Intersection*: the intersection operation between N sets groups together only those elements that belong to all N sets at the same time. The key operator for this action is the "and", denoted $\cap$. For example, the intersection operation between "portion of glacier below permanent snow altitude" and watershed A groups together all elements that belong to both of these subsets. In the case of the forest, the intersection between "deciduous" and "coniferous" produces a null result.

### 3.2.3. *Defining objects and collecting data*

The elementary operations presented here can be combined to create ad hoc cartographic objects. The spatial operators offered by GIS allow these actions to be applied to the geometric component of geographic information. While the geometric results of these operations are intuitive, the results of attribute table combinations are generally far less intuitive. In addition, urging users to be extra vigilant in checking the results of this combinatorial operation at attribute table level, we would like to remind you that operations based on set theory may lead to data aggregation or disaggregation, which calls for a number of mathematical precautions.

– *Partitioning and complementing operations*, by modifying the geometric component of the geographic information, require *a minima* an update of the attribute fields containing the measured data of this component, such as surface, perimeter and length, depending on the type of object. Second, if these operations are performed on a collection of objects, then the information previously stored in these objects is generally not modified. In the case of the forest, made up of several elements represented as surface objects, the structure of the semantic data is not modified (Figure 3.3). The configuration is different if these operations involve objects drawn and formalized a posteriori. Acquired or calculated quantitative data cannot generally be joined to the sets and subsets created. For example, in the case of the glacier, its constituent elements are not represented in the dataset and remain "in the black box". Partitioning the glacier surface according to the catchment area to which it belongs does not allow data acquired at the glacier level (e.g. climate and melting rate) to be added to this subset. Disaggregation operations are then necessary, if the assumptions underlying the change in scale are known (see intersection example below).

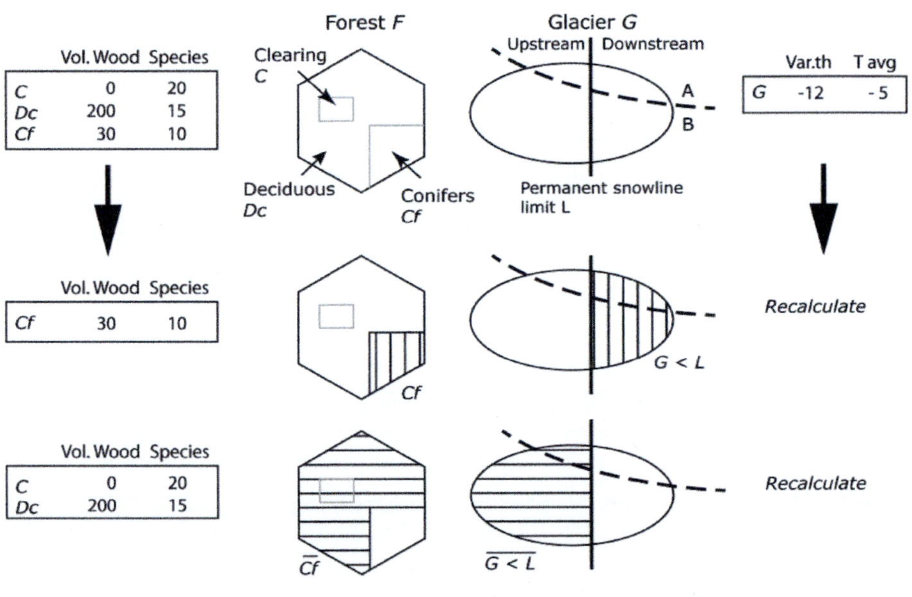


**Figure 3.3.** *Examples of attribute data management difficulties in the case of partitioning and union. For example, forest objects are characterized by the volume of wood produced (Vol.Wood in m$^3$) and the number of species present (species). Glaciers are characterized by variations in thickness (Var.th in m) and mean annual temperature (T avg °C)*

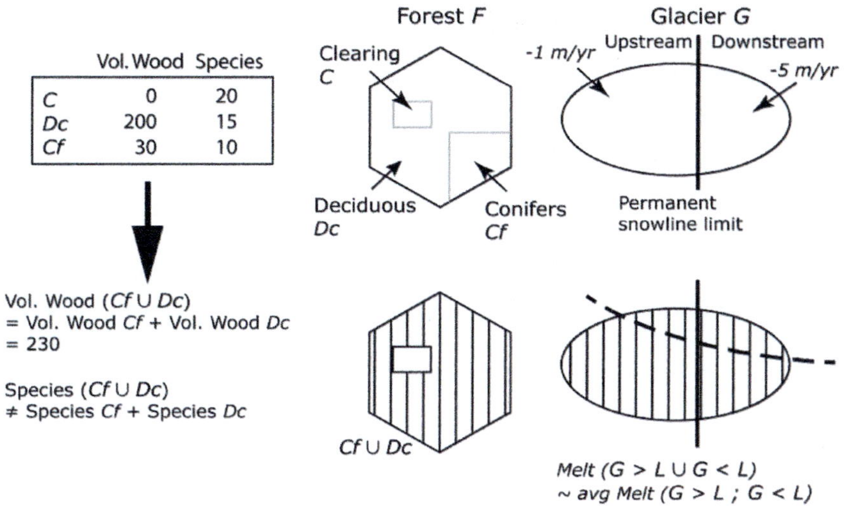


**Figure 3.4.** *Example calculations for unions*

– *Union*: the union operation can lead to a merging of the geometry of the objects, and thus to an aggregation of the data previously stored in the sets and subsets involved in the union. In the case of quantitative stock variables, the sum is generally the necessary mathematical aggregation operator. For example, if coniferous areas, with a wood volume of 30 $m^3$, and deciduous areas, with a wood volume of 200 $m^3$, are combined in a new set, then the latter will indeed have 230 $m^3$ of wood (Figure 3.4). However, this rule cannot be generalized: for example, it is not possible to add the number of fauna species present in the clearing zones to the number observed in the deciduous zone. In the case of quantitative rate variables, the sum has no mathematical significance and should therefore be avoided. The average, on the other hand, may have significance. In the case of the glacier, if a subset located below the altitude of permanent snow, where the loss of thickness is 5 m/year, is combined with the subset located above permanent snow, where the loss of thickness is 1 m/year, then an average can be calculated (incorporating, for example, a weighting, linked here to the surface area of the subsets).

– *Intersection*: intersecting N sets involves successive sets of combinatory operations to deconstruct the geometry of the original sets down to the desired level of granularity. This deconstruction requires disaggregation of the data, and it is generally not possible to associate the data acquired at the scale of the initial sets with the sets resulting from the intersections. Downscaling should only be used if there is a known link between the initial level and the final level, that is, if the sub-grid dynamics are known in such a way that it is possible to relate the macro variable to micro contextual variables. In the case of quantitative stock variables, the assumption that stock is distributed in proportion to surface area can be considered. For example, in the case of the glacier, the intersection between the "portion of glacier below the permanent snowline" and watershed A will be assigned a volume of ice corresponding to the ventilation of the stock estimated for the "portion of glacier below the permanent snowline" as a direct function of the proportion of the surface area of watershed A, relative to watershed B, in this zone of glacier below the permanent snowline (Figure 3.5). This hypothesis, though plausible, can only be used as an estimate, with a margin of error due to local variability in the geometry of the glacier and its bedrock. In many other cases, and particularly in the case of quantitative rate variables, downscaling necessitates the integration of new data. For example, the rate of thickness loss calculated over a wider area cannot be assigned to the subset created by the intersection without incorporating corrective factors derived from local context effects. In this case, topoclimatic data would be necessary (exposure to solar radiation in particular).

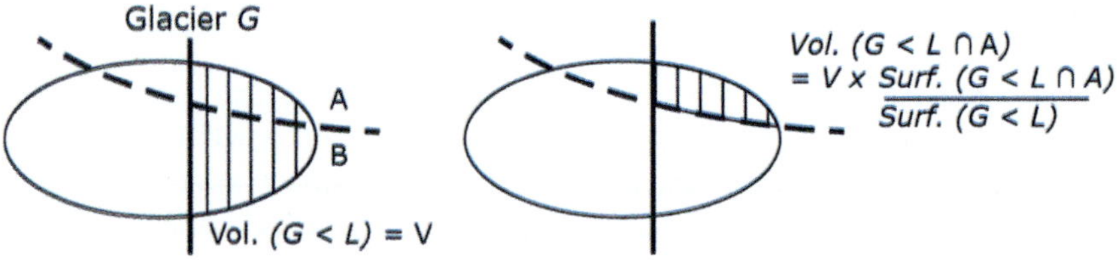


**Figure 3.5.** *Example calculations when sets intersect*

## 3.3. Dealing with disparate and incomplete data: examples from environmental geography

### 3.3.1. *Defining cartographic objects from incomplete data*

#### 3.3.1.1. *How to fill in the gaps?*

Mapping a state of the biophysical environment requires a reading that is as objective as possible and, for example, avoids the modifiable areal unit problem (MAUP, see Chapter 4), which can distort the reading of spatial structures. The main solution is to work with regular grids, whether raster or vector-based, to reveal patterns of organization: gradients, discontinuities and poles (or hotspots in today's terminology). In most cases, the challenge is to fill in the gaps between station observations in order to obtain a continuous representation of the phenomenon. An intuitive approach is to use an interpolation method to estimate the level of the phenomenon at each grid entity. This will be applied here in the context of an example of mapping the state of freeze-up in a massif. Other approaches are possible, based on set theory, notably when the data associated with biophysical entities need to be processed and stored in a grid with a sufficiently large mesh size to enable the use of the operators described above.

Whatever the method, the challenge is to choose a relevant indicator that reveals the state of the environment in relation to the issue under study. The aim is to build indicators that are robust, that is, reproducible and sufficiently stable to allow for comparisons of time and space between multiple geographical objects.

COMMENT.– The term "indicator" is also frequently used in spatial analysis demonstrations. Its etymological meaning is emphasized: "one who points out the significance of a phenomenon" (Brunet 1993). It therefore refers to a variable that is selected a priori as being representative of a phenomenon to be studied. While any statistical data or measurement derived from field observation can be an indicator, we prefer to use this term when referring to a conceptualized variable, designed to approximate a phenomenon or the property of a geographical object. In this case, the construction of an indicator is an essential part of the scientific process. The first step is to clarify and qualify the property or phenomenon under study, and to understand why raw, unprocessed data are insufficient when it comes to providing a reliable, objective measure. Indicators will then combine several elementary variables to propose a synthetic measure of the phenomenon and/or property being studied.

### 3.3.1.2. *Case study: mapping glaciation*

On a regional scale, glacier geometry cannot be represented: glaciers have been generalized into a subset of points, drawn in order to map the glacial equilibrium-line altitude (ELA). However, this cartographic translation cannot be carried out without difficulty, as the pattern varies over time, for each date being evaluated. Indeed, as a result of glacial confluence during glaciation and glacier splitting during retreat, some glaciers may join together to form a single glacier, while others may break up to form several. Similarly, a glacier may simply disappear or be born over the study period.

This instability of the glacier body raises a twofold question. First, how can one draw up diachronic maps of moving objects? Second, how can one quantify ELA variation using simple statistical parameters (mean, standard deviation, etc.)? Indeed, neighboring glaciers, resulting from the splitting up of a single glacier, will have similar ELAs and will weigh heavily on the calculation of the mean. These two problems can be solved by switching from a discrete representation of an ELA to a continuous one. This involves interpolating ELA values in order to construct a spatial field. Climatologists, seeking to recreate precipitation or temperature fields, have adapted various interpolation methods (Joly et al. 2003; Madelin 2013), which here largely inspire the work carried out on ELA. The principle is first to summarize the main features of spatial, regional variations in ELA. Spatial variation in ELA is related to latitude and longitude using geographically weighted regression (Cossart 2013). The fields obtained at different dates can then be compared with each other to produce diachronic maps. Finally, statistics for assessing variations in ELA elevation can be calculated, no longer from a corpus of "moving" individuals, but from all the pixels of the diachronic maps created in raster format.

**Choosing an indicator to map the state of glaciation**

The glacier equilibrium-line altitude (ELA), or permanent snow line, is the theoretical boundary between the glacier accumulation zone (the zone upstream of the glacier where snow accumulates and gradually transforms into ice) and the glacier ablation zone (the zone downstream where ice loss through melting or sublimation is dominant). At the end of the glacier's ablation period (summer in the Alps, for example), it is possible to observe the zone where snow accumulates permanently on the glacier's surface: the ELA then coincides with the snowline.

ELA varies over time, mainly according to two climatic parameters: precipitation and temperature. If temperatures rise, ice melting increases, the glacier's ablation zone expands and ELA increases. If precipitation increases, snow accumulation on the glacier increases, the accumulation zone expands and ELA decreases. Finally, an increase in ELA results in a retreat of the ice front, a contraction of the glacier, while a decrease causes an advance of the ice front, that is, an extension of the glacier.

There are several methods for estimating the ELA (Meierding 1982), most of them based on a geometric reconstruction of the extension of the glacier under study. It is commonly assumed that the accumulation zone corresponds to the upper two-thirds (in terms of altitude) of the glaciated area, with the ablation zone corresponding to the lower third. A map of a glacier's extent, compared with a topographic grid, can thus be used to reconstruct the altitudinal distribution of the glaciated area and determine the position of the ELA.

The altitude of the ELA, measured for different glaciers and at different dates, is fed into a spatiotemporal database, the basis for regional synthesis and mapping.

**Box 3.2.** *Equilibrium-line altitude: an indicator to be mapped*

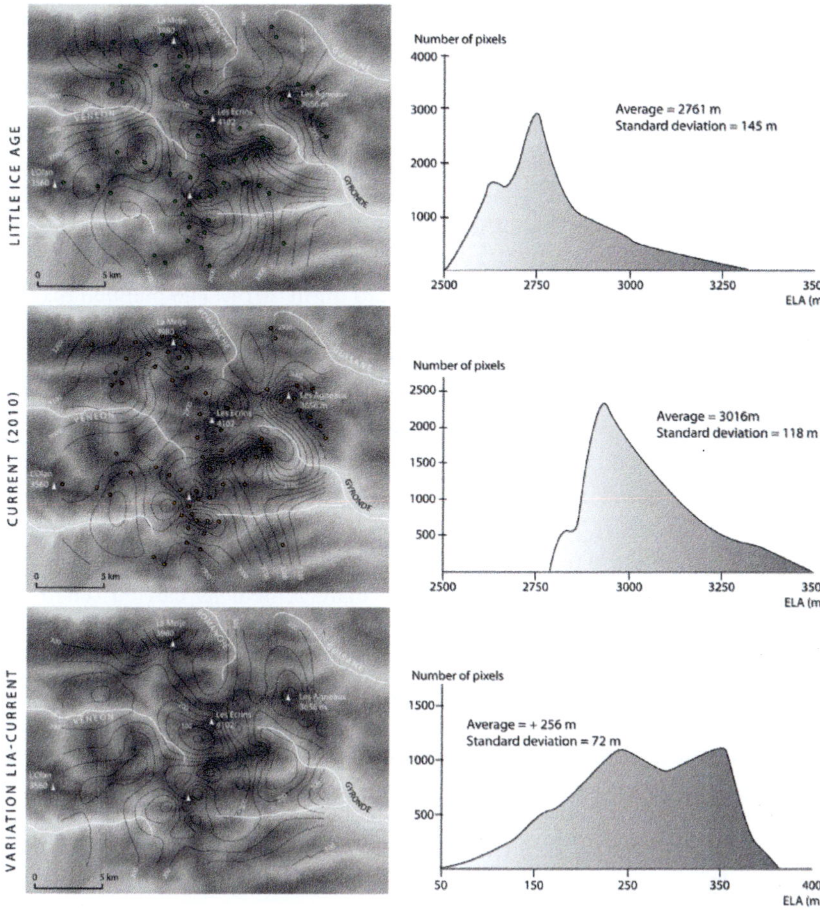


a) b)

**Figure 3.6.** *Mapping the state of glaciation using incomplete data. (a) Diachronic maps of ELA; (b) statistical distribution of ELA values*

This method is applied to the Écrins Massif (Figure 3.6). In 1880, 68 glaciers were identified; of these, four have disappeared, while the splitting up of the remaining 64 has led to an increase in the number of glaciers (72 in 2010). This example illustrates the importance of reconstructing the ELA elevation field to overcome the instability of the point segmentation. Since 1880, which corresponds approximately to the end of the Little Ice Age, there has been a clear upward trend in ELA (+180 m). Beyond these average values, estimated over the entire study area, a disparity is apparent in the way that glaciers react. Indeed, the diachronic maps clearly show that eastern glaciers have reacted with greater intensity than western glaciers.

#### 3.3.1.3. *Case study: environmental challenges in forestry*

Among the environmental debates associated with forests in France, which have emerged at the national level, those associated with the monocultures of exotic coniferous species have recently been in the media, particularly following the enactment of the recent *Loi pour l'Agriculture, l'Alimentation et la Forêt* "law for agriculture, food and forestry", which was voted on in 2014. The planned plantation of conifers is criticized in particular for altering biodiversity, and thus the ecological state of forests, with all that implies: soil acidification, simplification of the stratified structure and even the homogenization of ecological niches are just a few examples of these alterations (Ponge 2019). What is more, monocultures of exotic coniferous species reveal an opposition between local and central power since the 19th century, when the centralized service of forest engineers was created to govern over forest management in France (Corvol 1987; Simon et al. 2007), imbuing a social and cultural dimension to these forest environmental issues. These conflicts partly reflect the denigration by forest engineers of forest management as carried out by local farmers (Dodane 2009).

To complement these approaches, spatial analysis of conifer afforestation processes on a regional scale is required: this involves documenting transition patterns from which certain areas are converted to densely coniferous woodland. More precisely:

1) The rate of coniferous forest cover still requires some quantification and mapping.

2) The spatial patterns of conifer afforestation expansion need to be characterized in order to identify the underlying processes structuring this expansion.

We focus here on the Morvan massif, which is one of France's most densely forested regions, and where afforestation is the subject of debate and conflict (Ponge 2019; Marage 2020).

In order to answer these questions, the simple visualization of the expansion of deciduous and coniferous areas is totally inadequate (Figure 3.7). While it provides a numerical assessment, showing that the surface area of coniferous forest has increased significantly since 1990, reaching 78,660 ha in 2018, it is ineffective in identifying the spatial logic behind this expansion. The problem is twofold. First, the collection of diachronic maps representing surface areas and land use patterns is difficult to read. The cartographer can, of course, use intersection combinations to identify parcels that have undergone conversion(s). While these tools exist in GIS, they contribute to the dissipation of the geometry of the objects, making them unreadable at the scale of regional analysis. Second, coniferous forestation is a relative concept, that is, it corresponds to the proportion of resinous trees within a forest massif. This rate therefore evolves not only as a function of the absolute value of coniferous areas, but also as a function of deciduous areas, bearing in mind that the latter also have their own dynamics, linked to both ecological and anthropogenic factors.

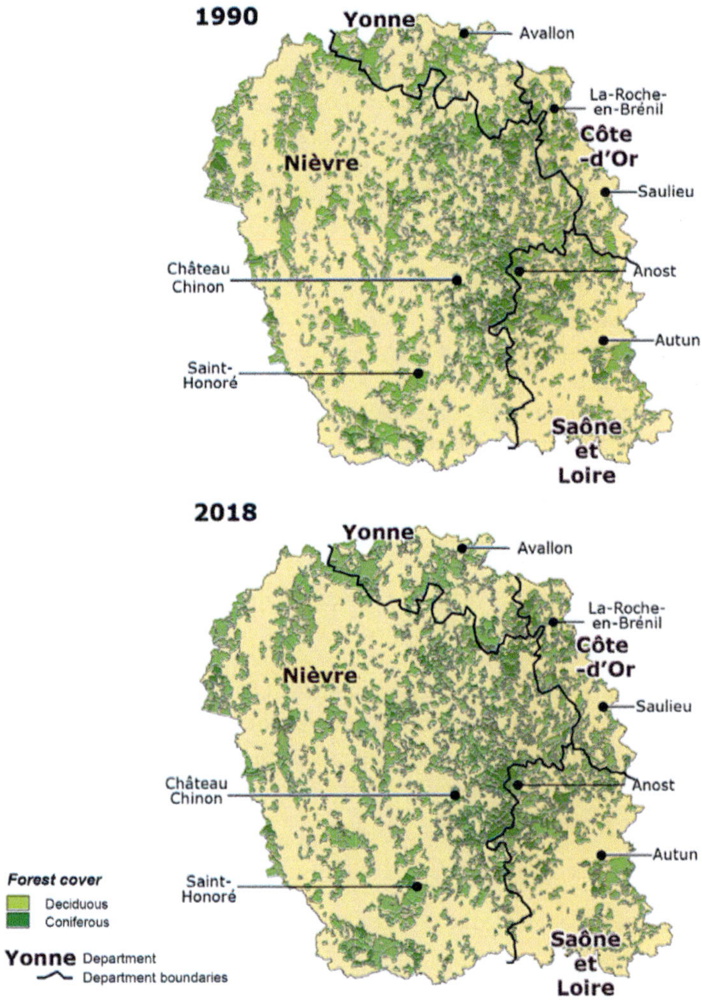


**Figure 3.7.** *Map showing the raw data on forest land use in the Morvan in 1990 and 2018 (source: Corine Land Cover)*

The solution to this dual problem is to construct a synthetic indicator that can be mapped at the desired scale. One way of understanding the co-evolution of coniferous and deciduous trees from an environmental point of view is to characterize the fragmentation of forest cover, in other words, the extent to which the expansion of coniferous areas is creating a sprawl of deciduous areas.

**Choosing an indicator to map forest fragmentation**

Fragmentation is a key concept in landscape ecology and corresponds to the degree of fragmentation of ecosystems and habitats. In this case, the indicator is based on the hypothesis that the emergence of a coniferous plot within a deciduous forest creates fragmentation, such that ecological connectivity may be impeded. Indeed, a coniferous plot leads to a loss of habitat due to the highly simplified structure of the monospecific plantations. Ultimately, it constitutes an obstacle to the free movement of species. Fragmentation is all the more damaging if it is accompanied by a loss of initial habitat area. Loss of surface area means that species no longer necessarily have sufficient critical surface area to satisfy their basic needs.

Fragmentation can be measured by calculating the distance between the center of gravity of each deciduous plot and the center of gravity of the five nearest coniferous plots. Lower distance values indicate a higher fragmentation pattern in the coniferous/deciduous forest. The distance values are interpolated in a raster grid (Figure 3.8). The grid facilitates comparison over time, and a synthetic map highlighting locations affected by increasing fragmentation (decreasing distance) compared with locations showing decreasing fragmentation (increasing distance) can be constructed for the period 1990–2018 (Figure 3.9).

**Box 3.3.** *Mapping forest fragmentation*

To fully map the extent of the environmental challenges, and in particular the threats posed by the spread of conifers on deciduous forests, we can add data relating to the stock of deciduous (in surface area) that could be milled. So as not to distort the reading of spatial structures, these data are broken down into a regular grid by the following operations:

1) intersection of deciduous polygons with mesh subdivisions;

2) calculation of the areas of the entities created;

3) combination of entities created within each mesh by aggregating their area (sum).

The same grid can be used for the different dates available to add information to the synthesis map (Figure 3.9).

In the case studied, forestation is accompanied by significant spatial fragmentation of the forest landscape. In fact, in the areas most affected by conifer afforestation (e.g. Haut-Morvan), the distances between the centers of gravity of deciduous and coniferous plots are very small (less than 1 km), reflecting a patchwork pattern. In other areas (notably the west), the distance is greater (5–7 km), suggesting that the forest is more made up of large, homogeneous patches (Figure 3.8). Since 1990, the general trend has been to exacerbate fragmentation: average distances have fallen from 3.1 to 2.6 km. This reduction is most marked in outlying areas (to the west and north), while fragmentation in Haut-Morvan remains fairly stable or is even slightly reduced (Figure 3.9).

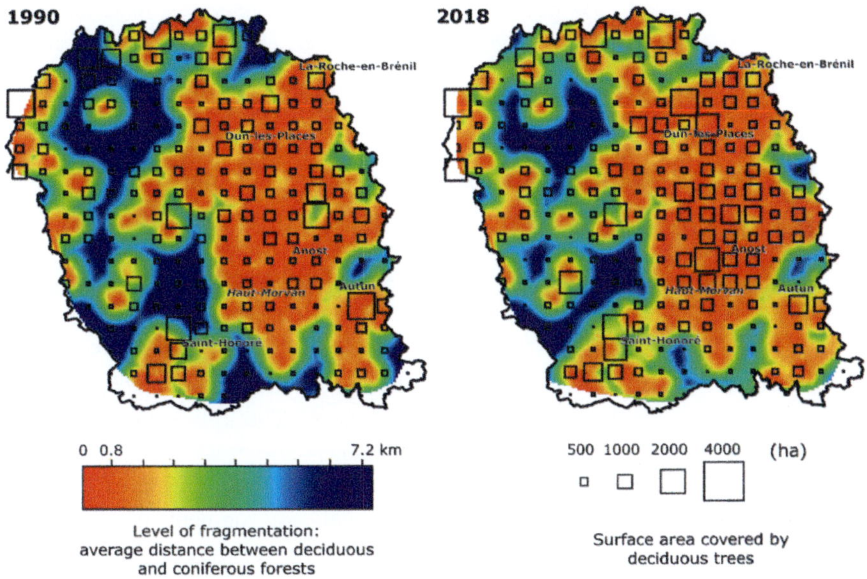


**Figure 3.8.** *Level of fragmentation within Morvan Forest*

The map shows that the area around Saint-Honoré is subject to increased fragmentation, while there are still large tracts of deciduous forest. In this area, the potential for the planned plantation of conifers (and concomitant fragmentation) is therefore high. Similar patterns can be observed to the north. Conversely, the Haut-Morvan is characterized by a stabilization or even slight reduction in fragmentation, while an increase in the area of deciduous forest is observed. This pattern, however, is occurring in an area already densely affected by conifer afforestation, which has probably reached a threshold in terms of thickening and fragmentation.

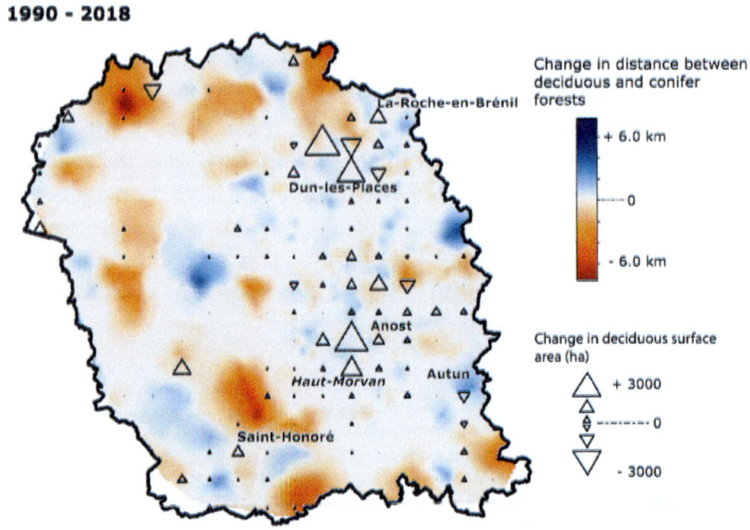


**Figure 3.9.** *Mapping the evolution of Morvan Forest fragmentation*

### 3.3.2. *Defining cartographic objects at reference scales*

#### 3.3.2.1. *Selecting the reference step*

Data derived from observation of the biophysical environment must be compared with territorial grids for two complementary reasons. The aim is both to provide decision-support elements for environmental managers and practitioners operating at the chosen scales and to identify spatial associations and, ultimately, possible causalities between processes in the biophysical environment and economic, demographic and social processes. In this case, the scale of reference must enable us to measure the intensity of spatial associations between phenomena of different natures.

Applied to the issues of forest cover in the Morvan, this approach involves comparing data on the state of forest cover with two variables linked to agricultural trajectory (agricultural hold versus agricultural decline) and demographic trajectory (rate of depopulation). The underlying hypothesis to be tested is that of increasing forest cover, associated with a negative agricultural or demographic trajectory. The test is carried out at the community level. The grids are fine enough to reveal different evolutionary trajectories depending on local contexts (proximity to urban centers, differentiated biophysical context between Haut and Bas-Morvan, etc.) (Cossart 2021).

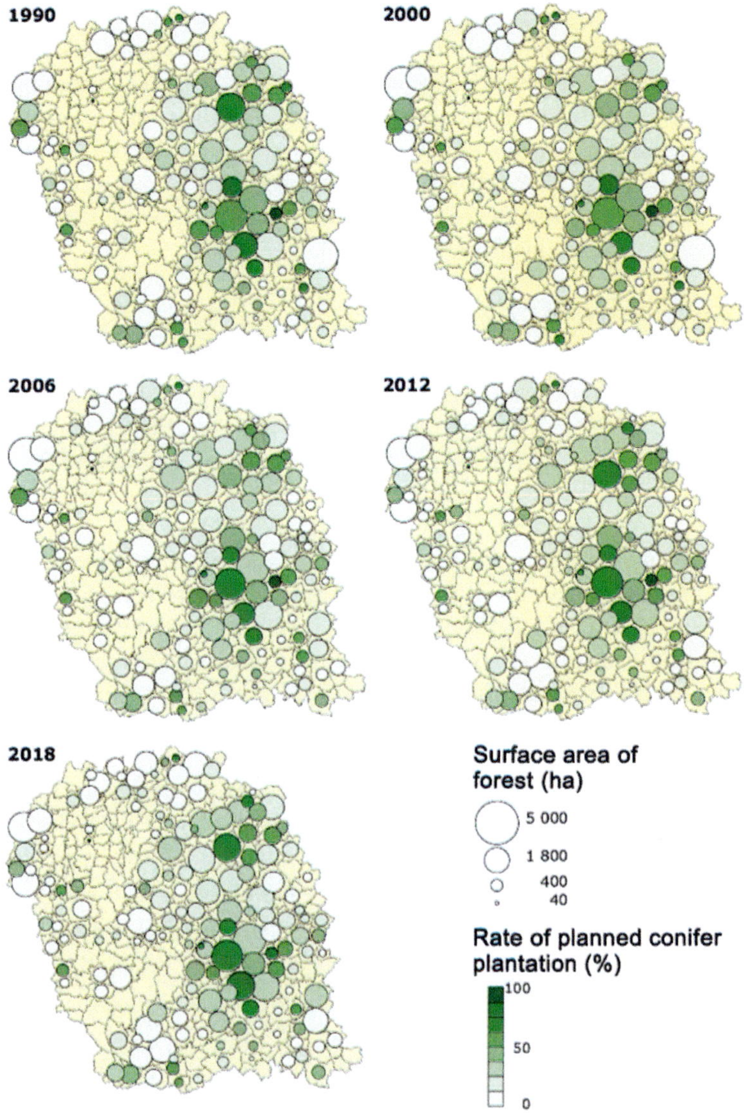


**Figure 3.10.** *Mapping the forest cover of the Morvan massif, from 1990 to 2018 (source: Corine Land Cover)*

In terms of data processing, the polygons corresponding to the coniferous and then deciduous forest zones are intersected by the polygons corresponding to the community grid. The surface entities thus created are then combined to form a single "coniferous forest" entity and a single "deciduous forest" entity for each community. The surface areas of these new entities are recalculated, with the results appended to the community data. For each community, we thus have the area covered by deciduous forest and coniferous forest, and this enables us to calculate the coniferous forestation rate (ratio between the area covered by conifers and the area covered by the forest as a whole). This calculation is carried out for several dates, from 1990 to 2018, in order to reconstitute the dynamics of the planned conifer plantation over a sufficient history (Figure 3.10).

#### 3.3.2.2. *Spatial associations and causality*

The collection of diachronic maps, while offering an initial representation of the dynamics of forestation, does not allow for a synthetic reading. This recurring problem in the representation of time (see Chapter 6) is exacerbated here by the fact that the variable represented (the rate of forest cover) depends both on the absolute value of the area covered by conifers and on the absolute value of the area covered by the forest as a whole. The latter is not stable and evolves according to the area covered by deciduous trees. To deal with this lack of simplicity in reading, a hierarchical ascending classification can be carried out: the chi-square metric is chosen here to carry out a classification that highlights types of evolving profiles independently of absolute values. Applied here to the surface area of coniferous forests, the classification method is detailed elsewhere (Cossart 2021), with the emphasis here on the resulting cartographic representation (Figure 3.11).

The classes are represented by the visual variable color, suggesting the intensity of the coniferous forestation dynamic: green for communities where coniferous forestation is decreasing and red for communities affected by worsening coniferous forestation (Figure 3.11). The darker the red, the higher the rate of afforestation. In order to visualize whether or not the changes affect a vast forest massif, the color is applied in proportional circles representing the surface area of the coniferous forest (the reference date here is 2018).

The spatial expansion of coniferous forest in the Haut-Morvan shows a classic pattern of spatial diffusion (Hägerstrand 1952). The source of the spread is in the Haut-Morvan, corresponding to Class 1 communities that have been precociously affected by conifer afforestation. Here, conifer cover is relatively stable, but at high levels. From this zone, spatial expansion was observed (mainly toward the north) at the beginning of the 2000s, with the emergence of a conifer belt around Dun-les-Places and La Roche-en-Brénil. At this point, the communities concerned reached a saturation point in the forestation process, materialized in the evolutionary profile by a plateau since 2000 (Class 3). Finally, the spread is now continuing in all outlying areas, where conifer afforestation was still negligible in recent decades (Classes 4 and 5).

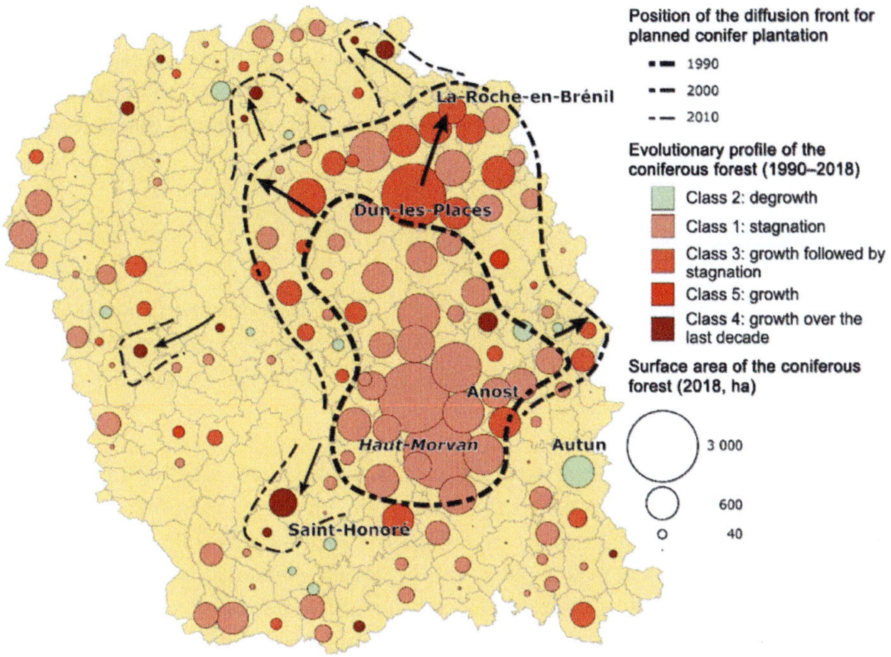


**Figure 3.11.** *Spatial increase of forest cover in the Morvan since 1990 (source: Corine Land Cover)*

Regardless of distance from the source area, the expansion of coniferous woodlands follows an S-shaped logistic curve, starting slowly, accelerating and then slowing down as a result of saturation. In the outlying communities, the process starts later (Classes 4 and 5). The evolutionary profile still places these communities in the acceleration phase of the process, which ultimately reaches lower proportions than in the other sectors, partly due to this initial delay. Conversely, in the area corresponding to the source of diffusion, the communities are at the end of the S-curve evolutionary model (Class 1): due to the already high level of conifer afforestation in 1990, there are few areas left to convert, and the opportunity for conversion remains low over the last three decades. Class 3 communities, on the other hand, show an intermediary pattern in which the entire logistic curve can be observed over the reference period.

In addition to this pattern of spatial diffusion, it is also possible to look for associations between the dynamics of tree-planting and agricultural or demographic trajectories. From this, a number of hypotheses can be put forward as to the causes of the emergence and the spread of planned conifer plantation. Diffusion patterns can be explained in part solely by the physical proximity of the areas progressively affected. In the present case, comparing these data on a local scale shows the rapidity of the changes observed and the facilitation of coniferous forestation in a context of agricultural abandonment.

## 3.4. Conclusion

We have recalled here some mathematical fundamentals derived from set theory, which underlie operations that are classically and easily performed in GIS software. The aim is to provide a few frameworks for these operations to show the range of possibilities in the construction of ad hoc cartographic objects.

In the field of environmental geography, it is all the more essential to reflect on the ways in which objects are constructed, since the challenge is to link biophysical data – which is often continuous, or in any case free of administrative boundaries – with socioeconomic data, aggregated in territorial divisions, which remains a complex problem. What is more, current scientific issues, and more broadly contemporary environmental challenges, focus on the dynamics of change that materialize in space and affect territories. In this context, the use of longitudinal observations is required and increases the difficulties involved in data management and representation.

The aim is to build cartographic objects that can be compared over time. This construction must go hand in hand with the search for an appropriate spatial granularity. Indeed, the grain size of the objects must enable spatial associations to be sought between disparate phenomena (e.g. biophysical, social and economic), evolutionary trajectories to be revealed (e.g. diffusion patterns) and the role of stakeholders involved at the relevant level to be compared.

The presentation of various examples, such as the case study on a glacier or a forest, highlights the multiplicity of methodological choices that can be made to meet the stated objectives and create suitable cartographic objects. The main pitfall is to test these choices by purely empirical means, whereas these must be based on explicit formal and mathematical frameworks. These conditions are all the more necessary as the objects created form the basis for the construction of indicators adapted to the objectives of cartographic representation, as explained in Chapter 4.

## 3.5. References

Brunet, R. (1972). Organisation de l'espace et cartographie de modèles. *Espace géographique*, 1(1), 43–48.

Claval, P. (1972). La réflexion théorique en géographie et les méthodes d'analyse. *Espace géographique*, 1(1), 7–22.

Cole, J.-P. and King, C. (1968). *Quantitative Geography.* Wiley, London.

Corvol, A. (1987). *L'Homme au bois. Histoire des relations de l'homme et de la forêt (XVII^e^-XX^e^ siècle).* Fayard, Paris.

Cossart, É. (2013). Influence of local vs. regional settings on glaciation patterns in the French Alps. *Geografia Fisica e Dinamica Quaternaria*, 36(1), 39–52.

Cossart, É. (2021). "Le Morvan est-il maudit ?" Analyse spatiale d'un territoire hyperrural. *Territoires en mouvement* [Online]. Available at: http://journals.openedition.org/tem/7276.

Dodane, C. (2009). Les nouvelles forêts du Massif Central : enjeux sociétaux et territoriaux : ces hommes qui plantaient des résineux pour éviter la friche. PhD Thesis, ENS, Lyon.

Durand-Dastès, F. (1993). Questions sur les modèles. *Géopoint*, 9–17.

Hägerstrand, T. (1952). The propagation of innovation waves. *Lund Studies in Geography*, 4, 1–20.

Joly, D., Nilsen, L., Fury, R., Elvebakk, A., Brossard T. (2003). Temperature interpolation at a large scale: Test on a small area in Svalbard. *International Journal of Climatology*, 23, 1637–1654.

Lambert, N. and Zanin, C. (2016). *Manuel de cartographie. Principes, méthodes, applications*. Armand Colin, Paris.

Madelin, M. (2004). Utilisation de l'interpolation spatiale pour estimer un alea gélif : application aux gelées printanières du vignoble marnais de 2003 et comparaison avec les dégâts observes. *Journées de climatologie 2004 Nancy : la représentation spatiale des variables climatiques*, 67–72.

Marage, D. (2020). Vivre de paysage dans le Morvan ou l'impensé de la gestion forestière morvandelle. *Projets de paysage* [Online]. Available at: http://journals.openedition.org/paysage/8182.

Mathian, H. and Sanders, L. (2014). *Spatio-temporal Approaches: Geographic Objects and Change Process*. ISTE Ltd, London, and John Wiley & Sons, New York.

Meierding, T.C. (1982). Late Pleistocene equilibrium-line altitudes in the Colorado Front Range: A comparison of methods. *Quaternary Research*, 18, 289–310.

Ponge, J.-F. (2019). Quel avenir pour la forêt française ? *Société Française d'Écologie* [Online]. Available at: https://www.sfecologie.org/regard/r86-mai-2019-jf-ponge-foret-et-sols/.

Simon, L., Clément, V., Pech, P. (2007). Forestry disputes in provincial France during the nineteenth century: The case of the Montagne de Lure. *Journal of Historical Geography*, 33(2), 335–351.

# 4

# Mapping and Identifying Geographic Configurations: The Example of Segregation

**Sylvestre DUROUDIER**
*Géographie-cités, Université Paris 1 Panthéon-Sorbonne, France*

## 4.1. Introduction

The representation of spatial forms, often through mapping, is a fundamental issue in the geographical approach to understand and explain societies. On the one hand, forms, also called spatial configurations, are generally complex, involving interactions between different dimensions at various geographical levels. However, on the other hand, what is captured by the map as a privileged tool for analyzing these forms depends closely on the measures chosen. These measures are part of a chain of procedures that starts with theoretical concepts, passes through data processing and ends with spatial representations. These methods therefore involve constrained or deliberate choices, in terms of the construction and visualization of results, which help to show or hide spatial forms. In so doing, these choices orient and subjectivize the geographical analysis of the phenomenon studied, whether or not they lead to misinterpretations, and whether or not they enable the level of geographical organization to be highlighted.

This chapter therefore tackles a problem that is simple in its formulation: how can spatial configurations be objectively analyzed? To shed light on this question, the chapter focuses on the thematic example of segregation. Indeed, understanding

*Handling and Mapping Geographic Information*,
coordinated by Claire CUNTY and Hélène MATHIAN. 

the spatial forms of segregation, defined as the uneven distribution of social groups in space, is fundamental to deciphering and objectifying the mechanisms behind the production of residential inequalities, both in terms of the factors responsible for these situations and the consequences they are generating.

Thematically, this approach is part of a long scientific tradition: the sociological urban ecology of Chicago School in the first half of the 20th century, then the geographical reformulations of the social area analysis approach in San Francisco from the 1950s onwards, before being taken up again via factorial ecology (Berry and Kasarda 1977), as well as by the French approaches to factorial urban ecology (Pruvot and Weber-Klein 1984) and the social division of space (Roncayolo 1996; Madoré 2004). More recently, these debates have been renewed with the rediscovery of factorial urban analysis by North American geographers (Reibel 2011): "methodological innovations for measuring, explaining and assessing the impacts of segregation and neighborhood change" (Wong et al. 2007). Very early on, this vast field developed well-known models for the social organization of urban spaces: concentric, sectoral and polynuclear. These models have been used time and again to read socio-spatial patterns, transposed indifferently onto other geographical contexts or other social dimensions. Therefore, everything seems to have been said about segregation analysis methods (Rhein 1994) in a vast literature in which two main approaches coexist: in one, segregation indices are used to compare cities with each other, mainly by North American works; on the other hand, multivariate analyses are applied to identify intra-urban spatial structures, which is the dominant approach in France. However, the permanence of these uses in this particular case calls for renewed reflection on the principles, presuppositions and challenges of approaches aimed at capturing and measuring spatial configurations and geographical organizations.

This chapter thus proposes a progressive discussion of the various theoretical and methodological issues involved in the analysis of spatial forms. The first part deals with the construction of geographical objects for analysis and the cartographic visualization of data. The second part starts with the definition of the categories of analysis to present the two main families of segregation measures: segregation indexes and multivariate analyses. The third section discusses these classic static approaches and presents approaches for a dynamic analysis of segregation: spatial indices and autocorrelation, discontinuities and neighborhood potentials.

This chapter focuses on ethno-racial segregation in Raleigh, North Carolina's capital and metropolitan statistical area (MSA) of 1.4 million inhabitants in 2020 (see Figure 4.1). This city has an ethno-racial profile typical of cities in the Old South: on the one hand, the historical dualism between a White majority (58% of the population) and a Black minority (18% of the population), and, on the other hand, growing ethno-racial diversification with Hispanic (12%) and Asian (7%) minorities. In order to study segregation in Raleigh, this chapter analyzes data from the 2020 General Census produced by the US Census Bureau. These data correspond to the ethno-racial affiliations declared by residents, aggregated into five categories that are common in North American studies[1]: Asian people, White people, Hispanic people, Black people and other affiliations. The case of Black people, abundantly documented in North American literature, serves as a common thread, and is punctually enriched by the other groups.

## 4.2. Mapping: rendering spatial configurations visible or invisible

The analysis of spatial configurations depends first on the ways in which geographical information is processed and mapped, according to a set of construction choices whose methods and pitfalls are well documented in cartography books and manuals[2]. Without going back over these aspects exhaustively, this first part discusses three decisive points for objectifying the forms of segregation in Raleigh: the definition of the space studied, the spatial units used and the making of the map.

### 4.2.1. *Defining the study framework*

The identification of spatial organizations is first and foremost linked to the definition of the space under study, as the perimeter chosen sets the interpretative framework for values, in this case those of segregation: are we talking about strong contrasts or, on the contrary, limited segregation?

In Raleigh, the most segregated neighborhoods for Black people are found not only in Wake County, but also in the chief municipalities of the outlying counties of Franklin and Johnson (see Figure 4.2). Therefore, while a focus on central Raleigh (see Figure 4.2(c)) gives the image of a sectorial bipartition, a broadening to all the counties of the metropolitan area (see Figure 4.2(a)) nuances this contrast both by the existence of a vast area whose neighborhoods have more than 30% Black residents, and by the segregation of this group in the peripheral municipalities.

1 For a detailed discussion of the definition of U.S. Census categories, see Le Goix (2016) and Duroudier (2018).

2 We refer, for example, to the books of Monmonier (1993) and Lambert and Zanin (2016).

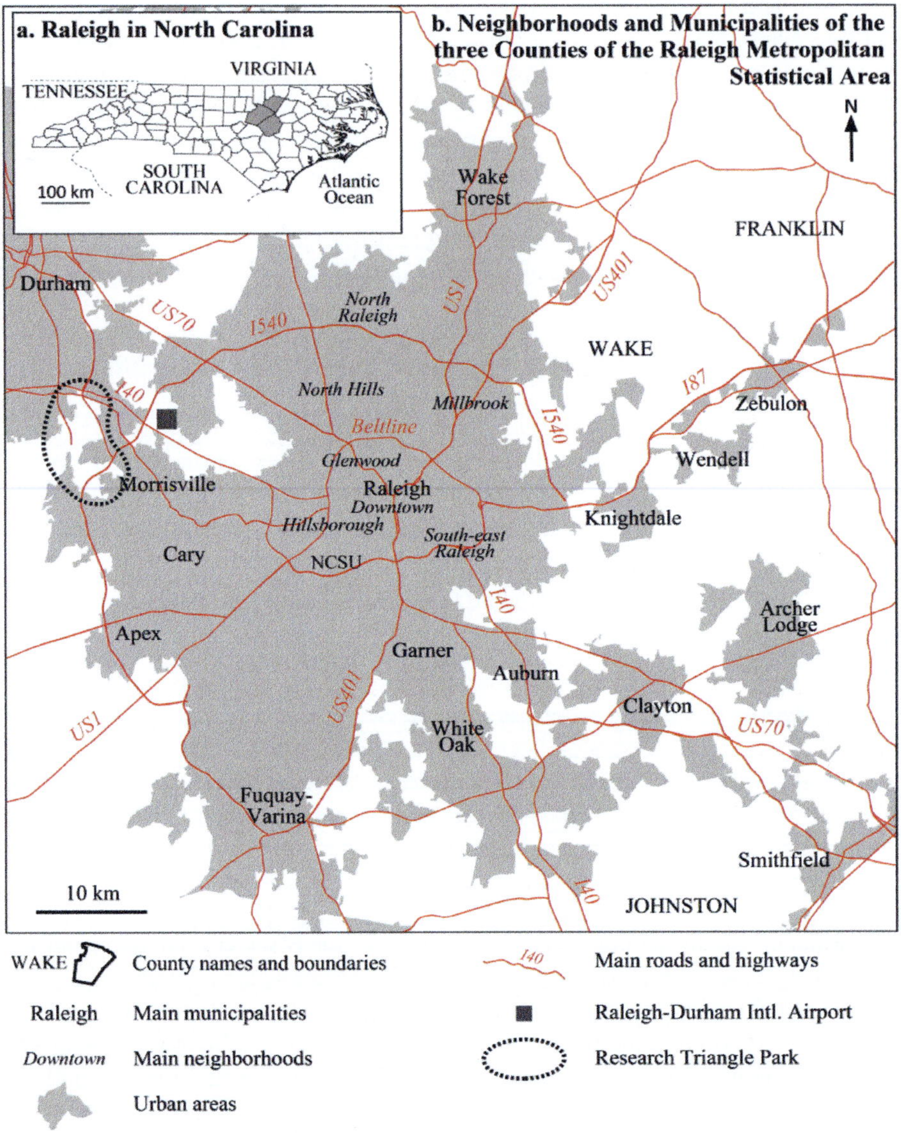


**Figure 4.1.** *Raleigh MSA, North Carolina (sources: US Bureau of Census, 2020; NHGS, 2021; Wake county, 2021. Map: S. Duroudier 2022)*

Several choices are then possible when it comes to defining the study area. For example, this chapter focuses solely on the three counties of the Raleigh metropolitan area (Figure 4.2(a)). Cartographically, this choice implies that the map stops at the limits of the space being studied, with the neighboring spaces being off the map. So if the geography of ethno-racial groups continues beyond the boundaries of Raleigh, this device has the effect of rendering the space under study "out-of-bounds", giving the false image of an insular space.

Another possibility is to restrict the map frame to the territory most intensely affected by the phenomenon. Here, this approach would tend to focus on Wake County (see Figure 4.2(b)) or the most central areas of Raleigh (see Figure 4.2(c)), where the historic Black and White segregated neighborhoods are located. More broadly, this approach is often used to show archetypal geographical figures such as the Black ghetto or community enclaves. On the other hand, it marginalizes in the discourse other interesting spaces, such as the mix that exists in many suburbs. Above all, this framing refers to the goal of the analysis: what socio-spatial configurations are we trying to capture?

### 4.2.2. *The scale of spatial configurations: a question of geographic mesh*

#### 4.2.2.1. *Geographical meshes and the modifiable areal unit problem (MAUP)*

A second important feature of pattern analysis is the choice of a mesh to aggregate data (Openshaw 1981). For intra-urban ethno-racial segregation in the United States, the US Census provides aggregated data in three nested meshes:

– *Blocks*, which are blocks or allotments of a few hundred inhabitants;

– *Block Groups* (BG), which are groupings of four blocks and generally comprise between 800 and 3,000 inhabitants;

– *Census Tracts* (CT), which are groupings of four block groups and generally comprise between 2,500 and 8,000 inhabitants.

Figure 4.3 shows the maps associated with each of these geographic meshes. The analysis of segregation patterns is a good example of the influence of mesh size on the interpretation of spatial organization. Numerous French and North American studies of ethno-racial, educational or socioeconomic segregation have shown how the design of meshes influences the profile of neighborhoods and the analysis of segregated forms (Fotheringham and Wong 1991; Taylor et al. 2003; François and Ribardière 2004; Ball and Petsimeris 2010).

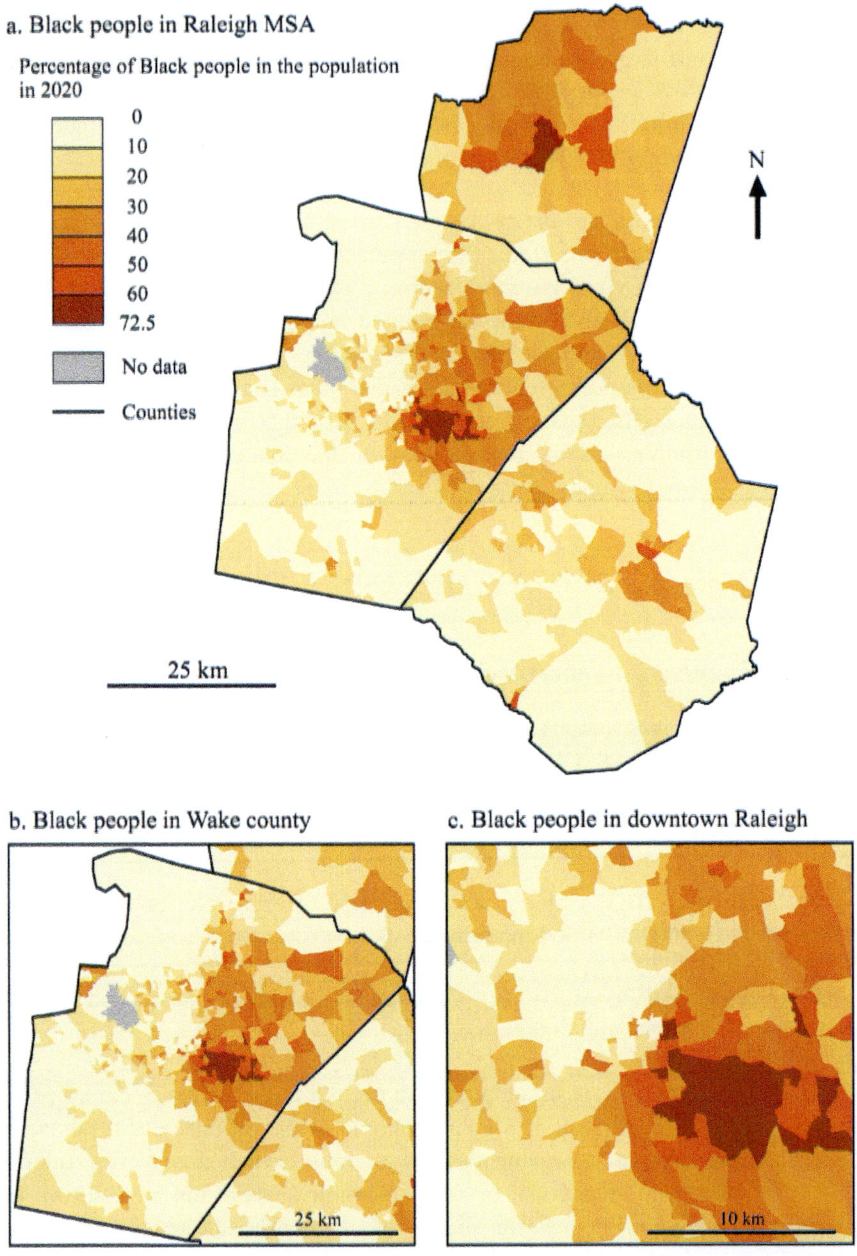


**Figure 4.2.** *Three framings of the map of Black people in the Raleigh MSA in 2020 (source: US Bureau of Census, 2020; Map: S. Duroudier, 2022)*

This question refers more broadly to the MAUP, that is, the effects of mesh size on the measurement of the unequal spatial distribution of a population or resource (Madelin et al. 2009). For example, a residential block occupied entirely by Black people, which is aggregated at the upper level with neighborhoods corresponding to the historic central ghetto, will reinforce the intensity of racial concentration in this neighborhood. Conversely, if this block is aggregated with spatial units characterized by a mixed ethnic composition, it will reinforce the image of increasing ethnic diversity.

#### 4.2.2.2. *Resolution of the mesh and scale of organization of the phenomenon*

The phenomenon observed varies according to the scale of the spatial partition, so that the mapping of Black people in three meshes (see Figure 4.3) shows a diversity of forms of segregation.

While the Raleigh MSA is organized around a main Black ghetto at the census tract scale, multiple highly segregated forms emerge at finer scales. At the block scale, the metropolitan area is broken down into numerous high concentrations of Black populations, particularly in the north, in Franklin County, whereas these spaces seem more mixed at the census tract scale. Therefore, the smaller the mesh, the greater the local fragmentation due to the internal homogeneity of the mesh that mechanically increases.

Conversely, the enlargement of spatial units has the effect of increasing internal diversity while making them more similar to each other. This aggregation effect depends on the structuring of the geographical space, so that blocks make it possible to capture local contexts of segregation, as is the case for Black people in some housing estates in western Wake County. Nevertheless, larger meshes can also capture other structures, such as a gradient on a metropolitan scale, particularly in the east-central part of the metropolitan area.

#### 4.2.2.3. *Mesh shapes and scale of organization of the phenomenon*

The analysis of spatial configurations is also sensitive to the shape of meshes. This problem, well documented in electoral geography through the question of gerrymandering (Boquet 2003), raises the question of which mesh is relevant to the nature of the studied phenomenon. The first option is to rely on political and administrative divisions. In terms of segregation, their interest lies in their direct articulation with public policies, for example, in the social housing, school investment, public transport or accessibility to public services.

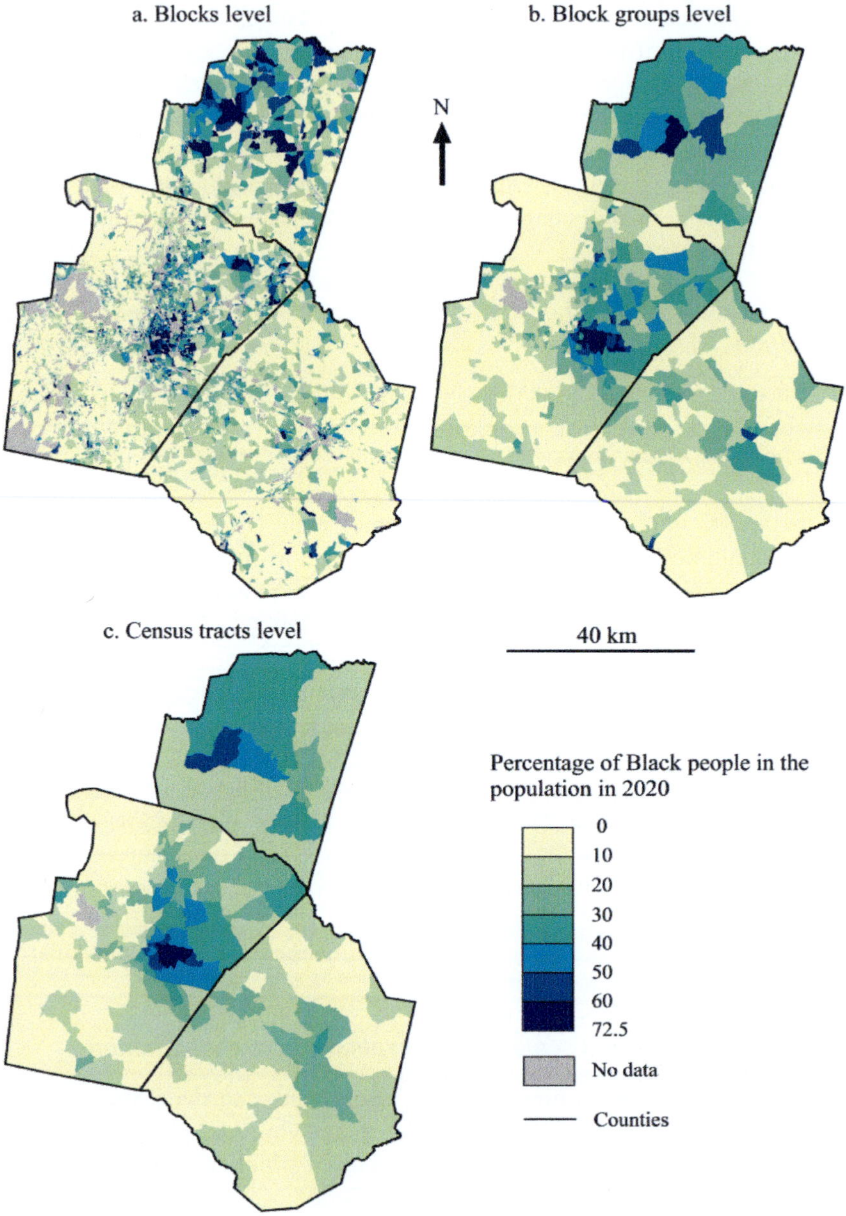


**Figure 4.3.** *Three representations of segregation: Black people in the Raleigh MSA in 2020 according to three US Census meshes (source: US Bureau of Census, 2020; Map: S. Duroudier, 2022)*

A second option is to use the statistical delineations proposed by the US Bureau of Census. In line with the objective of exhaustive coverage of the territory, these observation meshes are constructed according to identical criteria everywhere: generally based on the road network and on comparable numbers of resident population.

However, these meshes are not always neutral, since, for example, the construction of census tracts for the 2000 general census in the United States followed the objective of maximizing ethno-racial homogeneity within the spatial units, with the aim of facilitating the identification of segregated neighborhoods and the implementation of desegregation policies (Duroudier 2018).

A third possibility is to rely on geometric meshes. These seem to objectify geographical configurations to a greater extent, since the division of space follows the same geometric principle (squares, diamonds, hexagons), remedying surface inequalities between meshes (Openshaw 1981). For data that have already been aggregated, this means recalculating headcounts in new spatial units through a process of interpolation, which raises further methodological and theoretical issues in defining the shape and size of the mesh (see Figure 4.4).

#### 4.2.2.4. *What is the right partition of space?*

What is the appropriate partition of space to remedy the effects of a MAUP (size and shape) in the analysis of the spatial configurations of a phenomenon? On the one hand, this question does not arise because a MAUP can also be a revealing prism for geographical forms (Madelin et al. 2009). On the other hand, there is no definitive answer to this question, as the choice of level is made upstream of data processing. Whether we are talking about a social or environmental phenomenon, studies often opt for the finest possible mesh, in order to get as close as possible to the elementary units being observed. Such is the case with work on segregation, which seeks to get as close as possible to supposed homogeneous neighborhoods. However, this choice, at the interface between analytical objectives and methodological constraints, is not trivial, as the following example from Raleigh demonstrates.

First, the choice depends on the analytical objectives: is it to provide a picture of Black segregation in Raleigh? To identify these segregated neighborhoods in anticipation of the desegregation policies? To compare these segregated forms with other socioeconomic or demographic characteristics? These objectives call for the use of different partitions of space, such as geometrical ones in the first case study due to the abstraction of this type of mesh, or statistical meshes (CT and BG) as in the last case study, in order to be able to link it to other statistical data sets.

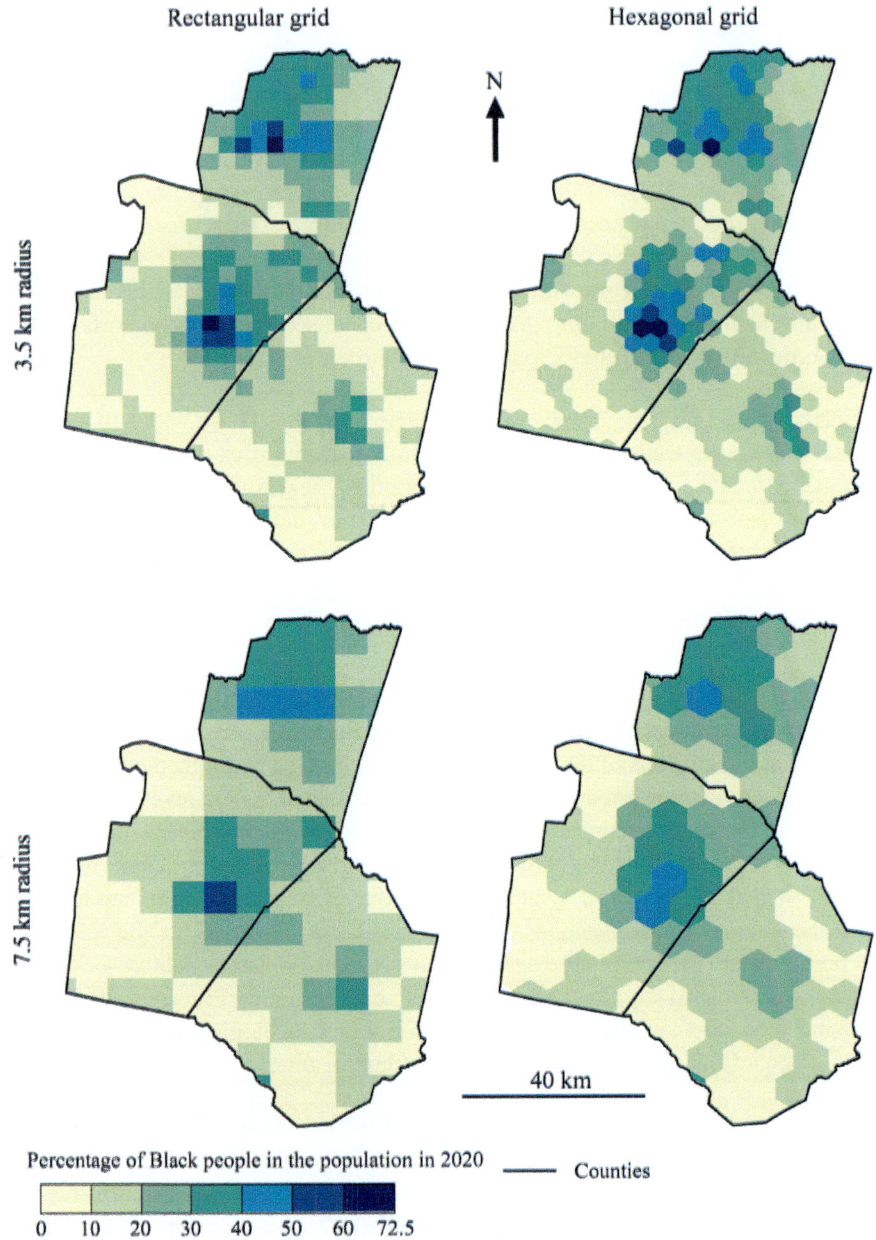


**Figure 4.4.** *Resolution and mesh shape: Black people in the Raleigh MSA in 2020 according to two types of regular meshes (source: US Bureau of Census, 2020; Map: S. Duroudier, 2022)*

Second, the choice of mesh is closely linked to the visualization provided by the map, as the forms must be legible and understandable (see Figure 4.3). In this respect, blocks allow us to appreciate not so much a spatial structure on a metropolitan scale as very local contrasts. Therefore, the choice of mesh size is often made after an exploratory phase, during which testing of cartographic visualizations reveals a variety of more or less relevant results, in relation to the definition of the area under study.

Third, the choice of mesh depends on the target audience. The level of technicality, the scale and the nature of the division will depend on whether the analysis is aimed at scientists, public authorities or local residents: for example, the mesh will be more political for municipalities, or at a medium level of aggregation for local residents.

In the remainder of this chapter, segregation measurements and maps will use the intermediate level of block groups, as it provides a more detailed picture of segregation than the census tracts typically applied in North American studies.

### 4.2.3. *How the map is created*

A third issue concerns the map-making process itself, since the choices made at this stage are crucial in rendering visible or invisible the spatial configurations of segregation. This stage is caught between the analytical objectives that determine the type of cartographic representation and the choices relating to the data represented and the visual parameters.

#### 4.2.3.1. *Spatial configurations and the nature of thematic variables*

In the broadest sense, the notion of segregation refers to the spatial exclusion of a group according to a discriminating criterion, a group which then finds itself in a situation of concentration within the space. Schematically, this concentration corresponds to two forms of the values taken by the variables. On the one hand, it is absolute, when we need to know in which neighborhoods, for example, the greatest number of Black people can be found. On the other hand, concentration is relative when the question concerns the identification of neighborhoods where the degree of Black homogeneity is greatest. The spatial forms of segregation associated with these two analytical objectives differ, but also they generally imply different modes of cartographic representation (see Figure 4.5).

First, absolute concentration is measured by the number of group members in each spatial unit. The proportional circle map (see Figure 4.5(a)) shows that Black people are most numerous in Raleigh's central and eastern neighborhoods. However, their presence is also diffuse in the other neighborhoods of the agglomeration, with the exception of Glenwood and the northwest, while block groups in outlying counties show lower concentrations.

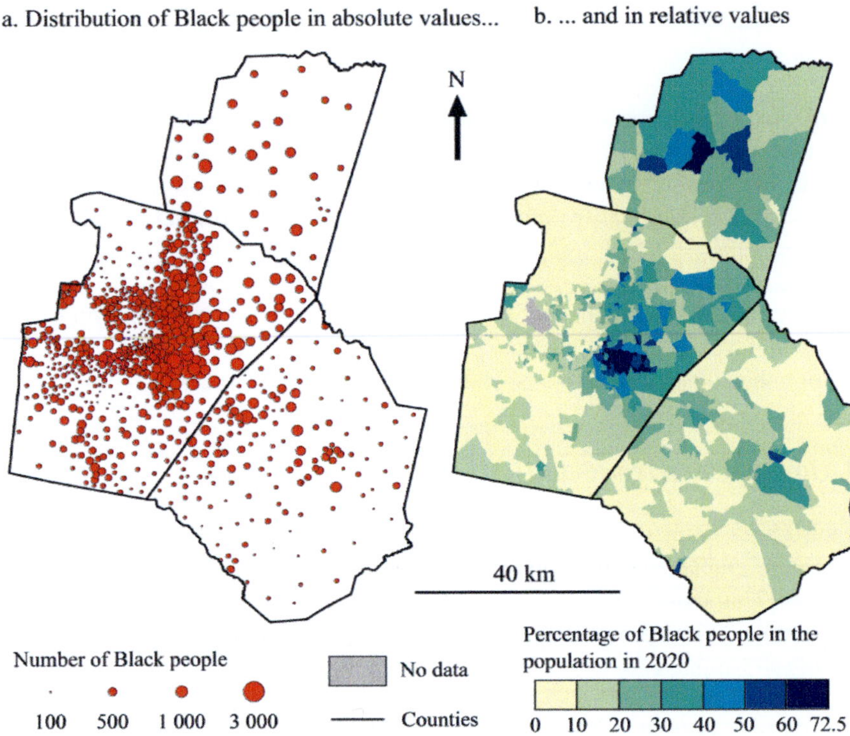


**Figure 4.5.** *Absolute and relative distribution of Black people in Raleigh MSA in 2020 (source: US Bureau of Census, 2020; Map: S. Duroudier, 2022)*

Second, relative concentration is analyzed by the group's percentage of the total local population. The choropleth map of Black percentages of the population (see Figure 4.5(b)) particularly highlights the historic ghetto in central Raleigh. In contrast, most of Raleigh's neighborhoods appear to be relatively unsegregated for Black people, while some rural areas are very homogeneous even though they have few Black residents.

In works on segregation, the focus is often on relative concentrations and percentage maps, while absolute numbers are rarely visualized. This may be due in part to the fact that percentages are a better way of capturing the geographical forms of homogeneity, particularly for places such as the Black ghetto. However, this is a partial picture, as the presence of Black people appears more diffuse in Raleigh. Thus, Figure 4.5 provides two complementary views of Black segregation in the Raleigh metropolitan area.

There are also other, rarer modes of representation that attempt to account for the joint logic of the distribution of numbers and relative weights.

A first solution is to depict the uneven distribution of ethno-racial groups using a dot map (see Figure 4.6). On these maps, each dot represents an equal number of people, usually a multiple of 10, in this case 100 Raleigh inhabitants. Since information is not available at the individual level, the points are randomly located within the block groups. In addition, each dot is characterized by an attribute such as ethno-racial affiliation, allowing for a multi-dimensional reading of segregation. So, while the density of points in the spatial unit reflects the absolute concentration of people, the homogeneity of neighborhoods is visually assessed by mixing the colors of the dots. The cartographic result is interesting: in addition to providing an image of diffuse mixing, it enables us to grasp the local contexts of segregation in greater detail: large-scale structures such as the partition between White neighborhoods in the northwest and Black and/or Hispanic neighborhoods in the southeast, as well as very local forms such as the concentration of Asian people in the very dense neighborhoods to the west of the NCSU campus.

A second solution is to map segregation using cartograms. These maps without Euclidean scale are the result of anamorphic techniques, which transform the geometry of spatial units on the basis of a quantitative attribute (see Chapter 7). In Figure 4.7, the Raleigh metropolitan statistical area is redrawn according to the Gastner, Seguy and Moore algorithm by (a) the total population in each block group and (b) the total number of Black people. The resulting maps provide a better understanding of local segregation contexts because, by giving greater visual weight to densely populated areas, they reveal the effects of proximity or distance between groups. According to total population, Raleigh seems to be divided into two main sectors in its center, one of which is a high concentration of Black people and the other is very homogeneous for White people, while the suburban rings are more diverse. On the contrary, the cartogram according to the number of Black people clearly illustrates the existence of a Black ghetto, characterized by the highest absolute and percentage concentration, while the peripheries are compressed due to their low numbers. The combination of choropleth mapping and geometric transformation techniques therefore has the advantage of placing the relative concentration of Black people in contexts of unequal densities. On the other hand, it makes analysis more difficult, as geographical landmarks are modified.

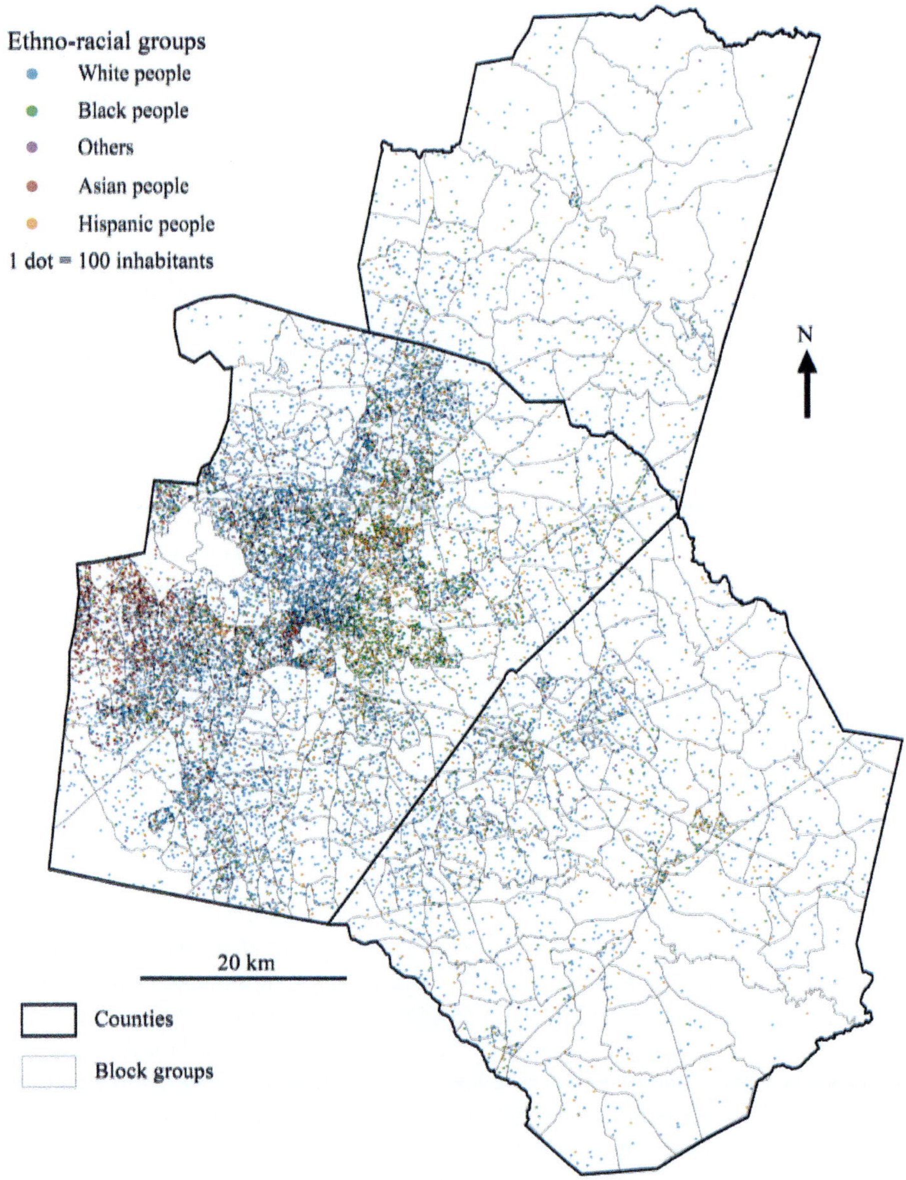


**Figure 4.6.** *Dot map of ethno-racial affiliations in the Raleigh MSA in 2020 (source: US Bureau of Census, 2020; The Racial Dot Map, Cooper Center, University of Virginia 2012; Map: S. Duroudier, 2022)*

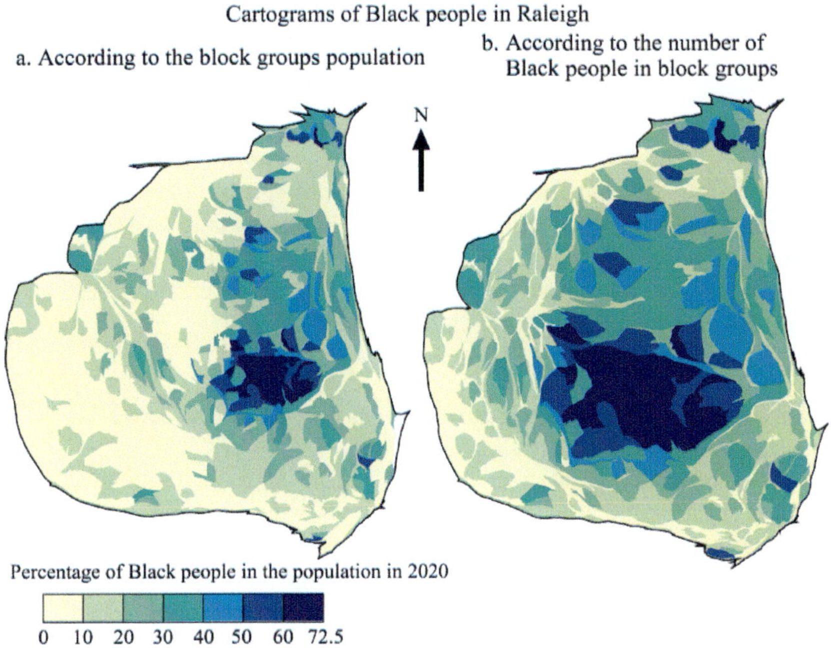


**Figure 4.7.** *Maps showing percentage of Black people in the Raleigh MSA in 2020 using Gastner et al.'s method with two distinct populations (source: US Bureau of Census, 2020; anamorphosis: package CartogramR, 2022; Map: S. Duroudier, 2022)*

### 4.2.3.2. *Data classification to reveal spatial patterns*

In the case of choropleth maps, the most frequently used, a crucial point in the map-making process, is data classification (discretization). This step, well known and documented in cartography manuals or applications (Béguin and Pumain 1994; Lambert and Zanin 2016), consists of cutting a continuous statistical distribution, such as the percentage of Black people, into a reduced number of classes to which spatial entities are assigned according to their values. However, depending on the classification methods used and the number of classes, the cartographic results differ greatly and guide the analysis. Figure 4.8 compares the variability of organizations produced by two methods and two choices of number of classes. Therefore, the parameters of classification lead to the highlighting or invisibilization of certain geographical forms, so that there is an eminently committed dimension to this stage, which requires explicit and justified positioning in the geographical analysis. For example, discretization by equal intervals illustrates the persistence of a Black "hyperghetto" as described by Massey and Denton (1993), whereas discretization by quantiles tends to invalidate it.

Another important parameter in map construction is the use of visual variables of color and value. First and foremost, it is a question of making the map legible: for example, by accentuating contrasts between levels of gray, by reinforcing proximities between colors, by facilitating immediate understanding through colors modeled on stereotypes, etc. With regard to ethno-racial segregation and discriminations based, among other things, on skin color, it would be easy to play on these colors: mapping Black people in a brown chromatic range (see Figure 4.2) or Asian people in yellow... Nevertheless, in turn, this can reinforce the stigmatization of these groups, making these choices inseparable from ethical questions and implying an explicit positioning vis-à-vis the choices made.

## 4.3. How to measure a phenomenon so as to reveal its forms

The previous section dealt with the analysis of spatial configurations through map-making, sidestepping questions of measurement, reduced simply to the weight of ethno-racial groups. However, on the one hand, the identification of spatial patterns is not limited to high concentrations, such as the most segregated neighborhoods. On the other hand, the definition of these forms is linked to questions relating to the measurement of phenomena and situated further upstream in the processing chain, for which segregation is a particularly demonstrative example: for example, the definition of statistical categories, the multifactorial aspect and the simultaneous consideration of several groups, or the level of aggregation of results (global or local). This section presents two main methodological approaches, which are particularly popular in segregation analyses and which produce contrasting results: indexes (here dedicated to measuring segregation) and multivariate analyses (within the framework of factorial urban ecology).

### 4.3.1. *The dependence of measures to the definition of categories*

First, measuring segregation in Raleigh requires defining the ethnic and racial groups studied. This is a fundamental point, as the perimeter of the categories has a direct impact on the counting of inhabitants, and therefore on the spatial forms observed. On the one hand, this question is part of broader debates on the relevance of the concept of race and ethnic statistics, complex issues that are not addressed here. On the other hand, given the persistence of racial discrimination, most North American studies use aggregate census categories without engaging in any real discussion of the effects of these definitions. The "race" variable proposed by the US Census is a very good example of the fact that these categories are constructions not devoid of discrimination or racist prejudice (Le Goix 2016, 2021).

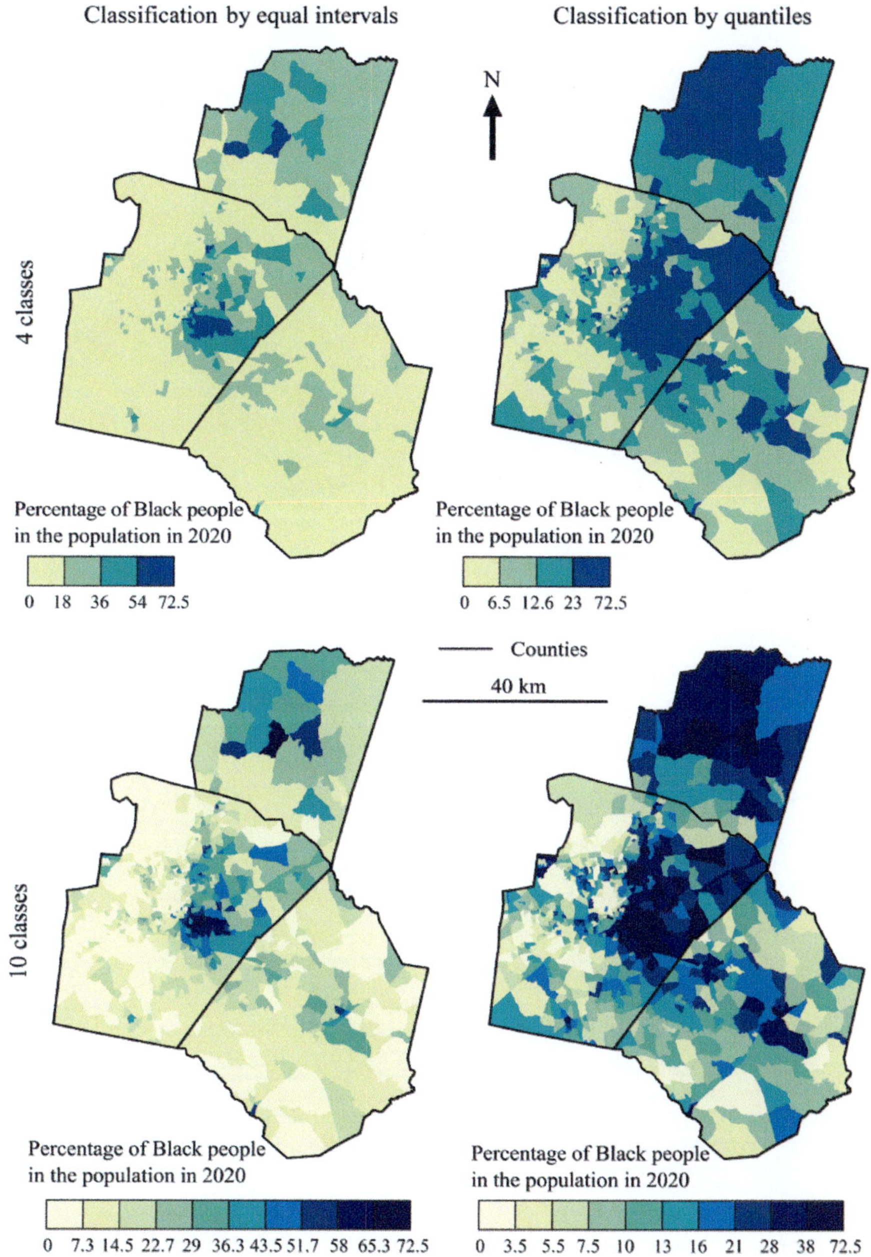


**Figure 4.8.** *Black people in the Raleigh MSA in 2020 according to several discretizations (source: US Bureau of Census, 2020; Map: S. Duroudier, 2022)*

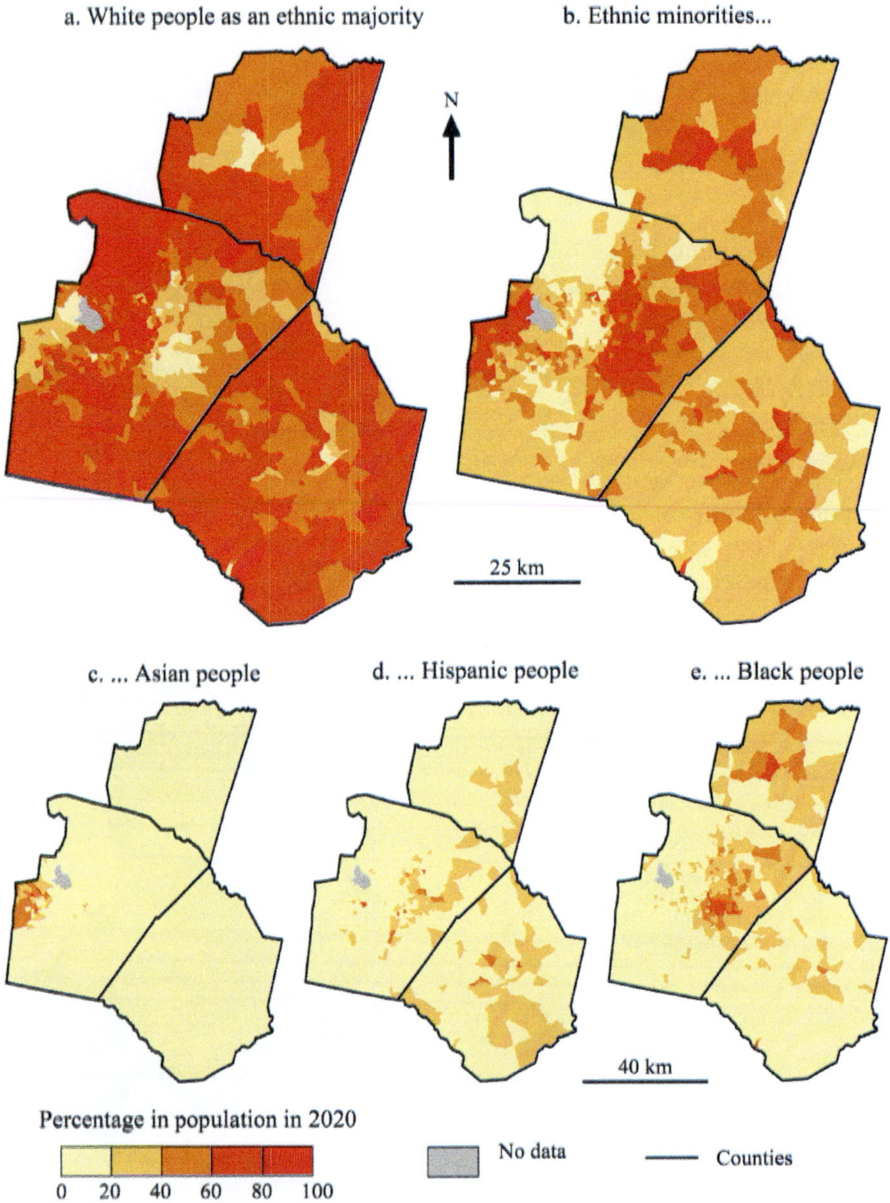


**Figure 4.9.** *Minorities and ethnic majorities in the Raleigh MSA in 2020 (source: US Bureau of Census, 2020; Map: S. Duroudier, 2022)*

Many North American studies, for example, focus on minorities defined as a single group whose segregation from a White majority group is assessed. This is primarily understood in terms of spatial assimilation theory, according to which the integration of ethnic minorities into the "Anglo-White" model requires access to the urban suburbs (Alba and Logan 1991; Charles 2003). Initially applied to immigrants, this interpretative scheme was transposed to Black segregation, then generalized to all ethnic minorities (Wright et al. 2005). It explains the preponderance of the majority/minority dialectic in measures of segregation.

As shown in Figure 4.9, this level of group definition helps to construct an opposition between White neighborhoods in homogeneous rural areas or in the suburbs of Raleigh's western half (see Figure 4.9(a)) and neighborhoods with a high concentration of ethnic minorities in the historic ghetto, more widely in the eastern half of the city, as well as in certain recent neighborhoods in western Wake County (see Figure 4.9(b)). Nevertheless, this level of reading masks very different logics in the segregation of ethno-racial groups (see Figures 4.9(c)–(e)): while Black people are organized around central neighborhoods, Asian people appear highly concentrated to the west toward the Research Triangle Park, while Hispanic people are located more diffusely throughout the city.

### 4.3.2. *The indices approach*

#### 4.3.2.1. *The variety of indexes and their abundance in segregation studies*

A first approach to forms of spatial organization involves the calculation of indices. The theme of segregation is exemplary for this approach, while being a singular field of study due to the profusion of so-called segregation indexes. The following paragraphs present the methodological developments carried out specifically on the theme of segregation; however, the discussion has a broader scope that also applies to other themes and indices.

There is a tradition of measuring segregation based on indexes, dominant in North American urban sociology and geography since the 1940s (Rhein 1994). These indexes are aggregated, meaning that each one provides a single value characterizing the intensity of segregation for the entire studied space. Since then, an abundance of segregation indexes has been proposed (some of which are presented in Table 4.1), whose significance, contributions and limitations have given rise to considerable debates (Rhein 1994; Apparicio 2000). Paradoxically, the vast majority of works have used only the dissimilarity index proposed by Duncan and Duncan (1955). In a canonical article, Massey and Denton (1988) suggested classifying these indexes into five distinct dimensions of segregation.

*Equality* measures the degree of concentration of a group's statistical distribution compared to its theoretical equal repartition into spatial units, using a calculation similar to the Gini Index. These indexes are the most widely used, not least because of their simplicity of calculation and interpretation, since the result can be understood as the proportion of the group that has to move to reach the equal repartition. More specifically, the dissimilarity index measures the gap between the distribution of one minority group in relation to another majority group, which made it popular in particular in an approach to the segregation of Black people from White people, before being extended in the 1990s to other ethno-racial affiliations.

| Dimensions | Indices | White | Black | Asian | Hispanic | Other |
|---|---|---|---|---|---|---|
| Equality | Segregation (Duncan) | 0.35 | 0.4 | 0.54 | 0.3 | 0.11 |
| | Dissimilarity (Duncan), in relation to White people | – | 0.44 | 0.55 | 0.36 | 0.17 |
| | Entropy (Theil) | 0.13 | 0.15 | 0.26 | 0.09 | 0.01 |
| Exposure | Isolation (Bell) | 0.65 | 0.3 | 0.27 | 0.19 | 0.05 |
| | Interaction (Bell), in relation to White people | – | 0.45 | 0.49 | 0.49 | 0.59 |
| Concentration | Absolute (ACO) | 0.4 | 0.65 | 0.9 | 0.77 | 0.86 |
| | Relative (RCO) to White people | – | 0.1 | 0.46 | 0.2 | 0.1 |
| | Delta (Duncan) | 0.51 | 0.59 | 0.77 | 0.52 | 0.53 |
| Clustering | Absolute (ACL) | 0.24 | 0.15 | 0.22 | 0.05 | 0.01 |
| Centralization | Absolute (ACE) | 0.52 | 0.61 | 0.6 | 0.54 | 0.55 |
| | Relative (RCE) to White people | – | 0.19 | 0.03 | 0.08 | 0.04 |
| Multigroup Indices | Dissimilarity | 0.35 | | | | |
| | Entropy (H Shannon) | 1.21 | | | | |

**Table 4.1.** *Ethno-racial segregation in Raleigh MSA in 2020 according to different block group-level segregation indices (source: US Bureau of Census 2020; Tivadar* et al. *2014)*

*Exposure* measures the probability that members of one group will encounter members of their own group (isolation) or another group (interaction) in their spatial unit (Apparicio 2000). It is interpreted as "degree of potential contact between members of the same group (one group) or between members of two groups (intergroup) inside spatial units" (Massey and Denton 1989). For the isolation index, which is the most widely used, the higher the value, the greater the potential interaction, but this value is dependent on the overall size of the group.

*Concentration* takes into account the surface area occupied by a group in all the city's spatial units. For Massey and Denton (1988), the higher the concentration in a small area, the more segregated a group, and conversely, a low value indicates low segregation and high social integration. Paradoxically, from a geographical point of view, we can assume that lower density leads to a lower probability of social interaction. Moreover, this dimension is little used, as the surface area is dependent on urban density and does not reflect a real segregation effect.

*Clustering* refers to the geographical proximity of segregated spatial units within the metropolitan area. This dimension, which directly raises the question of the spatial configurations of segregation, has been the focus of particular attention by geographers. They have proposed indices that integrate the distance or contiguity of spatial units, but these are rarely applied due to their complexity. Section 4.4 provides an in-depth discussion of the relevance of these so-called spatial indices.

*Centralization* measures the distance or proximity of a group to a given location in the metropolitan area. In North American interpretative theories of spatial assimilation, this place corresponds to the city center in a ghetto-suburb duality. In practice, the aim is often to capture the degree of ethnic concentration within a central urban area, such as the central municipality.

#### 4.3.2.2. *The limits of the indices approach: crossing indices and multi-group indices*

For each of the dimensions defined by Massey and Denton, several indices can be computed (see Table 4.1)[3]. Both the dimensions and the indices deliver contrasting, even contradictory, results in terms of the segregation of ethno-racial groups. For example, Black people appear less segregated than Asian people according to the equality or concentration indices, but they seem the most segregated according to centralization. At this stage, without a map, this suggests the over-representation of Black people in the most central neighborhoods of the metropolitan area, as well as the existence of highly segregated neighborhoods for other groups in more distant municipalities. These results lead us to formulate several precautions regarding the use of global indices of segregation. First, how should these indices, whose results differ, be articulated? Should one dimension be chosen, or should they be combined, assuming their complementarity? In fact, this

---

3 Calculations were performed with the OasisR package based on equations compiled in Apparicio (2000) and Tivadar et al. (2014).

divergence and the complexity of index refinements have led most North American studies to focus on the single dimension of equality, and more specifically the dissimilarity index.

Then there is the uni-group nature of most of these indices that measure only one ethnic group at a time, and the results are compared a posteriori (see Table 4.1). However, it is interesting to measure these segregations jointly, as the segregation of some automatically has consequences for the spatial forms of others. This is why Duncan's dissimilarity index compares the distribution of a minority group with that of a majority group, historically Black versus White. The interaction, concentration and relative centralization indices proceed in the same way, explaining the absence of a value for White people in Table 4.1, since they serve as a reference. However, these bi-group indices are insufficient to capture the spatial forms of segregation in U.S. cities, not least because of the widespread increase in other ethnic groups. With this in mind, various multi-group indices aim to measure an overall level of segregation for all categories. More specifically, entropy indices calculate the degree of statistical heterogeneity of spatial units, transposed here into the notion of ethnic diversity, as opposed to situations of homogeneity and segregation. Multi-group entropy indices are becoming increasingly popular in the United States as ethnic diversity increases and segregation generally decreases within cities (Duroudier 2016).

#### 4.3.2.3. *Global and local indices*

Finally, the focus on global equality indices is the source of recurring criticism that focuses on two distinct aspects:

– on the one hand, the use of these indices to compare levels of segregation between cities, while equi-partitioning does not imply the same effort for cities with 2% or 30% of a group, even if segregation can be measured with the same intensity (Quinn and Pawarasat 2003);

– on the other hand, the shortcomings of these so-called "aspatial" indices in revealing intra-urban forms of segregation. This issue is well known to North American geographers, who refer to it as the "checkerboard problem" (White 1983). For example, the same value for an equality index can correspond to very different spatial forms, such as a sectorial partition, a center-periphery opposition, or a locally fragmented checkerboard. Concentration, centralization and clustering indices, on the other hand, do not have these shortcomings, by incorporating geographical parameters (see section 4.4).

To fill this last gap, one approach is generally to use local indices, which allow local contributions to a global form to be assessed. There are two possibilities here:

either by considering the local contributions of global indexes, or by using local equality indexes such as the location quotient and the local entropy index.

The location quotient, or Isard Index, measures the over- and under-representation of a given group in relation to the group's theoretical equi-representation: for each spatial unit, if the index is less than 1, the group is under-represented, while the higher it is above 1, the greater the spatial concentration. In its mapping (see Figure 4.10(a)), the location quotient brings no new results compared to the choropleth map of percentages (see Figure 4.2). Its value lies in the statistical standardization of the data, enabling comparisons to be made between ethno-racial groups whose numbers are very unequal in magnitude. Nevertheless, it has the same shortcomings as segregation indexes in terms of the significance of these local deviations in relation to the benchmark used to calculate over- and under-representation.

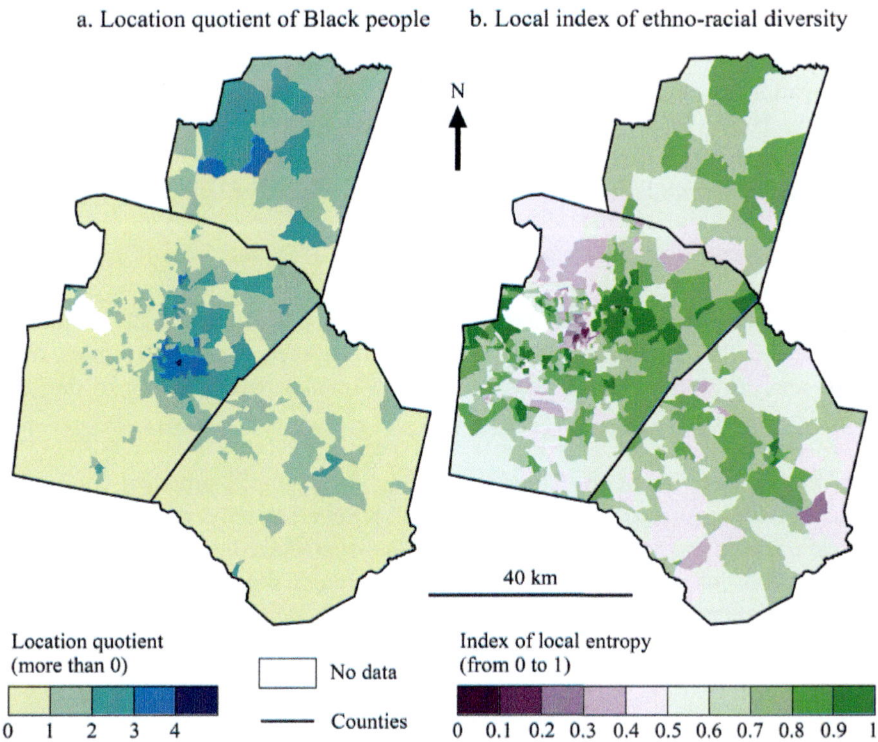


**Figure 4.10.** *Local segregation indexes for Black people in the Raleigh metropolitan area in 2020 (source: US Bureau of Census, 2020; Map: S. Duroudier, 2022)*

The entropy index is a local adaptation of the Piélou Index which, when applied to segregation, measures the level of ethno-racial diversity for each spatial unit of the metropolitan statistical area (see Figure 4.10(b)). The closer the values are to 0, the greater the ethnic homogeneity, while values close to 1 describe balanced proportions of groups. This multi-group index brings new results, as the historic Black ghetto, the archetypal form of American cities, tends to disappear behind two types of neighborhoods: on the one hand, the most homogeneous neighborhoods of Glenwood and north Raleigh, where White people remain homogeneous, and, on the other hand, the least segregated neighborhoods of the eastern half, where the ethno-racial diversity of inhabitants is more complete.

### 4.3.3. *Multivariate analysis approach: qualifying segregated neighborhoods*

A second approach to spatial organizations involves multivariate statistical analysis. These methods, often used in the social sciences in France, are not specific to segregation studies, so the discussion below goes beyond this thematic framework. However, the use of these methods to grasp segregated forms is ubiquitous, as they are part of a strong tradition going back to the currents of urban ecology and its reformulations in geography, in France and the United States.

Methodologically, this approach is distinguished by the use of multivariate statistics, that is, by jointly involving several variables, applied to a variety of social dimensions such as socioeconomic status, family and ethno-racial affiliations (Shevky and Bell 1955; Murdie 1969; Rhein 1994). In their general principles, these methods analyze vast tables of geographical information with the aim of deriving statistical patterns.

On the one hand, there are factorial analyses, for example, principal component analysis (PCA) or correspondence analysis (CA). Schematically, these methods use statistical covariations to produce new factors that synthesize the spatial associations and oppositions of variables. In terms of ethno-racial segregation in the United States, PCA often produces a first factor on which neighborhoods are opposed according to their level of concentration of White people and Black people, while a second factor shows segregated enclaves for smaller groups (Asian people, Hispanic people, others). On the other hand, classification methods (ascending or descending, hierarchical, mobile centers, dynamic clusters, etc.) measure the degree of similarity between spatial units in order to classify them into the most homogeneous groups possible.

#### 4.3.3.1. *Example of a hierarchical ascending classification of Raleigh neighborhoods*

Figure 4.11 shows the results of a hierarchical ascending classification computed on the five ethno-racial variables in Raleigh's 785 block groups in 2020. Incorporating almost 82% of the total variance, the eight-class partition reveals different types of neighborhoods.

– Profile 1: segregated and homogeneous White neighborhoods: on average, 82% of inhabitants in the 186 block groups are White;

– Profile 2: neighborhoods with a White majority close to the average metropolitan profile (244 block groups);

– Profile 3: 76 block groups with a White majority and Asian minority (average 18% of inhabitants);

– Profile 4: significant segregation of Asian people, on average the majority in 24 block groups;

– Profile 5: highly mixed neighborhoods with equal proportions of Hispanic people, Black people and White people (94 block groups);

– Profile 6: strong mix of Black people (average 30% of residents) and White people (46%) in 110 block groups;

– Profile 7: significant segregation of Black people, on average a majority in 60 block groups;

– Profile 8: two block groups stand out for the totality of their populations composed of people from other ethno-racial backgrounds.

Like this classification, spatial units can be mapped as belonging to a cluster, allowing us to read the geographic forms of segregation in Raleigh. First, there are different places of segregation that are not limited to the caricatured forms of the White suburbs and the Black ghetto. Indeed, while these are visible, they are complemented by a very highly segregated White neighborhood in Glenwood just northwest of downtown Raleigh, as well as segregated Black neighborhoods in some of the outlying counties. Above all, some neighborhoods to the west of the metropolitan area are clearly places where Asian people are segregated. Second, this typology provides a more nuanced picture of segregation in the Raleigh metropolitan statistical area, because while some profiles correspond to segregated neighborhoods, the majority of spatial units appear at different levels of ethno-racial diversity (e.g. Profiles 5 and 6).

#### 4.3.3.2. *Advantages and limitations of multivariate analysis*

These methods are therefore robust in identifying structuring socio-spatial profiles, justifying their use in a taxonomic approach to delineating homogeneous areas in urban spaces. In particular, factorial urban ecology has used these methods to forge classic theoretical models of urban organization (concentric, sectorial, polynuclear). The success of these methods in French geography since the 1970s can be explained by their conceptual association with regionalization approaches, as well as by their inductive character in the identification of spatially projected social hierarchies (Duroudier 2018). This approach has therefore been particularly concerned with reading the spatial configurations of segregation on the basis of social distances to elicit geographical figures.

However, factorial and classification methods have several limitations that make them complex, or at least more delicate, for identifying geographical forms of segregation. First, these methods are strictly statistical, so that their geographical dimensions only become apparent a posteriori via the mapping of results (Grasland 2004). In this respect, these methods make the often-implicit assumption that there are urban "regions" of social homogeneity, that is, that socially similar populations are spatially grouped together. Moreover, as with the choice of discretization, these methods involve parameters that need to be understood (standardization, aggregation methods, number of classes, etc.) and whose influence on the final result needs to be tested: for example, with regard to the weight of the other ethno-racial affiliation in the profiling of neighborhoods.

Last but not least, the complexity of the results of factorial analyses or classifications calls for a decisive effort in the graphic and cartographic visualization of results. The choice of a choropleth map, for example, implies a careful choice of colors to suggest spatial continuities and breaks. Therefore, the blue shades between Profiles 1 and 2 suggest a greater similarity between these two categories than with Profiles 3 and 4 in green shades, as well as for mixed neighborhoods, in red shades. One of the consequences of this play of chromatic contrasts is a subtle transformation in the analysis of segregation: from a continuous space traversed by gradients, this visualization suggests qualitatively distinct zones in terms of segregation or diversity. It thus invites us to draw unevenly segregated territories and questions the existence of urban boundaries and discontinuities.

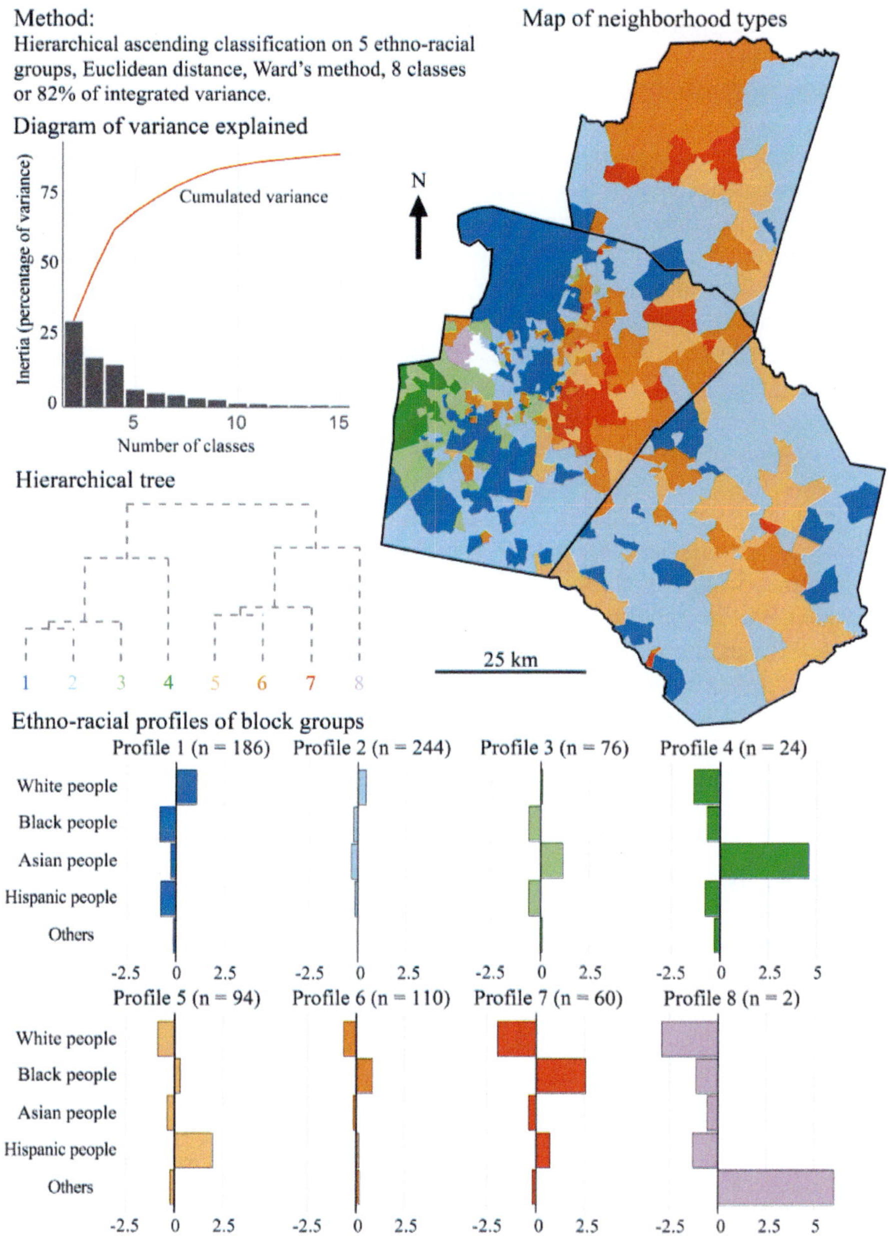


**Figure 4.11.** *Ethno-racial typology of Raleigh MSA neighborhoods in 2020 (source: US Bureau of Census; Map: S. Duroudier, 2022)*

## 4.4. Capturing spatial forms using dynamic approaches

### 4.4.1. *Why a dynamic approach to spatial morphologies?*

The spatial forms taken by phenomena such as segregation interact with a range of social and spatial factors that can be incorporated into measurements. However, the two main families of measures described above are static because, as the American geographer White (1983) observes, "calculations are based on parcels as discrete and independent units, ignoring the composition of nearby parcels". In other words, in terms of segregation, since the calculation does not take into account the position of spatial units in relation to each other, these approaches are insufficient to describe segregation as a process, that is, by considering the effects produced by spatial concentration and isolation, notably in terms of social integration or exclusion (Authier et al. 2006).

This observation is based on the hypothesis that there are neighborhood effects that both result from and contribute to the reproduction of socio-spatial inequalities. This hypothesis initially focused on the penalizing effects of the accumulation of disadvantages at neighborhood level on the socio-spatial trajectories of individuals, although this was subsequently extended to different social categories and contexts, as with the debate on residential diversity in France (Authier et al. 2006). In the United States, this type of hypothesis underpins the theory of the spatial assimilation of ethnic minorities in White suburban spaces, as per the induration analysis of the Black hyperghetto (Massey and Denton 1993).

This approach thus links the question of social integration, based on the strong ties of belonging to living spaces (neighborhood, municipality, school, etc.) and the weak ties of social relations to the question of concentration and isolation in space. In other words, the problem posed is that of the role of geographical proximity in social and spatial interactions. In this respect, various studies have shown that proximity induces interactions that thus have consequences for segregation: for example, on real estate dynamics in Los Angeles (Le Goix 2016) or on stigmatization and school avoidance practices in the Île-de-France French region (François 2002).

In order to analyze spatial forms, the difficulty hence lies in defining and taking into account proximity in measures of segregation. Yet this aspect is often overlooked (Rhein 2002). On the one hand, in theoretical terms, proximity refers to the notion of neighborhood: either the Durkheimian "*milieu*" (Authier et al. 2006), or the Chicago School "community" (Grafmeyer 1979). However, debates on the notion of neighborhood show both the difficulty of defining it and its inadequacy in capturing the interactions that occur at different scales (Fowler 2016). On the other hand, from a methodological point of view, proximity depends on the measures implemented. Furthermore, in the above approaches, spatial units are independent of each other in statistical analyses, which amounts to considering that proximity corresponds to the spatial unit and that it is within it that social and spatial interaction takes shape (Reardon and O'Sullivan 2004).

Neither neighborhoods nor spatial units can therefore capture the proximity involved in context effects. A third possibility relates explicitly to spatial analysis based on Tobler's (1970) First Law of Geography, according to which social interactions are a function of distance[4]. A recent example is the work of Sampson (2012): starting from a classical analysis of neighborhood effects through endogenous characteristics, he also considers the proximity of neighborhoods to one another to capture the role of proximity in social interaction. Like this study, the notion of neighborhood opens up to a range of approaches and methods within the field of spatial analysis. The remainder of this chapter presents three approaches that attempt to capture segregation in a dynamic way by integrating proximity roles: so-called spatial indices, the discontinuity approach and the measurement of spatial contexts.

### 4.4.2. *Approaches using spatial indices or spatial autocorrelation*

A first set of approaches, grouping together a wide variety of works, attempts to link geographical proximity and social similarity by directly integrating spatial parameters into the measurements. The first approach corresponds to the work carried out by North American geographers who, noting the shortcomings of classic equality and exposure indices, proposed new global indexes (White 1983; Morrill 1991; Wong 1993). These new indices, such as Morrill's or Wong's indices, incorporate spatial parameters that weigh the intensity of concentration.

A second approach derives from the methods of spatial autocorrelation analysis developed following Moran and Geary's work in economics in the 1950s and popularized in geography by Cliff and Ord (1973). This approach is a further formulation of Tobler's First Law on the assumption that two nearby locations are more similar than two distant ones. Autocorrelation indices thus attempt to measure spatial dependence, that is, the correlation between degree of similarity and spatial proximity. For the Moran Index, a positive autocorrelation indicates the existence of homogeneous regions (and suggests a dual structure), while a negative autocorrelation corresponds to a rapid alternation of different local contexts (rather a checkerboard pattern).

---

4 "Everything is related to everything else, but near things are more related than distant things" (Tobler 1970).

In these approaches using spatial segregation indices or spatial autocorrelation, the real issue is the parameterization of distance in the measurement, that is, the definition of what constitutes a spatial neighborhood. There are several possibilities:

– contiguity: neighbors are spatial units that share a common boundary;

– distance: neighbors are spatial units located within a defined radius (e.g. 10 km);

– number of neighbors: a defined number of spatial units (e.g. 15 block groups) is considered close.

In terms of results, the indicators of these two methods have two main elements in common. First, despite more sophisticated calculations, the results confirm the segregation logics already shown by the classic indices (see Table 4.2): Asian people and Black people are the most segregated. Moreover, for these groups, the high Moran index suggests the existence of a continuous zone of segregation, which the maps clearly show (see Figure 4.9). However, if the Moran index for White people is as high as that for Black people, the forms are trickier to interpret, as their map shows the existence of several zones of homogeneity for this most diffuse group in the Raleigh metropolitan statistical area.

Second, this highlights the inadequacy of these global indices, which only partially solve the "checkerboard problem", since several spatial configurations are possible for the same value.

| Types of indices | Indices | White people | Black people | Asian people | Hispanic people | Others |
|---|---|---|---|---|---|---|
| Spatial autocorrelation | Moran | 0.66 | 0.69 | 0.81 | 0.47 | 0.21 |
| | Geary | 0.32 | 0.31 | 0.21 | 0.51 | 0.78 |
| Segregation indices | Wong | 0.31 | 0.37 | 0.54 | 0.28 | 0.10 |
| | Morrill | 0.22 | 0.32 | 0.51 | 0.24 | 0.09 |

**Table 4.2.** *Spatial segregation indexes and spatial autocorrelation indexes for ethno-racial affiliations in Raleigh MSA block groups in 2020 (sources: US Bureau of Census 2020; Tivadar* et al. *2014; Geoda 2021)*

An interesting solution lies in the use of Local Indicators of Spatial Autocorrelation (LISA) developed by Anselin (1995). For each spatial unit, these indicators measure the level of differentiation between, on the one hand, resemblance to the metropolitan profile and, on the other hand, resemblance to neighboring spatial units.

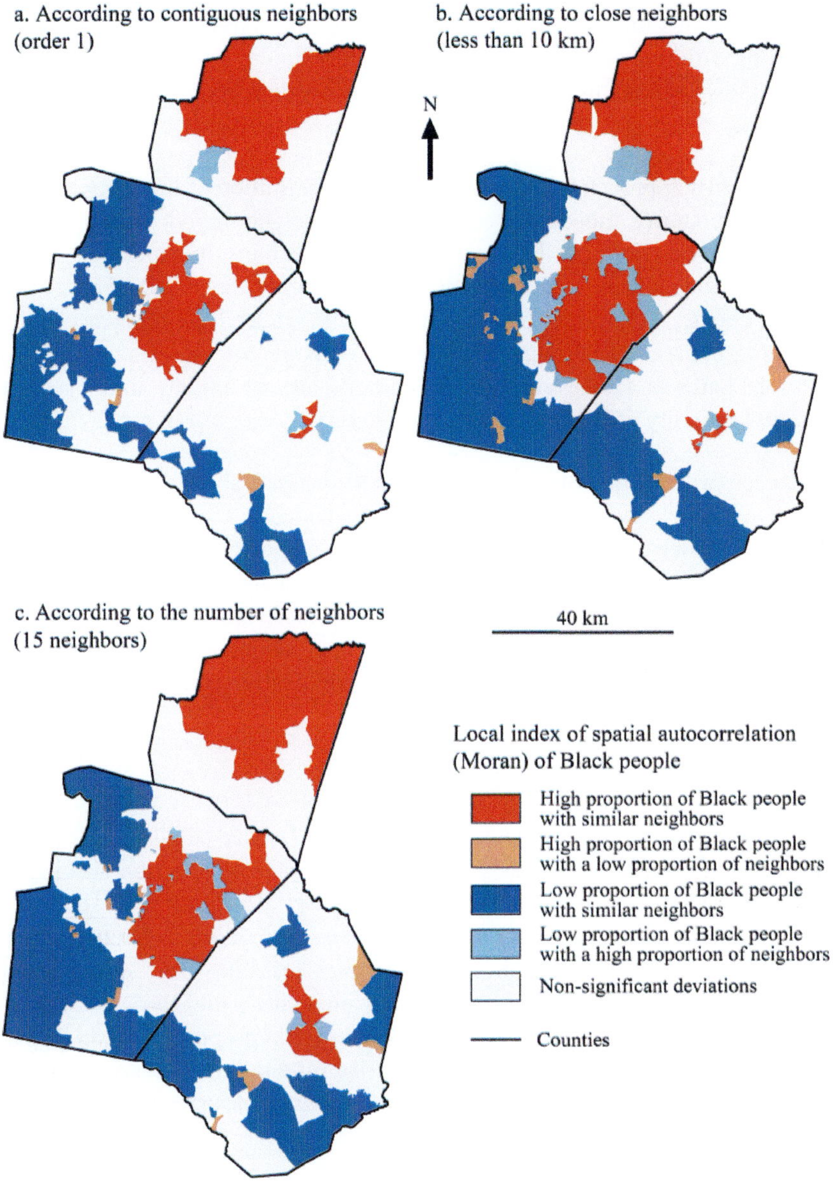


**Figure 4.12.** *Local Indicators of Spatial Autocorrelation (LISA) for Black people in the Raleigh MSA in 2020 according to different neighborhood measures (source: US Bureau of Census, 2020; calculations made in Géoda, 2022; Map: S. Duroudier, 2022)*

The value of this measure lies in the production of a typology (see Figure 4.12) distinguishing different types of configuration according to the local level of values (high in red, low in blue) and the level of neighborhood similarity (high in dark, low in light).

Moreover, this approach tends to accentuate the definition of homogeneous areas in segregation patterns. Indeed, by considering the neighborhood isotropically, that is, by integrating neighbors in the same way in all directions, this measure smoothes out rapid variations in group concentration and reinforces the visual effects of large segregated zones. This isotropic neighboring is indeed a relevant tool for characterizing the texture of segregation: either homogenous regions or fragmented checkerboard patterns. However, not all neighborhoods are similar, in the sense that there are discontinuity effects that produce ruptures in socio-spatial interaction.

In the case of Black people, the three types of distance show the image of a vast central zone of segregation, with high percentages and a homogeneous neighborhood. However, the metropolitan context changes significantly: while in the case of a neighborhood by kilometric distance, this area stands out as a strong Black ghetto against the rest of the agglomeration where there are few Black people, the other two types of neighborhoods reveal more contrasting situations where Black people can be absent, somewhat present, or are very locally segregated.

### 4.4.3. *Approach based on discontinuities*

A second approach, linking spatial proximity and social similarity, focuses on discontinuities, broadly defined as a particular spatial configuration where social differentiation is associated with spatial proximity. This approach lies at the crossroads of systems analysis and spatial analysis. According to Brunet (1965), a discontinuity is a rupture between spatial systems resulting from endogenous or exogenous dynamics and contributes to calling integration into question. Based on this conceptual framework, Decroly and Grasland (1992) proposed to read the levels of geographical patterns of fertility in Europe from the notion of discontinuity: this is defined as an exceptionally strong spatial break measured by the statistical dissimilarity between contiguous regions.

This approach is therefore doubly relevant to understanding the geographical forms of segregation. On the one hand, this socio-spatial separations approach responds directly to debates in numerous works on the tangibility of segregated neighborhood boundaries (François 1995; Duroudier 2018): for example, the significance of boundaries in the schematic models of the Chicago School, in the analysis of the American Black hyperghetto (Massey and Denton 1993), or the boundaries suggested by factorial analyses (see section 4.3.3). On the other hand, this approach is closely linked to debates on neighborhood effects and spatial autocorrelation, for if social similarity between adjacent neighborhoods determines the nature of spatial interaction, then discontinuities potentially generate barrier effects, discriminations and a reinforcement of inequalities. Therefore, reconciling two elements of segregation (spatial concentration and separation), this discontinuity approach is useful for delineating segregated forms by their most salient contours, in cases of both vast areas of segregation and urban fragmentation (Duroudier 2018).

This approach is more complex, however, as reading the geographic configurations of segregation relies on three methodological components (Duroudier 2018): measuring dissimilarity, the mode of cartographic representation and setting a threshold for discontinuities.

First, the intensity of separation is measured for each ethno-racial group as the dissimilarity on each segment between contiguous spatial units. The measurement of discontinuities requires the choice of a dissimilarity index from among many possibilities. Absolute dissimilarity corresponds to the difference between the values of the two spatial units: if they count 25% and 50% Black, they have a dissimilarity of 25 points. By contrast, relative dissimilarity calculates the ratio between the two spatial units: in the previous case, this ratio is 2, as in the case of two spatial units with 2% and 4% Black people. As per the maps (see Figure 4.13), relative dissimilarity is more relevant for measuring threshold effects in less segregated contexts, which is the case in Raleigh's western neighborhoods. Absolute dissimilarity, on the other hand, shows the most intense differences on the contours and within the neighborhoods with the highest concentration of Black people.

The next step in analyzing discontinuities patterns is to project the dissimilarity values onto a map (see Figure 4.14). The intensity of the dissimilarity is proportionally represented by the thickness of the lines between two spatial units. However, to interpret the meaning of discontinuities, that is, to identify on which side of the breaks segregated or non-segregated neighborhoods lie, one option is simply to superimpose the choropleth map of Black people in Raleigh, thus referring to the elements discussed above.

A final point concerns the discontinuity threshold: at what point does a dissimilarity value for Black people become strong enough to be significant in terms of segregation, discrimination and exclusion? The question is important because the spatial configurations highlighted depend on this threshold (see Figure 4.12): the lower the threshold, the more fragmented the situation will appear, as is the case for a 10-point difference in Black people in Raleigh. On the contrary, a higher threshold will restrict the analysis to salient geographical shapes, and in this case, for Black people with a difference of 20 points, it shows very well how the outer contours of the Black ghetto are structured. However, there is no definitive answer to the question of the threshold of discontinuity, as thresholds are relative to local segregation contexts. Within Raleigh itself, a 10-point difference in Black people may be noticeable in a very homogeneous White context and insignificant in segregated neighborhoods.

This spatial approach based on discontinuities is interesting from a theoretical point of view, as it takes into account neighborhoods while at the same time addressing the issue of urban boundaries. By analyzing the spatial composition of discontinuous segments, this approach yields rich results that complement more conventional representations of segregation: for example, it allows us to identify segregated enclaves, continuous mixed-use zones, or fragmented spaces alternating locally between different neighborhoods.

### 4.4.4. *Approach based on population potential*

Finally, a third approach, which combines proximity and similarity, corresponds to population potential calculation methods. This approach is interesting because it considers each entity in its spatial vicinity, which varies in scope and resemblance. The underlying assumption of this approach is to account for the probable interaction between individuals, which is proportional to the population potential and decreases with the distance separating them (Grasland 1990).

In principle, this method performs a spatial smoothing of the initial information by computing, for any point in space, a potential population concentration value from the values of entities in a defined spatial neighborhood. There are various methods for computing a population potential, in this case the Stewart method (Stewart and Warntz 1958).

The results are sensitive to the parameters chosen in terms of neighborhood scope and nature (see Figure 4.14). The neighborhood radius corresponds to the size of the window in which entities located around a spatial unit are considered as neighbors. This radius, defined according to interaction hypothesis, implies very different smoothing: while a wide neighborhood of 5 km highlights the high concentration of Black people in the center of Raleigh, a smaller window of 1 km shows the existence of several concentration nuclei, including those in suburban areas.

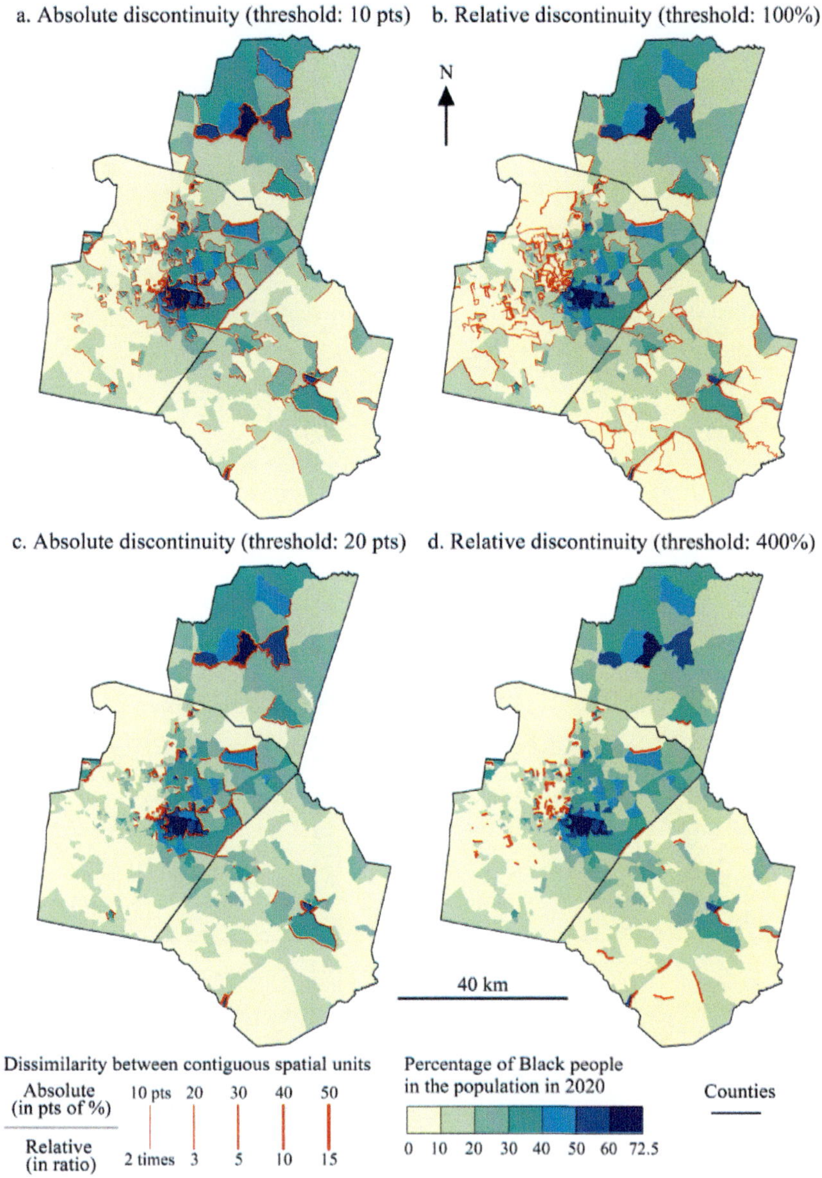


**Figure 4.13.** *Absolute and relative discontinuities of Black people in the Raleigh MSA in 2020 according to different thresholds (source: US Bureau of Census; Map: S. Duroudier, 2022)*

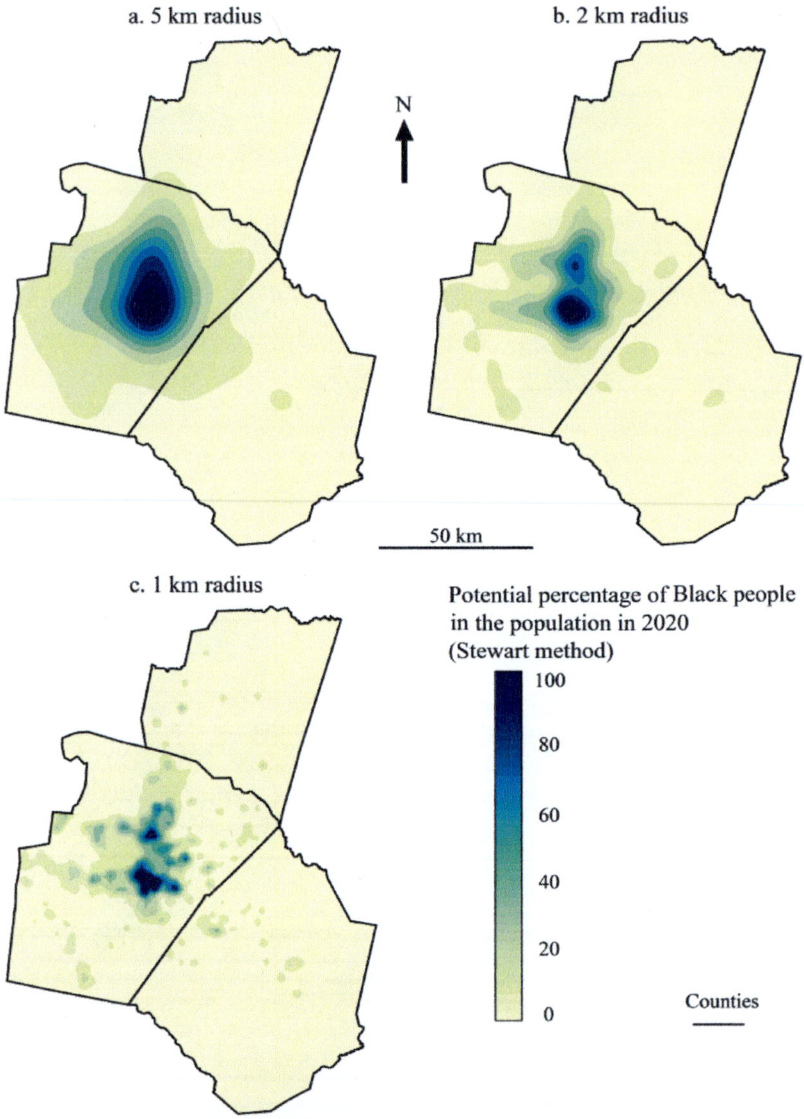


**Figure 4.14.** *Potential Black population in the Raleigh MSA in 2020 according to different neighborhood radii (source: US Bureau of Census, 2020; Map: S. Duroudier, 2022)*

The second parameter corresponds to the nature of the neighborhood: for example, does proximity have the same effect everywhere within a defined radius (disc neighborhood), or does its effect decrease progressively with distance (Gaussian neighborhood)? Depending on the nature of the neighborhood considered, and with the same radius, the geographical forms of segregation differ, particularly at the local level.

This approach is interesting for two reasons. On the one hand, the smoothing of information and the resulting visualization emphasize gradual variations in space. However, these maps also reveal abrupt variations in group concentration and discontinuities, when the isolines are closer together.

On the other hand, this approach is interesting for grasping multiscale spatial organizations of segregation: Figure 4.14 shows that Black people are organized around a central ghetto on the scale of the Raleigh metropolitan area, while on a more local scale it is possible to identify two main poles and a multitude of more segregated neighborhoods.

## 4.5. Conclusion

In the manner of Raymond Queneau, this chapter is somewhat of a *stylistic exercise* regarding the analysis of spatial configurations, here applied to those of segregation, in particular the segregated configurations of Black people in the Raleigh metropolitan statistical area. This overview shows that the analysis of geographical forms is the result of a set of theoretical, epistemological and methodological choices in order to achieve a fine-tuned map of the organization of space. In this respect, the figures presented in this chapter show that cartographic aspects are central to the interpretation of segregation.

The example of segregation also shows that classical statistical methods (indicators and multivariate analysis) are sufficient if the organization we are looking for is defined statically only as a concentration in space. However, if the forms sought are more diverse, or if the approach to the phenomenon is dynamic as a process that includes hypotheses in terms of spatial interaction and social integration, it is preferable to opt for more spatial approaches (autocorrelation indices, discontinuities, smoothing). Similarly, the parameters of the different measurement methods are important, as results can vary considerably depending on their granularity, whether it is the threshold for discontinuities, the number of types in a classification or the choice of the most relevant index. Therefore, while some parameterizations point to Raleigh's historical configurations and corroborate the historical model of the North American city, others identify more complex spatial configurations: for example, more ethno-racial fragmentation as well as mixed-use.

Finally, it is clear that the elements discussed in this chapter are part of a chain for the collection and transformation of geographic information, from the formulation of a question to data processing, and its representation and visualization. In this context, methodological choices are often the result of a compromise between the objective of researching socio-spatial morphologies, the empirical significance of the forms observed and the complexity of measurements and results in the context of debates renewed by the evolution of data, methods and social issues.

## 4.6. References

Alba, R. and Logan, J. (1991). Variations on two themes: Racial and ethnic patterns in the attainment of suburban residence. *Demography*, 28(3), 431–453.

Apparicio, P. (2000). Les indices de ségrégation résidentielle : un outil intégré dans un système d'information géographique. *Cybergeo: European Journal of Geography*, 134 [Online]. Available at: https://journals.openedition.org/cybergeo/12063.

Anselin, L. (1995). Local indicators of spatial association_LISA. *Geographical Analysis*, 27(2), 93–115.

Authier, J.-Y., Bacqué, M.-H., Guérin-Pace, F. (eds.) (2006). *Le quartier : enjeux scientifiques, actions politiques et pratiques sociales*. La Découverte, Paris.

Ball, S. and Petsimeris, P. (2010). Mapping urban social divisions. *Forum: Qualitative Social Research*, 11(2).

Béguin, M. and Pumain, D. (1994). *La représentation des données géographiques : statistiques et cartographie*. Armand Colin, Paris.

Berry, B. and Kasarda, J. (1977). *Contemporary Urban Ecology*. Macmillan, New York.

Boquet, Y. (2003). Cartes et élections : la construction des circonscriptions électorales aux Etats-Unis, entre égalité et équité. *Espace, Populations, Sociétés*, 3, 483–499.

Charles, C.Z. (2003). The dynamics of racial residential segregation. *Annual Review of Sociology*, 29, 167–207.

Cliff, A. and Ord, J. (1973). *Spatial Autocorrelation*. Pion, London.

Decroly, J.-M. and Grasland, C. (1992). Frontières, systèmes politiques et fécondité en Europe. *Espace, populations, sociétés*, 10(2), 135–152.

Dumolard, P., Dubus, N., Charleux, L. (2003). *Les statistiques en géographie*. Belin, Paris.

Duncan, O.D. and Duncan, B. (1955). A methodological analysis of segregation indexes. *American Sociological Review*, 20(2), 210–217.

Duroudier, S. (2016). La ségrégation raciale des villes américaines diminue-telle toujours ? *Urbanités* [Online]. Available at: https://www.revue-urbanites.fr/les-villes-americaines-la-segregation-raciale-des-villes-americaines-diminue-t-elle-toujours/.

Duroudier, S. (2018). Ségrégations et discontinuités dans les villes intermédiaires des États-Unis. PhD Thesis, Université Paris Diderot, Paris.

Fotheringham, A. and Wong, D. (1991). The modifiable areal unit problem in multivariate statistical analysis. *Environment and Planning A*, 23(7), 1025–1044.

François, J.-C. (1995). Discontinuités dans la ville. L'espace des collèges de l'agglomération parisienne (1982-1992). PhD Thesis, Université Paris I Panthéon-Sorbonne, Paris.

François, J.-C. (2002). Contexte théorique de la notion de discontinuité géographique. *Cybergeo: European Journal of Geography* [Online]. Available at: http://journals.openedition.org/cybergeo/347.

François, J.-C. and Ribardière, A. (2004). Qu'apporte l'échelon infra-communal à la carte des inégalités de richesse en Ile-de-France ? *M@ppemonde*, 75, 1–8 [Online]. Available at: http://mappemonde-archive.mgm.fr/num3/articles/art04305.pdf.

Grafmeyer, Y. (1994). Regards sociologiques sur la ségrégation. In *La ségrégation dans la ville : concepts et mesures*, Brun, J. and Rhein, C. (eds). L'Harmattan, Paris.

Grasland, C. (1990). Potentiel de population, interaction spatiale et frontières : des deux Allemagnes à l'unification. *L'Espace géographique*, 19/20(3), 243–254.

Grasland, C. (2004). Discontinuités statistiques et discontinuités spatiales. L'exemple des inégalités de richesse par habitant en Italie (1951-1991). In *L'analyse statistique des données spatiales. Acte des Journées d'étude statistiques de Marseille-Lu-miny*, d'Aubigny, G. and Lejeune, M. (eds). Dunod, Paris.

Lambert, N. and Zanin, C. (2016). *Manuel de cartographie : principes, méthodes, applications*. Armand Colin, Paris.

Le Goix, R. (2016). *Sur le front de la métropole : une géographie suburbaine de Los Angeles*. Publications de la Sorbonne, Paris.

Le Goix, R. (2022). Socio-spatial segregation in cities. In *Cities at the Heart of Inequalities*, Cottineau, C. and Pumain, D. (eds). ISTE Ltd, London, and John Wiley & Sons, New York.

Madelin, M., Grasland, C., Mathian, H., Sanders, L., Vincent, J.M., (2009). Das MAUP: Modifiable Areal Unit-Problem oder Fortschritt? Archive ouverte HAL [Online]. Available at: https://halshs.archives-ouvertes.fr/halshs-00424589/document.

Madoré, F. (2005). La ségrégation sociale dans les villes françaises réflexion épistémologique et méthodologique. *Cahiers de géographie du Québec*, 49(136), 45–60.

Massey, D. and Denton, N. (1988). The dimensions of residential segregation. *Social Forces*, 67(2), 281–315.

Massey, D. and Denton, N. (1993). *American Apartheid: Segregation and the Making of the Underclass*. Harvard University Press, Cambridge.

Monmonier, M. (1993). *Comment faire mentir les cartes : du mauvais usage de la géographie*. Flammarion, Paris.

Morrill, R. (1991). On the measure of geographic segregation. *Geographic Research Forum*, 11, 25–36.

Murdie, R. (1969). *Factorial Ecology of Metropolitan Toronto, 1951-1961*. University of Chicago Press, Chicago.

Pruvot, M. and Weber-Klein, C. (1984). Écologie urbaine factorielle comparée : essai méthodologique et application à Strasbourg. *L'Espace géographique*, 13(2), 136–150.

Quinn, L. and Pawasarat, J. (2003). Racial Integration in Urban America: A Block Level Analysis of African American and White Housing Patterns. Technical Paper, Employment and Training Institute, University of Wisconsin-Milwaukee.

Reardon, S. and O'Sullivan, D. (2004). Measures of spatial segregation. *Sociological Methodology*, 34(1), 121–162.

Reibel, M. (2011). Classification approaches in neighborhood research: Introduction and review. *Urban Geography*, 32(3), 305–316.

Rhein, C. (1994). La ségrégation et ses mesures, In *La ségrégation dans la ville : concepts et mesures*, Brun, J. and Rhein, C. (eds). L'Harmattan, Paris.

Rhein, C. (2002). Intégration sociale, intégration spatiale. *L'Espace géographique*, 31(3), 193–207.

Roncayolo, M. (1996). *Les grammaires d'une ville. Essai sur la genèse des structures urbaines à Marseille*. Éditions de l'EHESS, Paris.

Sampson, R. (2012). *Great American City: Chicago and the Enduring Neighborhood Effect*. University of Chicago Press, Chicago.

Shevky, E. and Bell, W. (1955). *Social Area Analysis*. Stanford University Press, California.

Taylor, C., Gorard, S., Fitz, J. (2003). The modifiable areal unit problem: Segregation between schools and levels of analysis. *International Journal of Social Research* Methodology, 6(1), 41–60.

Tivadar, M., Schaeffer, Y., Torre, A., Bray, F. (2014). OASIS – Un Outil d'Analyse de la Ségrégation et des Inégalités Spatiales. *Cybergeo: European Journal of Geography* [Online]. Available at: http://journals.openedition.org/cybergeo/26579.

Tobler, W. (1970). A computer movie simulating urban growth in the Detroit Region. *Economic Geography*, 46, 234–240.

White, M. (1983). The measurement of spatial segregation. *American Journal of Sociology*, 88(5), 1008–1018.

Wong, D. (1993). Spatial indices of segregation. *Urban Studies*, 30(3), 559–572.

Wong, D., Reibel, M., Dawkins, C. (2007). Introduction. Segregation and neighborhood change: Where are we after more than a half-century of formal analysis. *Urban Geography*, 28(4), 305–311.

Wright, R., Ellis, M., Parks, V. (2005). Re-placing whiteness in spatial assimilation research. *City and Community*, 4(2), 111–135.

# 5

# Map and Statistical Model to Explore Spatial Heterogeneity

**Mohamed HILAL and Julie LE GALLO**

*CESAER, INRAE, L'Institut Agro,*

*Université Bourgogne Franche-Comté, Dijon, France*

## 5.1. Introduction

The map is both a powerful medium and an instrument of research and expertise in its own right, particularly in the field of regional planning. Thanks to its evocative power and didactic virtues, it remains a privileged medium for illustrating and analyzing data, provided its requirements are met. In particular, it must be based on rigorous semiological rules and appropriate statistical data modeling tools. Conceived as an approach in its own right, the map takes part in all stages of geographic analysis, the results of which are not limited to a simple inspection or visual comparison of data (Muller 1977; Badariotti 2004). The map can be used to ask questions, formulate and test hypotheses, search for relationships, elaborate findings and present them in a form accessible to the reader (Courville 1995; Çöltekin et al. 2018).

This chapter shows how cartographic expression and statistical modeling of geographical data can be combined and complement each other to describe, explore and explain the spatial heterogeneity observed, in this case, in the real estate market in a French *département*. Using administrative data, previously transformed into a geographic dataset, it is possible to produce maps describing the initial data, a model explaining the variations of a datum in space and, finally, a cartographic

*Handling and Mapping Geographic Information*,
coordinated by Claire CUNTY and Hélène MATHIAN. 

representation of indicators derived from this model, revealing the local specificities of the area studied. Statistical modeling, used in conjunction with the map, aims to use appropriate estimation and inference methods to link field observations, collected by notaries, with a theoretical framework, in this case that of hedonic regression (Lancaster 1966; Rosen 1974), to estimate and explain the municipal distribution of house prices.

Statistical modeling of geographical data may appear simple on the surface, but it is fraught with pitfalls. Indeed, many of the conditions and assumptions of conventional statistical analysis are not met when data processing concerns economic, social or natural objects, one of whose primary attributes is location. For example, the most important condition in statistical modeling is the requirement of a random distribution of observed samples. However, spatial data almost always violate this requirement. The term describing this particular data configuration is spatial autocorrelation (see Introduction and Chapter 4), which is at the top of the list of issues to be addressed in the statistical modeling of spatial data (Cliff and Ord 1973; Ord 1975; Paelinck and Klaassen 1979; Anselin 1980; Le Gallo 2002; Oliveau 2010). Other, closely related characteristics also need to be taken into account, such as the spatial aggregation or modifiable areal unit problem (Yule and Kendall 1950; Openshaw and Taylor 1979), questions of scale and edge effects, etc. (see Introduction and Chapter 4). While these characteristics add complexity to the analysis of spatial data, they also allow for a richer analysis than the simple statistic distribution of values (mean, variance, etc.). Indeed, the analysis of spatial distributions requires a detailed exploration of the statistical distributions of variables through the prism of concepts such as distance, contiguity, neighborhoods and interaction.

This chapter is organized into three sections, each corresponding to a key phase in the statistical and spatial modeling process. Section 5.2 details the construction of the data, starting with the "land value requests" (DVF) files downloadable from the data.gouv.fr website. These raw open-source data have a complex structure. Pre-processing is therefore essential to extract individual mutations and associated variables, which can be used as is or aggregated to a more appropriate scale of analysis to answer a question. In section 5.3, we propose a set of statistical and cartographic analyses of the data aggregated at the municipal level, enabling us to appreciate the spatial variability of the main variables in the database. The presentation includes an analysis of global and local spatial autocorrelation. A hedonic model of house prices is presented in section 5.4. It shows how, in addition to the usual diagnostics of econometric regressions, maps of residuals, that is, the differences between observed values and the values estimated by the model, help to specify the model by informing the search for and choice of the model's explanatory variables. The econometric analysis is divided into two parts: on the one hand, we quantify the overall relationship between price and goods characteristics, and, on the other hand, we present the results obtained for a model that spatially differentiates the intensity of the relationship. The final section concludes by emphasizing the important role of the map in this approach to spatial data modeling, with the example of the spatial dynamics of house prices.

Finally, the entire chapter is part of a reproducible framework (Giraud and Lambert 2017; Orozco et al. 2020). Therefore, the study uses databases and cartographic backgrounds collected from open-source repositories, as well as the R Programming language, to be carried out in script form[1], which enables data import, statistical and cartographic processing, and illustrations (tables, graphs and maps).

## 5.2. From raw open-source data to statistical data

This section details the steps involved in extracting administrative data and building a geographic information matrix. It illustrates the need to clarify the scope of the data and to document the ways in which they are established. It also makes it possible to use the data for statistical and cartographic processing.

### 5.2.1. *Open-source data with a complex structure*

#### 5.2.1.1. *DVF, first and foremost a legal and tax object*

The DVF database, opened in 2019 by the *Direction Générale des Finances Publiques* (DGFiP), provides access to land sales carried out over the last five years in mainland France and the overseas departments and territories, except for Mayotte and Alsace-Moselle. Updated every six months, the database covers all transfers effected for valuable consideration (sale, exchange, expropriation or auction). The data do not contain any personal information (name or individual characteristics) of the sellers or buyers of a property. It solely provides information on:

– the sale price and transaction date of a built (apartments, houses, business premises) or unbuilt property (land, farmland, etc.);

– property description: number of rooms, surface area, lot number, etc.;

– property location (cadastral references and plot location coordinates).

1 All the codes are available at: www.forgemia.inra.fr/mohamed.hilal/map_dvf_iste.

Data recording is based on a logical model derived directly from the legal object, known as a document or deed of disposal. This object records one or more operations modifying rights over a property. A deed can thus combine, in an almost unlimited number of ways, real estate assets such as a plot of land (built or unbuilt), premises (apartment, house, business premises and outbuilding), a co-ownership lot or a volume lot.

**DVF data production chain**

Once the data have been transmitted by the notaries, the DGFiP carries out a number of computerized processes to update the tax authorities' information systems. "Fidji" processing (*Fichier informatisé des données juridiques immobilières*), a computerized real estate legal data file, is used to update the real estate file. "Majic" (*Mise à jour des informations cadastrales*), a cadastral information update, is used for tax purposes, updating information on built and non-built-up properties, mainly for the purpose of calculating local taxes (property, housing, business, business tax and household waste tax). These two processes feed the BNDP (*Base nationale des données patrimoniales*), which covers the whole of France, with the exception of the departments of Alsace, Moselle and Mayotte.

Extracted from the BNDP, the DVF file contains only the variables stipulated by the Decree of December 28, 2018, on the electronic publication of information on property values declared on the occasion of real estate transfers. All transfers are recorded, with the exception of non-expensive transfers (inheritance, bequest, donation, etc.) and property transfers not covered by a notarial deed (shares in non-trading property companies and certain public sales).

**Box 5.1.** *Creation of DVF data*

#### 5.2.1.2. *Data extraction and creation of a geographic information matrix*

At the time of writing, six DVF years (from 2016 to 2021) are available for download from data.gouv.fr[2]. The data files contain raw data that need to be transformed into a statistical table or a matrix of usable geographic information.

The main difficulty encountered when processing DVF files is linked to the fact that the same mutation can be entered across several lines within the file (Boulay et al. 2021; Casanova Enault et al. 2019). Indeed, depending on its degree of complexity, each deed or publication is broken down into a variable number of lines containing the same fields and mixing information linked to the different legal or fiscal operations recorded therein. A deed is transcribed on more than one line if it

2 Direct link to the DVF files used in this chapter: www.files.data.gouv.fr/geo-dvf/latest/csv/.

includes at least several dispositions, or several plots within a disposition, or several premises within a plot, or several tax subdivisions within a plot (recreational grounds, woods, orchards, meadows, etc.). The 40 fields or variables present in each file describe, in a nested and repeated manner, the deeds (date and date-number of the transfer), the dispositions (price) and the properties (type of premises, type of crop, address, etc.).

The creation of a mutation database in a format suitable for statistical analysis is a two-stage process. The first breaks down the raw DVF file into different tables, so as to aggregate information by layout and type of property (mainly premises and land). This preliminary processing is greatly facilitated by the use of the Cerema (2022) data model (see Figure 5.1). It extracts[3] four tables containing mutations, premises, land and property coordinates.

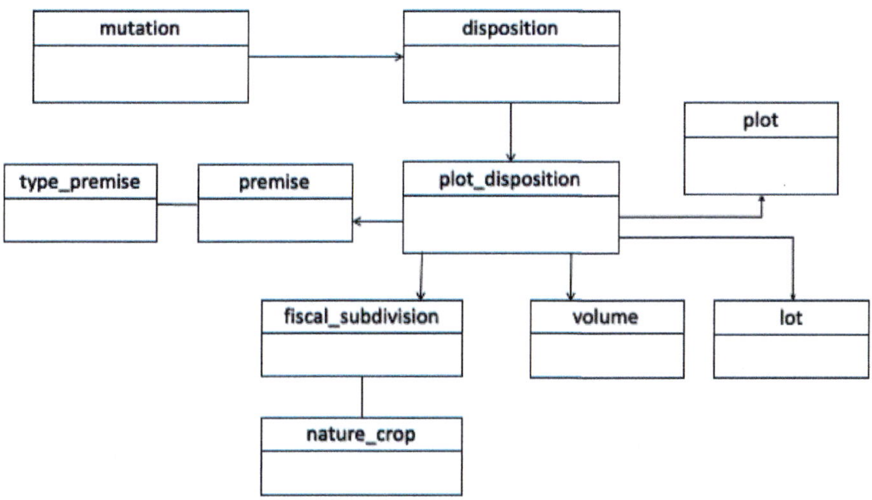


**Figure 5.1.** *DVF data model (source: adapted from CEREMA, Direction territoriale Nord-Picardie, March 2015)*

The second step produces a database alignment (one observation = one mutation) by concatenating the mutation, premises, land and property coordinates tables. Four additional variables are also created and added: the year, extracted from the

3 For detailed information on extractions, please refer to R code, available as a supplement to this chapter.

mutation date; the surface area, corresponding to the actual surface area of houses and business premises, or the habitable surface area of apartments; the total number of rooms; the type of property; and possibly the combination of properties present in the mutation (house, apartment, business premises, land without premises, combinations of at least two different properties).

### 5.2.2. *Object of study and scale of analysis*

The six raw DVF files downloaded contain 16,306,439 lines. After processing, the mutation database contains 6,969,281 observations and 36 variables. Table 5.1 gives a breakdown of the number of lines, dispositions and mutations per year. Faced with such a vast amount of data, it is necessary to target a specific phenomenon. In our example, the dataset collected is intended to be used to analyze the real estate market. However, it contains observations describing different market segments, each with its own logic of price formation, depending on the nature of the property, its destination, the profile of sellers and buyers, etc. The value of each property differs from one segment to another. The value of houses is different from that of apartments, although both types of occupancy are mainly bought by households, to live in or for investment purposes. Business premises meet specific needs (professional offices, craft workshops, warehouses, industrial or commercial buildings, etc.). Similarly, land can be used for a variety of purposes (agriculture, amenity, construction, etc.). Depending on the nature of the property and the type of use, it is possible to identify different market segments with different characteristics and land values. By way of illustration, we have chosen to focus solely on the sale of houses, restricting our geographical scope to the Côte-d'Or *département*. Furthermore, in the raw DVF files processed, the years 2016 and 2021 present two to three times fewer observations than the other years. Because of this apparent shortfall, we use only house transfers from 2017 to 2020.

The mutation database is accompanied by two geographic reference systems based on the land register. The first information layer describes the locations of parcels affected by land and property transfers. It consists of a set of points corresponding to parcel centroids. The second layer refers to the single locators of mutations corresponding to the centroid of the parcel with the largest premises, or the largest surface area in the absence of premises. The map (see Figure 5.2) shows the locators of all parcels affected by a mutation and the unique locators of house mutations[4].

4 All maps are created in R using the mapsf package (Giraud et al. 2022).

| | Number of lines | Number of mutations | Number of dispositions |
|---|---|---|---|
| 2016 | 1,617,368 | 694,209 | 701,727 |
| 2017 | 3,383,409 | 1,447,459 | 1,458,159 |
| 2018 | 3,332,865 | 1,433,542 | 1,442,894 |
| 2019 | 3,612,746 | 1,545,836 | 1,555,256 |
| 2020 | 3,149,482 | 1,338,164 | 1,344,832 |
| 2021 | 1,210,569 | 510,071 | 512,292 |
| Total | 16,306,439 | 6,969,281 | 7,015,160 |

**Table 5.1.** *Counting the number of lines, dispositions and mutations per year (source: DVF 2016-2021)*

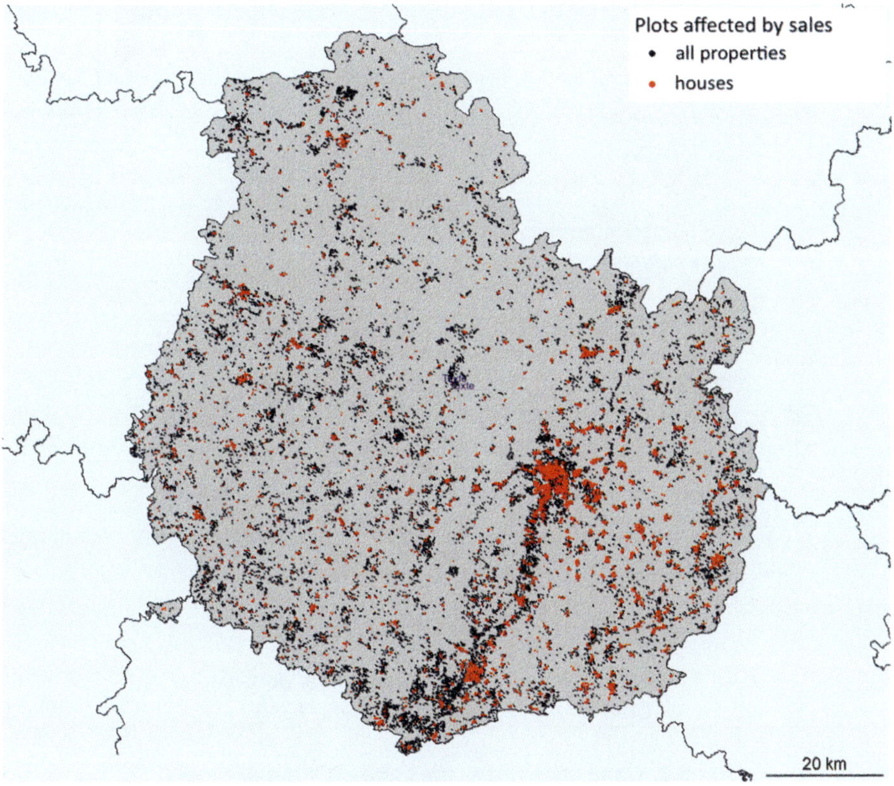


**Figure 5.2.** *Map showing the plots affected by sales in Côte-d'Or 2017–2020 (sources: DVF & IGN)*

**Brief presentation of the case study area**

Côte-d'Or *département* is located in central-eastern France, in the heart of the Bourgogne Franche-Comté region. It has a population of 533,220 (2018) and covers an area of around 8,800 km². The eastern part of the département is highly urbanized. Organized around Dijon and Beaune to the south, it concentrates most of the population, employment and facilities. The rest of the department is much less densely populated (Figure 5.3). It is structured around a few small towns, such as Châtillon-sur-Seine, Montbard, Semur-en-Auxois, Venarey-lès-Laumes, Saulieu, etc., which play a central role (Hilal et al. 2020).

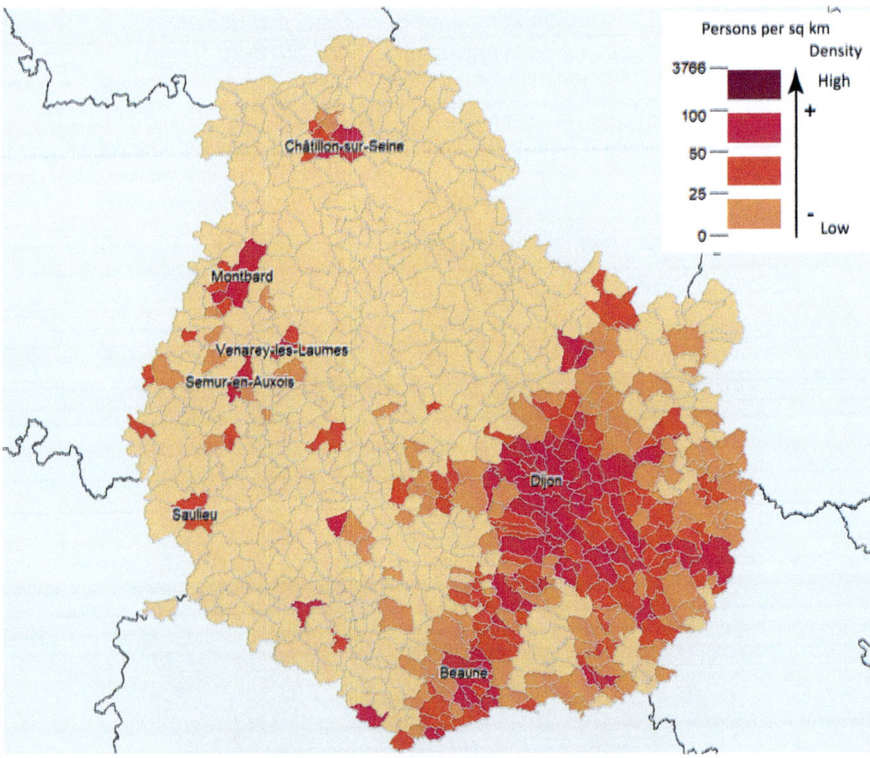


**Figure 5.3.** *Map illustrating municipal densities of Côte-d'Or (sources: INSEE and IGN)*

**Box 5.2.** *Case study area: the Côte-d'Or département*

Spatial heterogeneity is one of the fundamental characteristics of most geographical processes (i.e. non-stationarity), whether social or natural. For example, it is observed by many authors in the spatial or spatiotemporal dynamics of property prices (Guérois and Le Goix 2009; Baumont and Legros 2013; Chasco and Le Gallo 2015). It is explained by a difference in context, which can manifest itself on different spatial or temporal scales, generating a fluctuation in prices or any other observed characteristic. In the example developed here, it would be entirely possible and sensible to analyze the heterogeneity of house prices at the scale of mutations (individual data). However, this choice would not necessarily be relevant to this chapter. Apart from the resources mobilized, it would require gathering a set of variables established by observation or processing at the cadastral parcel level, which would go beyond the illustrative framework. As a result, we have preferred to analyze real estate spatial dynamics on a municipal scale. Although it is possible to observe sub-municipal heterogeneity, particularly within large cities, we feel that the scale of analysis we have chosen is appropriate, given the economic and geographical coherence of the municipalities. The data have therefore been aggregated over the entire period, giving priority to consolidating the spatial dimension of the data rather than the temporal dimension.

## 5.3. Preliminary explorations of spatial variations

In this section, we present a set of exploratory statistical and cartographic analyses of data aggregated to the municipality, with a dual objective. On the one hand, we wish to assess the spatial structure of each variable in order to identify, in a descriptive way, possible differentiating factors linked to the spatial dimension of the data: geographical situation (location), proximity to communication routes or major orographic complexes (adjacency), accessibility to major cities (distance), cardinal positions, etc. On the other hand, we test a few hypotheses of spatial organization, with a focus on the spatial organization of the data. Second, we test a number of spatial organization hypotheses, looking for possible territorial effects, such as EPCI (*établissements publics de coopération intermunicipale*) membership, public intermunicipal cooperation establishments, or simply spatial effects, such as neighborhood effects. In fact, the existence of such effects must be verified and taken into account in the development of the explanatory model presented in the remainder of this chapter. To this end, several maps will enable us to explore the spatial structure of the variables and test the hypothesis of an administrative effect in the spatial organization observed.

### 5.3.1. *Maps to explore the spatial structure of each variable*

Table 5.2 shows some of the descriptive statistics commonly used to describe quantitative variables. Synthetic quantities summarize the collection of values associated with each variable of interest. This is a simple way of describing and quantifying the characteristics of observations. These statistics can be accompanied by count tables and histograms for a more detailed analysis of the statistical distribution of values (see Figure 5.4). These elements can be used to answer various questions about the data: Are the observed values centered or clustered around one or more values? Are they spread over a wide or narrow range of values? Are they randomly distributed or do they follow a known statistical distribution? etc.

| | N | Mean | Median | Standard deviation | Min. | Max. |
|---|---|---|---|---|---|---|
| Average number of mutations | 688 | 25 | 11 | 79 | 1 | 1,811 |
| Average price (€) | 688 | 139,013 | 129,572 | 67,672 | 3,500 | 446,172 |
| Average surface area of homes (m²) | 688 | 129 | 123 | 38 | 22 | 416 |
| Average price per m² (€) | 688 | 1,169 | 1,053 | 527 | 159 | 3,852 |
| Average number of rooms | 688 | 4.3 | 4.3 | 0.9 | 0.5 | 11.0 |
| Average lot area (m²) | 688 | 3,967 | 1,484 | 20,176 | 182 | 429,133 |

**Table 5.2.** *Municipal descriptive statistics of homes sold (source: DVF 2017-2020)*

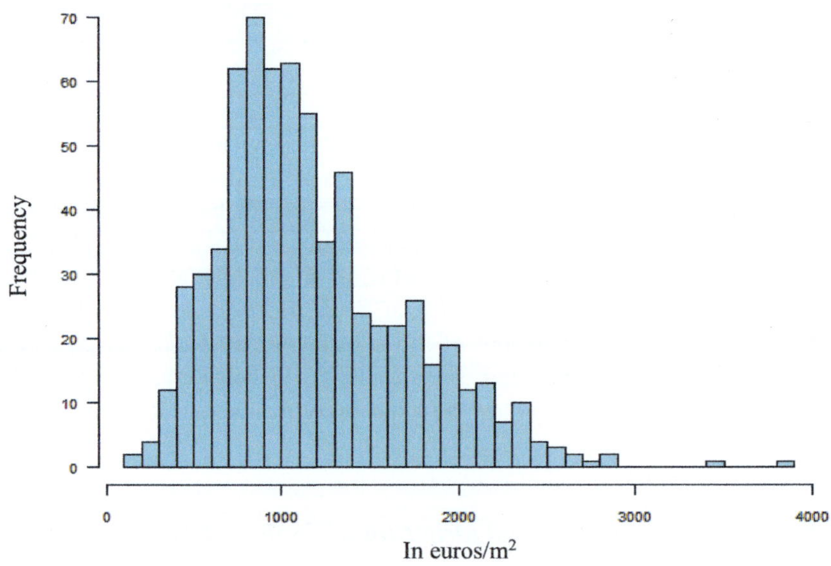


**Figure 5.4.** *Histogram showing the statistical distribution of average sales price of houses sold*

A histogram, like the one in Figure 5.4, provides a direct, global reading of the statistical distribution of the variable of interest, in this case the average selling price per m$^2$. The x-axis shows the price values, expressed in euros per square meter, divided into equal classes (here 30), and the ordinate shows the frequency corresponding to the number of times a value falls within a price class. The price distribution is unimodal and spread to the right (positive asymmetry). It is centered around 1,000, as indicated by the median value of 1,053 in the descriptive statistics table, and the majority of values are between 100 and 2,800, with a few rare values between 3,400 and 4,000. This description can be completed by investigating whether the distribution follows a known statistical law, such as a normal or exponential distribution, etc.

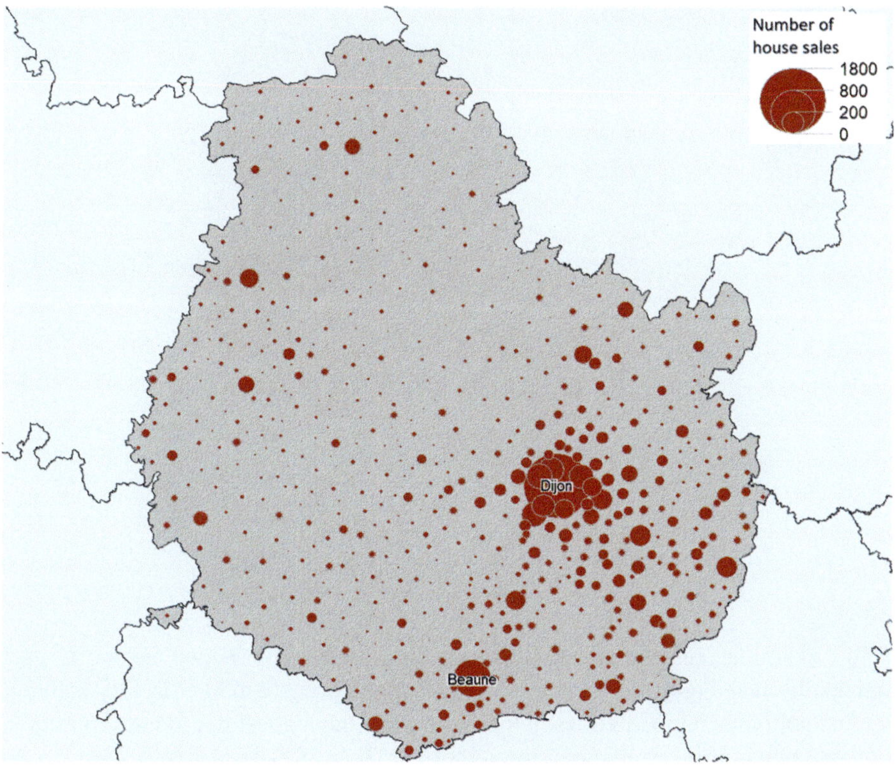


**Figure 5.5.** *Number of house sales by municipality 2017–2020 (sources: DVF and IGN)*

The spatial dimension, which is inherent in the data collected, is absent from the descriptive statistics presented here. The use of maps makes it possible to identify, for the variables used, the existence of a spatial structure linked to specific geographical situations (proximity, distance, neighborhood, etc.).

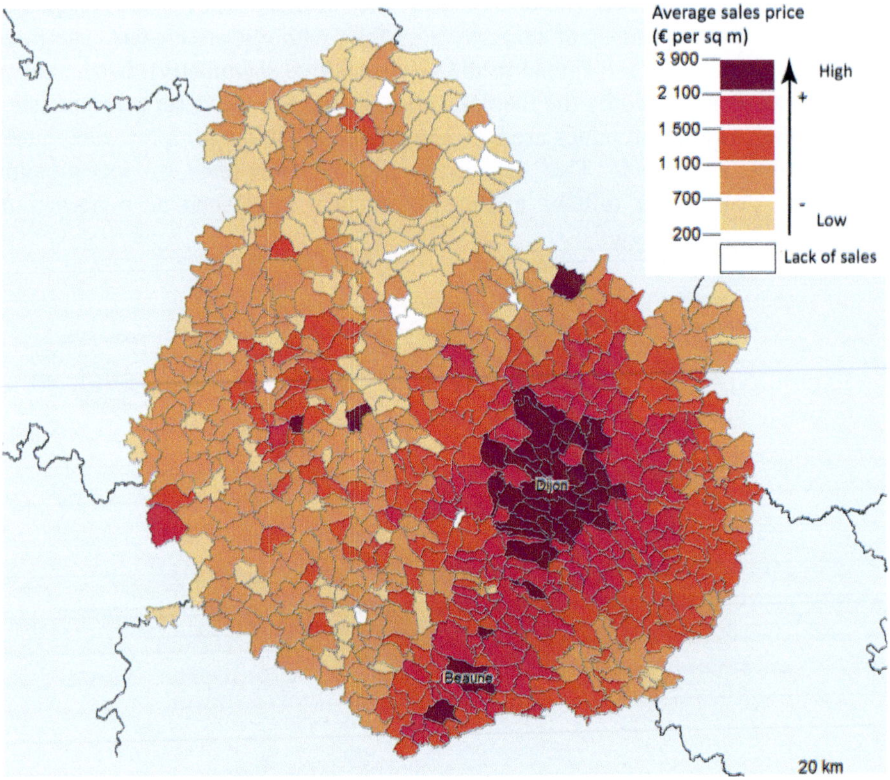


**Figure 5.6.** *Average sales price of houses sold 2017–2020 (sources: DVF and IGN)*

For example, on the map showing the number of house sales in each municipality (see Figure 5.5), we can see that the real estate market is very active in the most populous municipalities, which also have the highest house concentrations: Dijon and neighboring municipalities, as well as Beaune. Sales are also strong in the towns on the north-south axis between these two cities, and to the east of this axis. In the west of the *département*, a number of small towns and rural villages are also experiencing a more dynamic market. In addition, the map showing house sales prices (see Figure 5.6) shows that the average cost per square meter is highest in the east of the *département*, not only in Dijon and nearby municipalities, but also in Beaune and its suburbs. Municipalities on the Dijon-Beaune axis have slightly lower average prices, but they are still high compared to other municipalities in the *département*. Prices fall the further their distance from these areas.

The spatial distribution of observed values, for both sales volume and price (see Figures 5.5 and 5.6), is therefore not spatially uniform. The real estate market is dynamic in the region's major cities and in the densest areas, corresponding to the suburbs, while the average price per $m^2$ is higher in the cities and decreases with distance.

### 5.3.2. *Testing the hypothesis of an administrative or spatial effect in spatial organization*

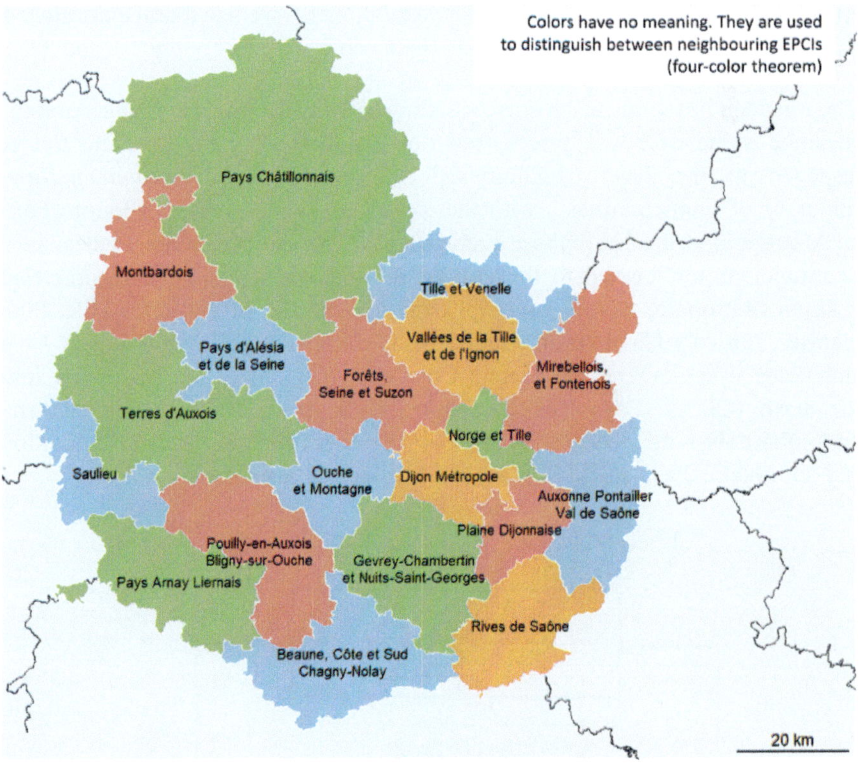


**Figure 5.7.** *Map showing EPCIs (public intermunicipal cooperation establishments) in Côte-d'Or (source: IGN)*

The intermunicipal differences observed may be the result of uneven population distribution or other underlying factors, themselves unevenly distributed across municipalities (job accessibility, levels of public facilities and services, natural amenities, etc.). In order to gain a better grasp of the spatial dimension of the variables under study, we use two methods to test two hypotheses of spatial organization: the existence, on the one hand, of an administrative effect linked to the grouping of municipalities into EPCIs and, on the other, of a spatial neighborhood effect or spatial autocorrelation.

#### 5.3.2.1. *Testing the hypothesis of an administrative effect*

We hypothesize that the EPCIs (see Figure 5.7), within which development and solidarity projects are drawn up, offer a specific living environment and conditions that may explain some of the real estate dynamics. The use of superimposed box-plots makes it possible to describe the distribution of average municipal values for each EPCI, and to compare the distributions between them (see Figures 5.8 and 5.9).

The nineteen box-plots, each corresponding to an EPCI, are ordered according to the average value of house prices (red dot inside the rectangles). The top box corresponds to the Pays Châtillonnais CC (*communauté de municipalities*), "community of municipalities", with the lowest average value; the bottom box is Dijon Métropole, with the highest average value. House prices are highest in the EPCI municipalities located to the east of the *département* on the Dijon-Beaune axis: Dijon Métropole, the communities of municipalities of Norge et Tille, Plaine dijonnaise, Gevrey-Chambertin and Nuits-Saint-Georges, and finally the Communauté d'agglomération Beaune Côte et Sud. Conversely, prices are lower in the north (CC du Pays Châtillonnais) and the west (CC du Pays d'Arnay, du Montbardois, de Pouilly-en-Auxois/Bligny-sur-Ouche, du Montbardois), Bligny-sur-Ouche, Pays d'Alésia et de la Seine, Terres d'Auxois).

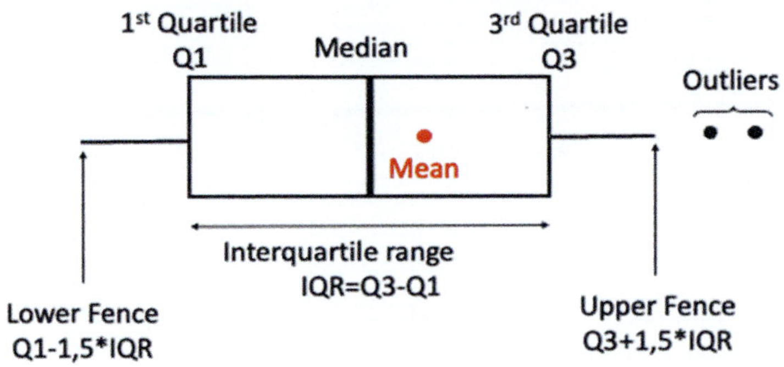


**Figure 5.8.** *Box-plot*

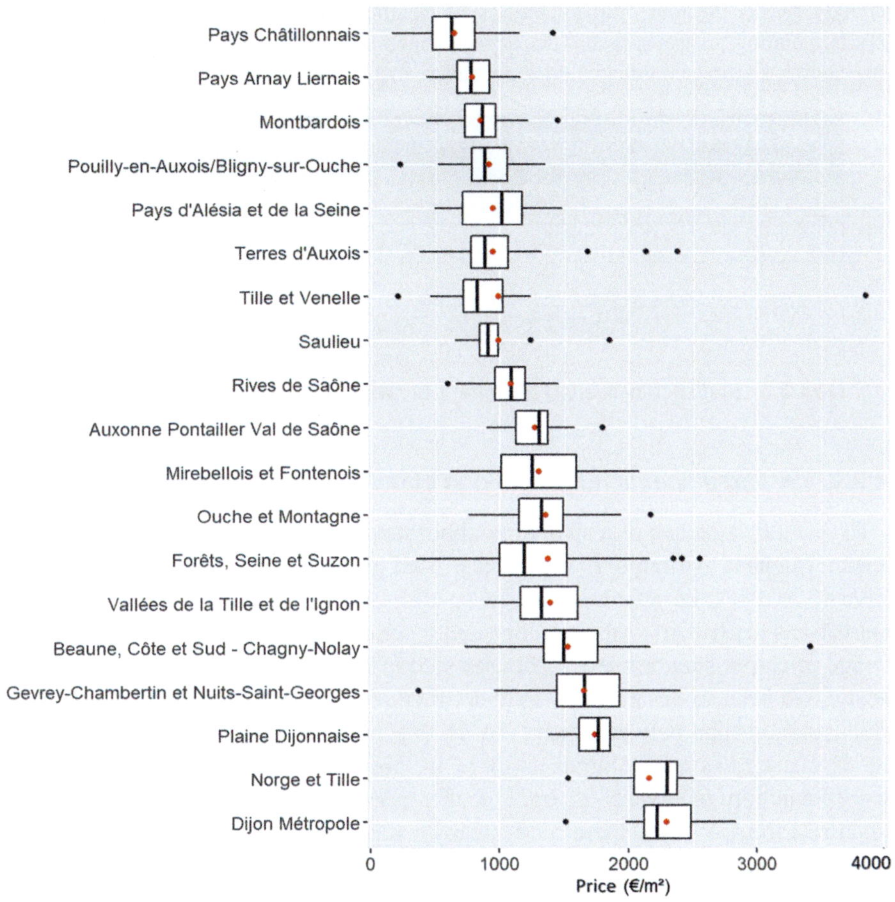


**Figure 5.9.** *Box-plot representing the relationship between average municipal house price per square meter and EPCI*

**Inter–intra analysis of variance**

Table 5.3 shows the results of an analysis of variance confirming the previous visual impressions. The null hypothesis tested is that of joint equality of average house prices between all EPCIs, while the alternative hypothesis assumes that there are at least two EPCIs between which average prices are significantly different. The value of the test statistic associated with this set of hypotheses is F = 60.121, which is higher than the critical value of the Fisher–Snedecor distribution with 18 and 669 degrees of freedom. The test result can also be appreciated directly by considering the critical probability, here displayed at 0 (in fact, a critical probability so low that it is rounded off to 0). This strongly rejects the joint null hypothesis of equality of house price averages between EPCIs, pointing to strong spatial disparities, which had already been appreciated via the map showing average municipal price per square meter.

| Degrees of freedom | Sum of squares | Averages | F-values | Pr(>F) |
|---|---|---|---|---|
| 18 | 117,769,458 | 6,542,748,0 | 60.121 | 0 |
| 669 | 72,804,441 | 108,825.8 | – | – |

**Table 5.3.** Analysis of variance table

**Box 5.3.** *Statistically testing the effect of belonging to a territorial network*

### 5.3.2.2. *Testing a spatial neighborhood effect hypothesis*

To test the existence of a spatial neighborhood effect, it is possible to calculate a statistic, such as Moran's *I*, commonly used to measure spatial autocorrelation (Oliveau 2010). Like any measure of spatial autocorrelation, it requires a set of observations described by their values and locations, in our example, the municipal average price per square meter of houses sold in Côte-d'Or. To calculate the Moran statistic, we need to describe the spatial structure of the entities, that is, specify the spatial relationships between them. To do this, we use a matrix of spatial weights with as many rows and columns as there are observations. This is a square matrix, the construction of which is one of the essential steps in calculating spatial autocorrelation. While there are many approaches to establishing spatial autocorrelation, the simplest method is to define the neighborhood using contiguity or adjacency criteria to construct a binary matrix, (0-1), where the value is 1 if the observations (in this case, municipalities) share a common boundary and 0 otherwise.

The Moran index describes the degree of overall spatial association of house prices. The null hypothesis tested is that the spatial distribution of these values is the result of a random process. In our example, the observed Moran index is equal to 0.725. The significance of this index is calculated using a permutation test (via Monte Carlo simulations). The simulation generates a very low critical probability, close to 0. Consequently, we reject the hypothesis of a random distribution of house prices.

The presence of spatial autocorrelation has important implications for subsequent statistical analyses. It reveals the existence of one or more non-random spatial processes linked, directly or indirectly, with the variable under study, here the unit price of houses sold in Côte-d'Or, generating an association between values observed in nearby locations. Processes can be social, economic, political, environmental, etc. Examples include ease or difficulty of access to workplaces, shops and services, the perceived or actual quality of schools or neighborhoods, the presence or absence of recreational areas, noise, pollution, and so on.

These processes have a spatial dimension that is not random. This needs to be taken into account, at the very least through formal modeling, in statistical analyses. We shall come back to this in the section on hedonic models. In this respect, the Moran index is a good tool for summarizing the spatial structure of the sample and providing the degree of spatial clustering of observations. However, this global index says nothing about the localization of similar-value groupings. On the other hand, positive overall spatial autocorrelation is not incompatible with a few municipalities in the sample that are very different from their neighbors.

To take this a step further, we need a local measure of spatial autocorrelation which, rather than providing an aggregated value, focuses on the relationships between each observation and its neighbors. This is what local indices of spatial association, known as LISA, accomplish (Anselin 1995). The construction of a Moran diagram helps us to understand the logic (see Figure 5.10). It is a scatterplot that crosses the values at one location with the mean values in the vicinity (defined above). The variables are centered and reduced so that they can be compared, and it can be shown that, in this case, the slope of the straight line fitting the scatterplot is equal to the value of Moran's $I$ statistic. The positive slope reflects positive overall spatial autocorrelation.

This figure allows disaggregation of the overall spatial autocorrelation: each of the two variables can be divided into two groups. Municipalities whose price is higher than the departmental average have positive reduced centered values, and those whose price is lower have negative values. The diagram is thus divided into four quadrants, depending on whether a given observation displays a value above (+) or below (–) the average of the original variable (average municipality price), or whether its spatial offset (average of adjacent municipality prices) displays a value above (+) or below (–) the average of the original variable. On the diagram, the four quadrants are identified by the labels "High-High", "Low-Low", "High-Low" or "Low-High".

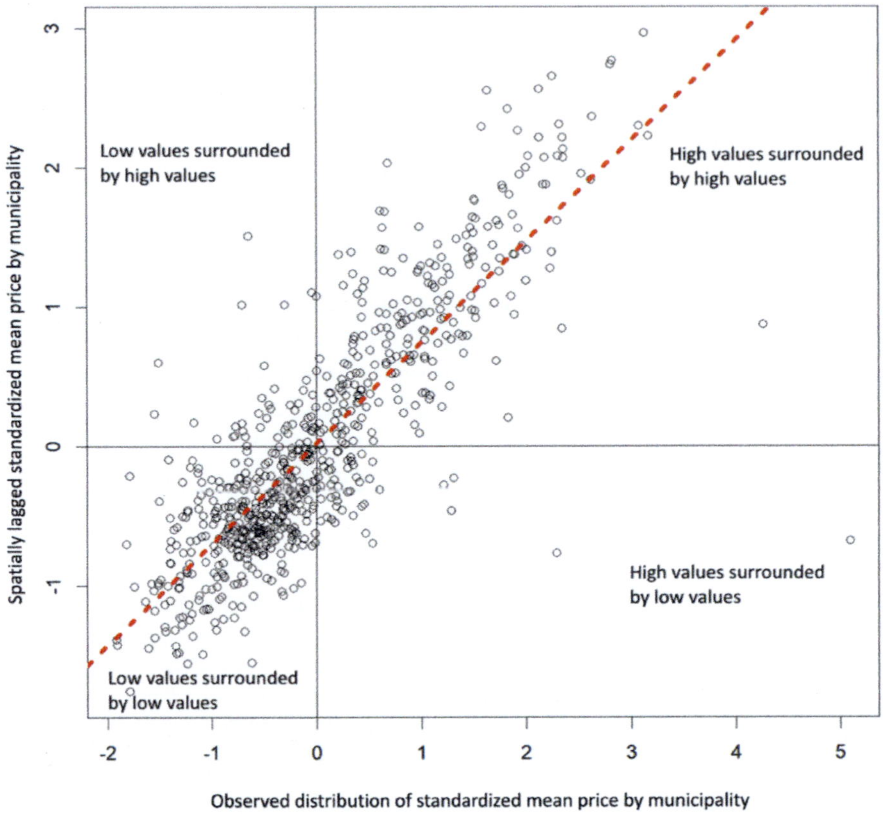


**Figure 5.10.** *Moran scatterplot (source: DVF 2017-2020)*

The statistical significance of local indices is also calculated using a permutation test (Monte Carlo simulations). Significant local Moran indices are plotted on the map in Figure 5.11. White indicates municipalities where indices are insignificant. Red and blue indicate positive and significant spatial autocorrelation of the “High-High” (high value surrounded by high values) or “Low-Low” (low value surrounded by low values) type, respectively. The groupings identified contrast, on the one hand, the north of the *département*, with more specifically the rural and forested municipalities of the eastern part of the Châtillonnais, where the average house price is significantly lower than the *département* average and, on the other, the municipalities of the Dijon urban area – and to a lesser extent that of Beaune, where the average house price is significantly higher than the *département* average.

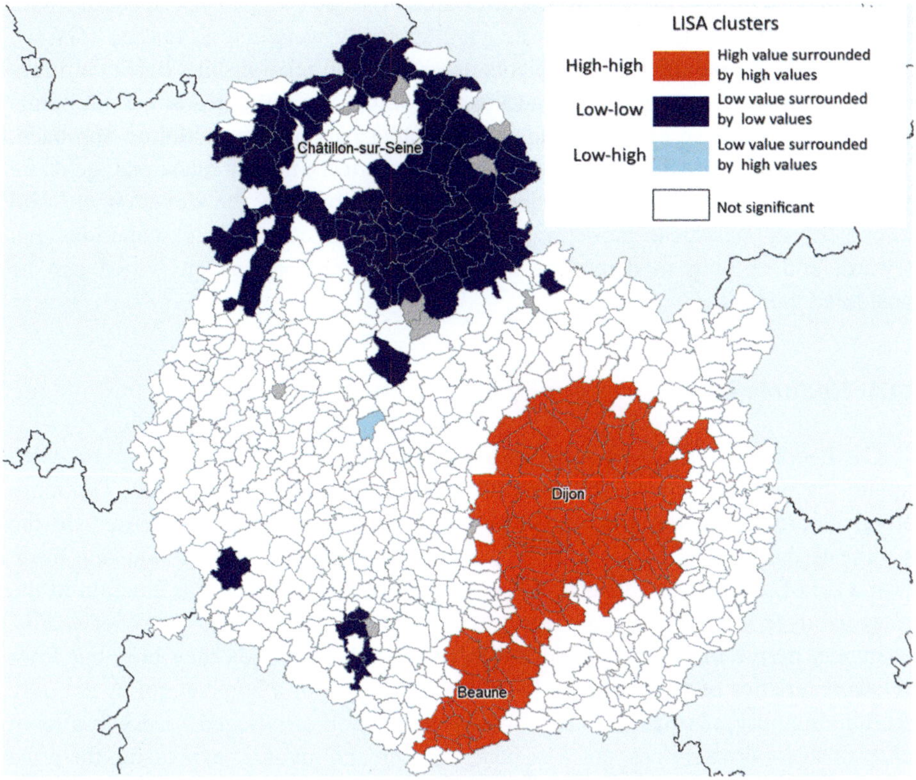


**Figure 5.11.** *Homogeneous concentration zones detected based on the local Moran Index of average price per m² (sources: DVF and IGN)*

## 5.4. Analyzing relationships statistically and rendering a map

The aim of this last section is to account for the fluctuations of a variable as a function of other variables. To do this, we draw on the hedonic method widely used by economists and geographers. After a brief presentation of the approach, we test two econometric models in which house prices are explained by their attributes and location. The first model, estimated by ordinary least squares (OLS) regression, quantifies the overall relationship between price and property characteristics. However, as we shall see, it is not certain that this relationship is spatially stable. Indeed, an attribute valued in an urban municipality may be valued less or not at all in a rural municipality and vice versa. The hypothesis on the spatial stationarity of the price–attribute relationship is probably not valid; in other words, there is a phenomenon of spatial heterogeneity. To take account of the spatial variability of the relationship between prices and attributes, it is possible to use a model with spatially variable coefficients, such as geographically weighted regression (GWR). In this model, the parameter coefficients describing the relationships between prices and attributes form continuous surfaces whose values are estimated at certain points in space. Maps are an essential and indispensable part of this modeling approach. They accompany statistical processing, enabling us to formulate questions, generate hypotheses and new processing ideas. They contribute to the development and enrichment of statistical modeling, supporting or refuting the explanations put forward, and helping to determine which of the explanations envisaged can be considered the best.

### 5.4.1. *Hedonic regression principles*

The hedonic approach is relatively old and has been used in many fields to explain the price of cars, wine, computer equipment, labor and housing. The more formal theoretical underpinnings (Lancaster 1966; Rosen 1974) are based on the hypothesis that the price of a complex, heterogeneous good such as housing, made up of a set of different but inseparable elements sold as a whole, is a function of the utility a buyer will derive from each of the good's components. In other words, consumers derive their satisfaction not directly from the goods they buy, but from the characteristics of those goods. Therefore, the buyer of a home acquires not only a certain quantity of square meters, but also a location, privileged access to a town center or to certain services, etc. The hedonic approach involves expressing the price the consumer is willing to pay for each type of good as a function of the quantity or quality of the various attributes. A hedonic model will therefore seek to explain the price of a home in terms of its structural characteristics (intrinsic), such as the number of rooms, the year of construction, the number of bathrooms, etc., and in terms of the characteristics of its environment (extrinsic), such as distance from the town center, the absence or presence of pollution, the quality of the neighborhood, the view, etc.

Estimating a hedonic model involves studying the relationship between the price of a good (explained variable) and all its different characteristics (explanatory variables). This relationship is analyzed by means of a multiple linear regression known as hedonic regression. For example, the price of a house can be summarized using a hedonic price function as follows:

$$P_i = \alpha + \sum_k X_{ik}\,\beta_k + \sum_l Z_{il}\,\gamma_l + \varepsilon_i$$

The price of a house ($P$) is a function of its attributes $X_{ik}$, for example, living area and number of rooms, land area, etc., and of its location described by characteristics $Z_{il}$, such as distance from the urban center, neighborhood quality, presence of green amenities, etc. The error term $\varepsilon_i$ is a random variable that captures all the variables influencing house prices that are not taken into account in $X$ and $Z$. The variation in house price resulting from the marginal variation in one of its characteristics is called the hedonic price, implicit price or price differential. Regression results indicate the extent to which house values will change for a small change in each characteristic, with all other characteristics remaining constant.

The relationship between price and characteristics can be established at an individual level, as in the model presentation, or at an aggregate level. In the latter case, price and characteristics are aggregated (Baltagi et al. 2014). In our case, aggregation takes place at the municipal level.

### 5.4.2. *Model with no spatial attributes*

The first model is estimated using the OLS method, including only the property attributes available in the municipal database: the average municipal value of houses sold is explained by the average surface area of the property, the average number of rooms and the average surface area of the land. As the quantity of each of the explanatory variables does not vary linearly with price, we opted for log-linear regression, replacing the variables price, average surface area of homes and average surface area of lots by their logarithms. The variable average number of rooms remains unchanged. Based on the estimation results presented in Table 5.4, the tests indicate that the model is correctly specified and that collinearity is low (VIF < 10)[5]. The coefficients (in the "Estimate" column) associated with the three explanatory variables are statistically significant (in the "p-value" column). All other things being equal, a 10% increase in the average surface area of a house leads to a 7.8% rise in the average price per m$^2$. Similarly, an additional room, with constant living area and lot size, increases the price by 0.14%. This corresponds to an additional €19,462 for a municipality with an average sales price of €139,013 (departmental average).

5 In regression, the variance inflation factor (VIF) assesses whether the explanatory variables are correlated with each other (collinearity or multicollinearity), which could reduce the reliability of the model. A VIF value greater than 10 indicates a high degree of multicollinearity: the variation in the variable will appear greater and more influential than it actually is. If the value is closer to 1 or lower, then the model is much more robust, as the variable is not influenced by its correlation with other variables.

| Explanatory variables | Estimate | Estimated standard deviation | Student's $t$ statistic | $p$-value | VIF |
|---|---|---|---|---|---|
| (Constant) | 8.00 | 0.33 | 24.48 | 0.00 | – |
| Average living area (log) | 0.78 | 0.08 | 9.79 | 0.00 | 1.83 |
| Average number of rooms | 0.14 | 0.02 | 5.70 | 0.00 | 1.76 |
| Average lot size (log) | –0.09 | 0.02 | –4.50 | 0.00 | 1.06 |

**Table 5.4.** *Model estimation results without spatial attributes (source: DVF 2017–2020)*

The quality of the regression, measured by the coefficient of determination ($R^2$), is 33%. It is significant and positive, but relatively low. The model residuals map shows that the estimated errors, represented by a harmonic double gradation, are not randomly distributed between municipalities (see Figure 5.12). On the contrary, their distribution shows groupings similar to those identified on the LISA map (see Figure 5.11), contrasting municipalities in the urban areas of Dijon and Beaune with rural municipalities in the north of the department.

Observing such systematic spatial structures in the residual map indicates the omission of important spatial variables in the model, in this case distances to urban centers (large cities or rural towns).

In order to improve this initial estimate, variables describing geographical locations are introduced in a second model. We focus on the accessibility of centers where shops, public and private services, educational establishments, medical professions and facilities, etc., are located. Four levels of centers are identified (Hilal and Piguet 2022). Local centers have only basic commerce and services, while major centers are the best equipped. In between are the intermediate and structuring centers, with an ever-increasing range of services on offer.

We hypothesize that the accessibility of these centers can have an effect on house prices, with, all other things being equal, higher values in the centers and decreasing values further away. Prices should also rise as the level of services in the centers increase. Road travel times between municipalities and the nearest center of each level are calculated using Odo-matrix (Hilal et al. 2018). The maps in Figures 5.13(A) and (B) show the spatial variation in accessibility of the four levels of equipment and service centers.

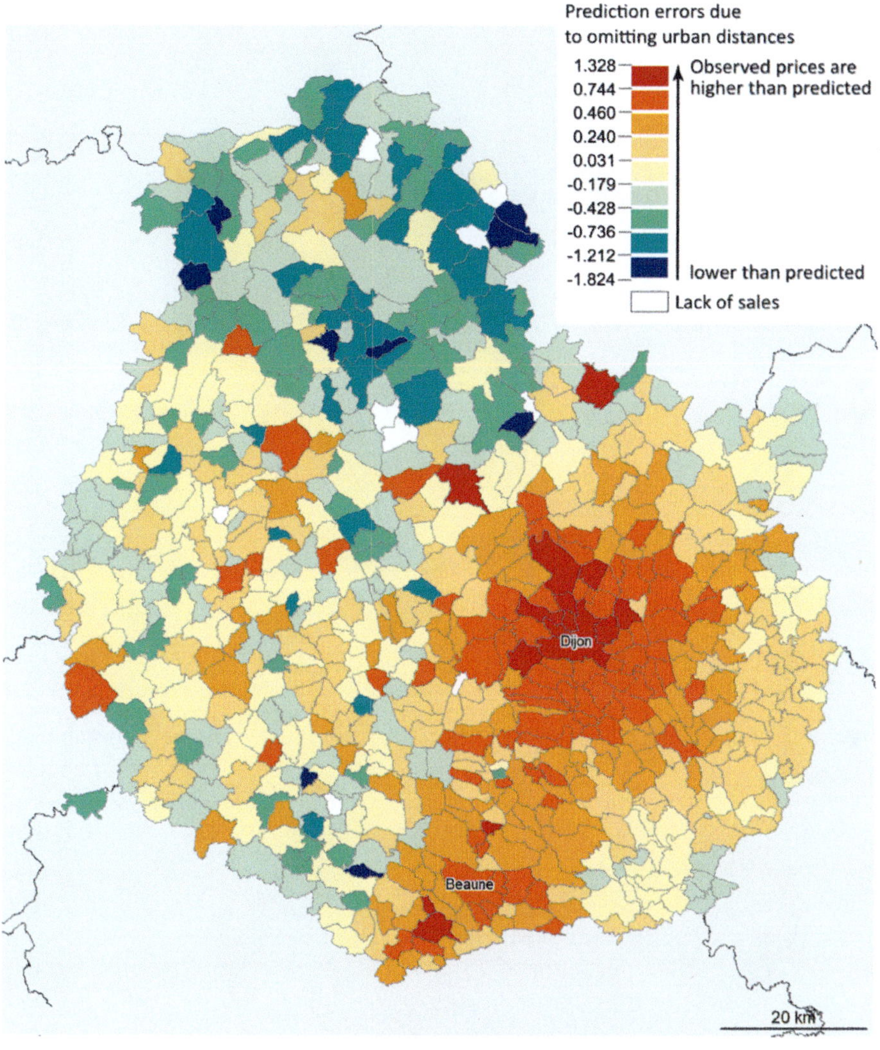


**Figure 5.12.** *Map showing residuals for the model with no spatial attributes using OLS (sources: DVF and IGN)*

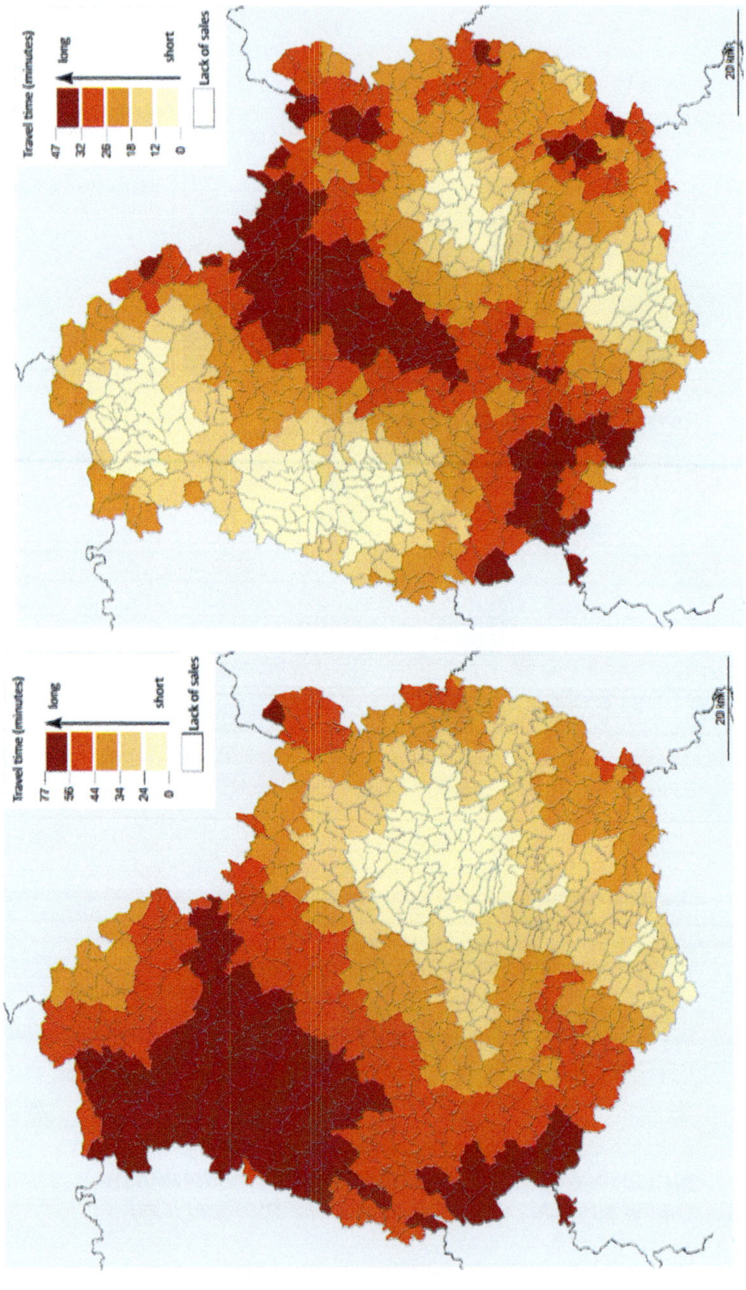


**Figure 5.13A.** *Spatial variability of accessibility of the four levels of centrality: (a) Access to the major center (Dijon); (b) access to urban centers (sources: INREA/ANCT & IGN)*

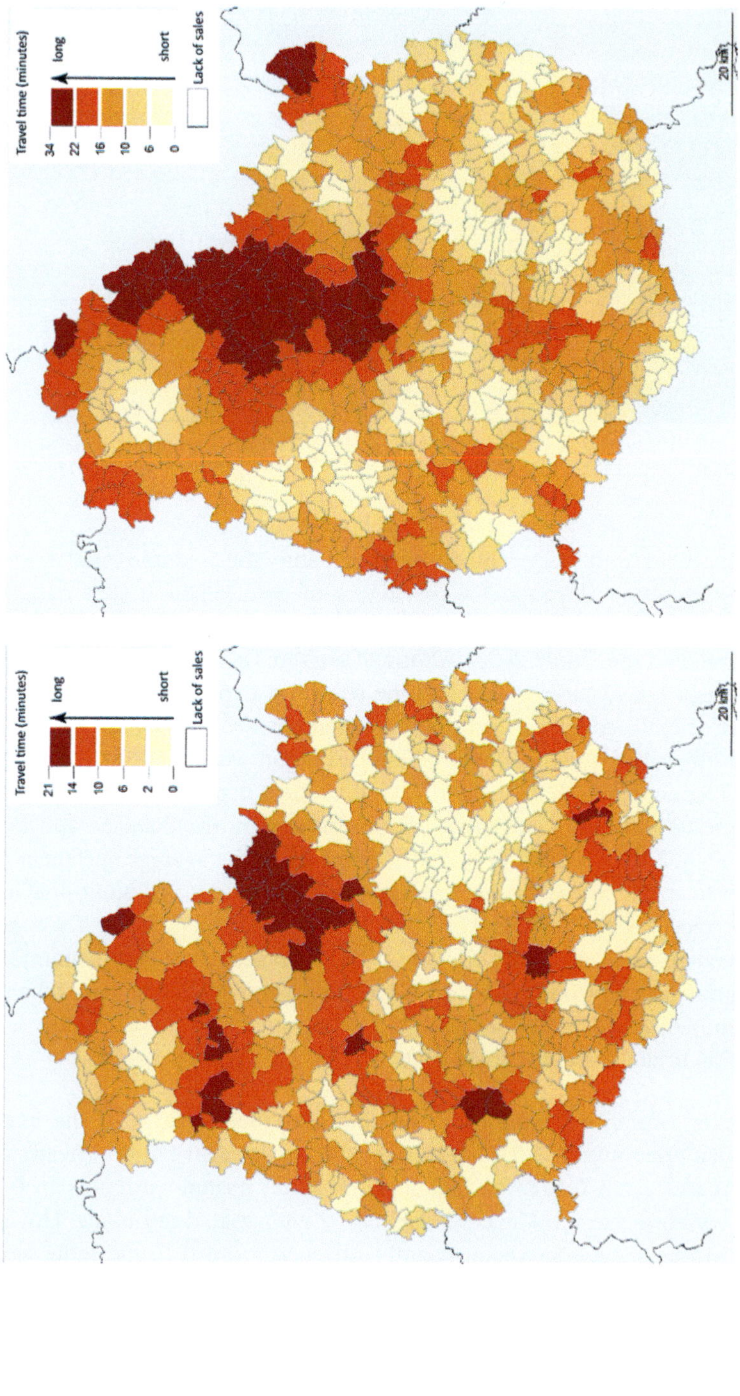


**Figure 5.13B.** *Spatial variability of accessibility of the four levels of centrality: (a) accessibility of intermediate centers; (b) accessibility of local centers (sources: INREA/ANCT & IGN) (continued)*

| Explanatory variables | Estimate | Estimated standard deviation | Student's *t* statistic | *p*-value | VIF |
|---|---|---|---|---|---|
| (Constant) | 9.30 | 0.23 | 40.20 | 0.00 | – |
| Average living area (log) | 0.57 | 0.06 | 10.16 | 0.00 | 1.90 |
| Number of rooms | 0.09 | 0.02 | 5.16 | 0.00 | 1.85 |
| Average lot size (log) | 0.07 | 0.01 | 4.83 | 0.00 | 1.31 |
| Accessibility to local centers | 0.00 | 0.00 | 1.05 | 0.29 | 1.48 |
| Accessibility to intermediate centers | –0.01 | 0.00 | –4.91 | 0.00 | 1.48 |
| Accessibility to urban centers | –0.01 | 0.00 | –6.26 | 0.00 | 1.36 |
| Accessibility to major centers | –0.02 | 0.00 | –23.29 | 0.00 | 1.40 |

**Table 5.5.** *Model estimation results with spatial attributes (source: DVF 2017-2020)*

Introducing these four accessibility variables into the hedonic model greatly improves the estimates. The model's coefficient of determination ($R^2$) rises from 33% to 69%. All coefficients are statistically significant, with the exception of local center accessibility (see Table 5.5). Although in the first model without spatial attributes, average house surface area had the strongest explanatory power (highest *t*-value), in the second model, accessibility to Dijon has the best explanatory power. Therefore, all other things being equal, a house in Dijon worth €500,000 would sell for €350,000 15 minutes from the center. The theoretical depreciation of €150,000 is obtained by multiplying the coefficient of the variable by the distance and by the starting price (–0.02 × 15 × 500,000). A 10% increase in the average surface area of a home leads to a 5.7% rise in the average price. Similarly, an additional room increases the price by 0.07%, corresponding to an additional €12,511 for a house paying the average price observed in Côte-d'Or. The accessibility of intermediate and urban centers plays a significant role, with a less steep decline in prices than that estimated from Dijon. A house worth €200,000 would sell for €170,000 within 15 minutes of an urban or intermediate center.

The value of Moran's *I* measuring the spatial autocorrelation of the average municipal house price was 0.725. Its value for the spatial distribution of estimation residuals decreases from 0.578 for the model without spatial attributes to 0.300, when spatial variables are introduced in the model with spatial attributes. However, the value of Moran's *I* remains significantly different from 0 in the latter model, meaning that the residuals are spatially autocorrelated. Figure 5.14 shows a map of these residuals. Some spatial groupings of high and low values can still be observed.

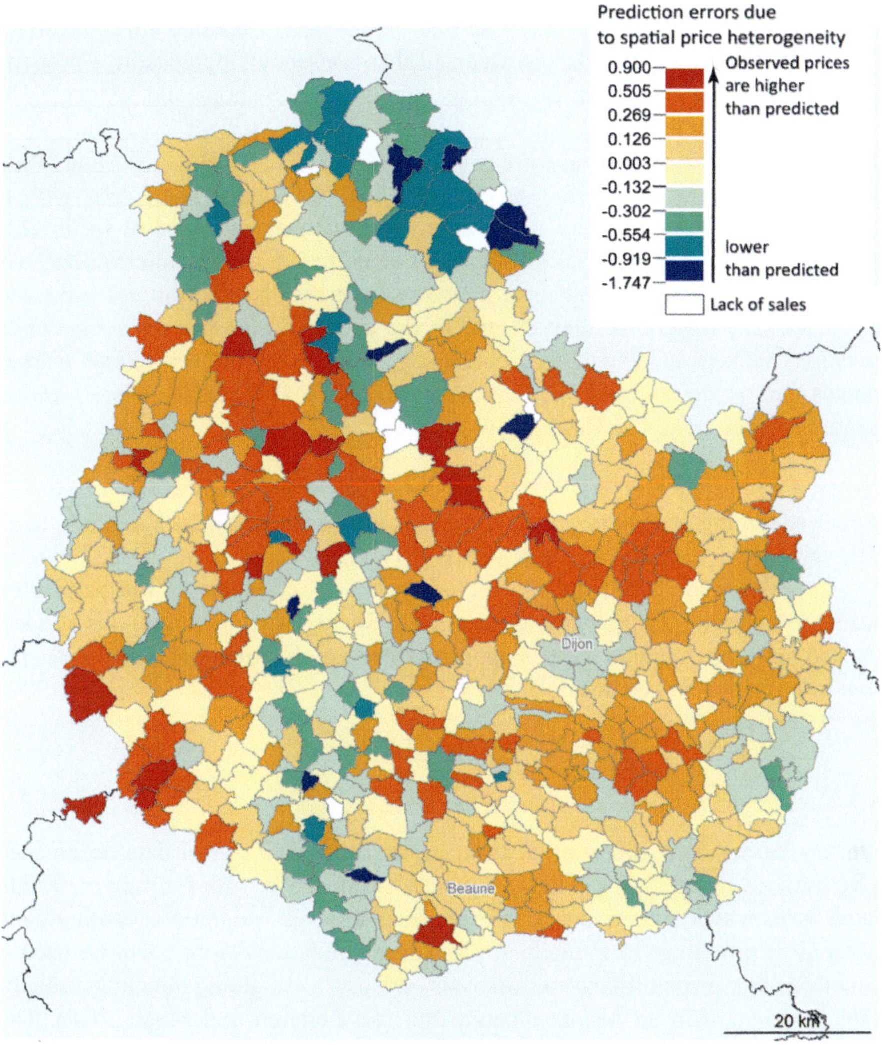


**Figure 5.14.** *Map showing model residuals with spatial attributes, estimated using OLS (sources: DVF and IGN)*

The presence of significant persistent spatial autocorrelation in the residuals may indicate the possible omission of spatial price heterogeneity. The next section examines this point.

### 5.4.3. *Spatial model estimated by GWR*

In the two previous models, we assumed that households purchasing a property would be willing to pay more for a larger house close to shops and services. The models use as explanatory variables property attributes (surface area, number of rooms, etc.) and proximity to amenities measured by the distances to the four levels of amenities and service centers. These are applied to the entire study area, with an overall estimator that does not necessarily take adequate account of local price variations. In fact, regardless of proximity to amenities, not all municipalities have the same real estate prices, and some may be systematically more or less expensive than others. The difference may be due to the age of the property (obsolescence, renovation, vintage effect) or the spatial segmentation of the real estate market, variables that we do not observe (Dambon et al. 2022). This phenomenon is known as spatial heterogeneity (Le Gallo 2004).

In its simplest form, spatial heterogeneity means that certain model parameters can vary according to location. To capture this heterogeneity, it is possible to use a model with local coefficients such as GWR (Fotheringham et al. 1998; Lu et al. 2014; Comber et al. 2022). In this regression, the coefficients of the parameters describing the relationships between prices and attributes form continuous surfaces estimated in each municipality using data from neighboring municipalities. The model to be estimated becomes:

$$P_i = \alpha_i + \sum_k X_{ik}\,\beta_{ik} + \sum_l Z_{il}\,\gamma_{il} + \varepsilon_i$$

In this model, the calibration of local parameters assumes that data observed in the vicinity of municipality $i$ have a greater influence on the estimate than data located further away. Consequently, each observation is weighted according to its geographical proximity to municipality $i$. Several functional forms can be used to account for this proximity. In our analysis, we use a weighting function, called a kernel function, with an adaptive bandwidth (de Bellefon and Floch 2018). This means that the extent of the kernel is determined by the number of neighbors maximizing the predictive power of the model (minimization of the CV validation score).

The output of the GWR model includes, in addition to the estimation results, diagnostic information, a summary of the variability of local parameter estimates (minimum, lower quartile, median, upper quartile, maximum) and the coefficients of the globally estimated parameters (see Table 5.6). As expected, the explanatory variables of the GWR model show spatially differentiated behaviors identified by parameter variation. For example, all other things being equal, a 10% increase in house surface area leads to a price rise of between 1.1% and 10.8%, depending on the municipality. The same applies to the size of the plot of land, where doubling the size generates an increase of between 0.6% and 12.2% depending on the municipality, or to the price paid for an additional room, which varies between €3,475 and €30,722. Coefficients relating to access to amenities and service centers also show spatial variations in price decreases with distance.

| Explanatory variables | Min. | First quartile | Median | Third quartile | Max. | Global |
|---|---|---|---|---|---|---|
| (Constant) | 6.43 | 8.69 | 9.78 | 10.68 | 11.33 | 9.30 |
| Average living area (log) | 0.11 | 0.24 | 0.44 | 0.61 | 1.08 | 0.57 |
| Number of rooms | –0.03 | 0.05 | 0.09 | 0.14 | 0.22 | 0.09 |
| Average lot size (log) | 0.01 | 0.04 | 0.06 | 0.08 | 0.12 | 0.07 |
| Accessibility to local centers | –0.02 | –0.01 | –0.00 | 0.01 | 0.03 | 0.00 |
| Accessibility to intermediate centers | –0.05 | –0.01 | 0.01 | 0.01 | 0.02 | –0.01 |
| Accessibility to urban centers | –0.02 | –0.01 | –0.01 | 0.00 | 0.02 | –0.01 |
| Accessibility to major centers | –0.03 | –0.03 | –0.02 | –0.01 | –0.00 | –0.02 |

**Table 5.6.** *Descriptive statistics for the coefficients of the model with spatial attributes estimated using GWR (source: DVF 2017–2020)*

The parameters of the GWR model, unlike those of the OLS model, can be shown on a map. The model has an acceptable quality of fit, with local $R^2$ values ranging from moderate (0.47) to good (0.83). On the map in Figure 5.15, the quality of estimates is best in urban and peri-urban municipalities in the east of the *département*. It is less good in municipalities throughout the western part, corresponding roughly to the prefectural district of Montbard, with the lowest quality in the rural municipalities of Auxois and Châtillonnais. As for the residuals, they show no particular spatial pattern (see Figure 5.16), indicating that the spatial autocorrelation of residuals detected earlier was indeed due to unmodeled spatial heterogeneity.

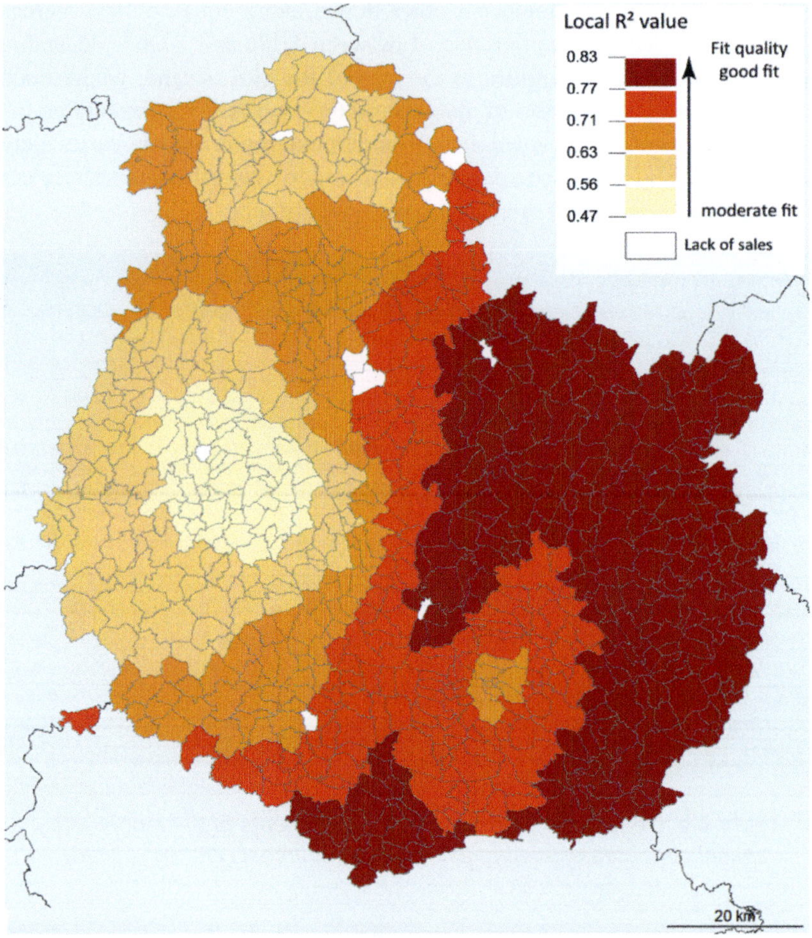


**Figure 5.15.** *Map showing local coefficients of determination for the model with spatial attributes estimated using GWR (sources: DVF and IGN)*

Depending on the significance and intensity of the relationship between the average house price and the explanatory variables, the maps (see Figures 5.17(A)(C) and 5.18(A) and (B)) identify specific hedonic price gradients that differ from one variable to the another one. Therefore, the three variables describing house characteristics show particular spatial contrasts. Average house surface area plays a significant role in the average selling prices observed in 71% of the department's municipalities, all of which lie north of a line running from the Vingeanne to the Beaunois, with a main focus of high values in the Châtillonnais and a secondary focus in the Auxois. The coefficients associated with average number of rooms and average lot size are significant in 62% of municipalities, but with very different spatial configurations. The average house price is partly determined by the average number of rooms in the southern half of the département, excluding the Dijon plain and the Saône Valley. A focus of high, positive coefficients appears in the Beaunois region. The spatial pattern of coefficients associated with average land area is completely different. Indeed, the municipalities with significant coefficient values are located in the north and west of the Côte-d'Or, with several foci of positive and high values on the periphery of the département, north of the Châtillonnais, Pays d'Arnay and Sud Beaunois.

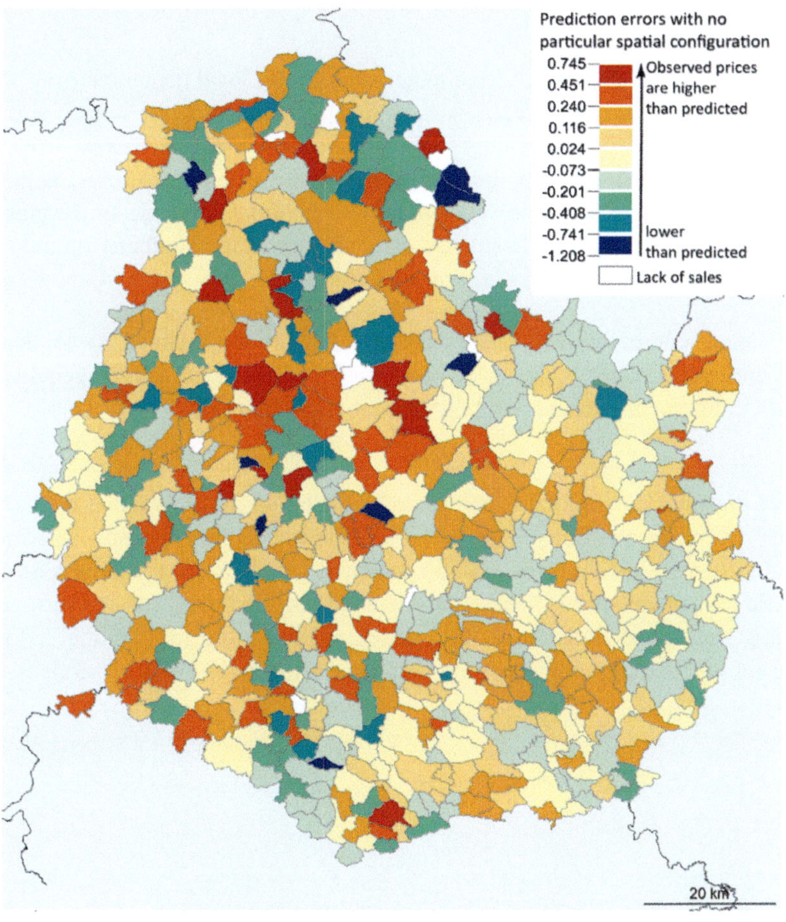


**Figure 5.16.** *Map showing residuals from the model with spatial attributes estimated using GWR (sources: DVF and IGN)*

Accessibility is expected to have an inverse effect (negative coefficients) on price. Prices should fall as commuting times increase. Accessibility to the major center has a significant effect on average house prices in 89% of municipalities.

As expected, prices fall with distance from Dijon. However, price decline is not isotropic. The decline is most marked in two foci located to the north and south of the Burgundy capital. Accessibility to structuring centers is a factor in house prices in 60% of municipalities, with a slightly steeper decline in prices in the Vitteaux, Liernais and Arnay-le-Duc areas. The coefficients associated with distance from intermediate centers are significant in 53% of municipalities. The decline in house values with distance is much more marked in the Châtillonnais, in the north of the département.

Finally, house prices are sensitive to accessibility to local centers in only 18% of municipalities. These municipalities are located more to the west, and their coefficients are positive, meaning higher prices even further away from local centers. They form a ring on the periphery of the northern Auxois, suggesting a possible effect of the Montbard TGV station, one of the gateways to Burgundy for Parisian secondary residents. However, distance from the Montbard station, tested in a complementary model, does not improve the model.

The maps, combined with statistical modeling, have enabled us to describe, explore and explain the house prices observed in Côte-d'Or. Heterogeneity is an important component in characterizing the real estate market.

The observed behavior between prices, house characteristics and spatial attributes does not correspond to the usual statistical or administrative breakdowns, since the estimated coefficients are not distributed according to known perimeters. The spatial disparities observed reveal municipal groupings that can be assimilated to specific real estate sectors. These perimeters, identified thanks to the cartographic processing associated with the GWR model, integrate the joint spatial effects of housing supply, demographic dynamics and access to employment or services.

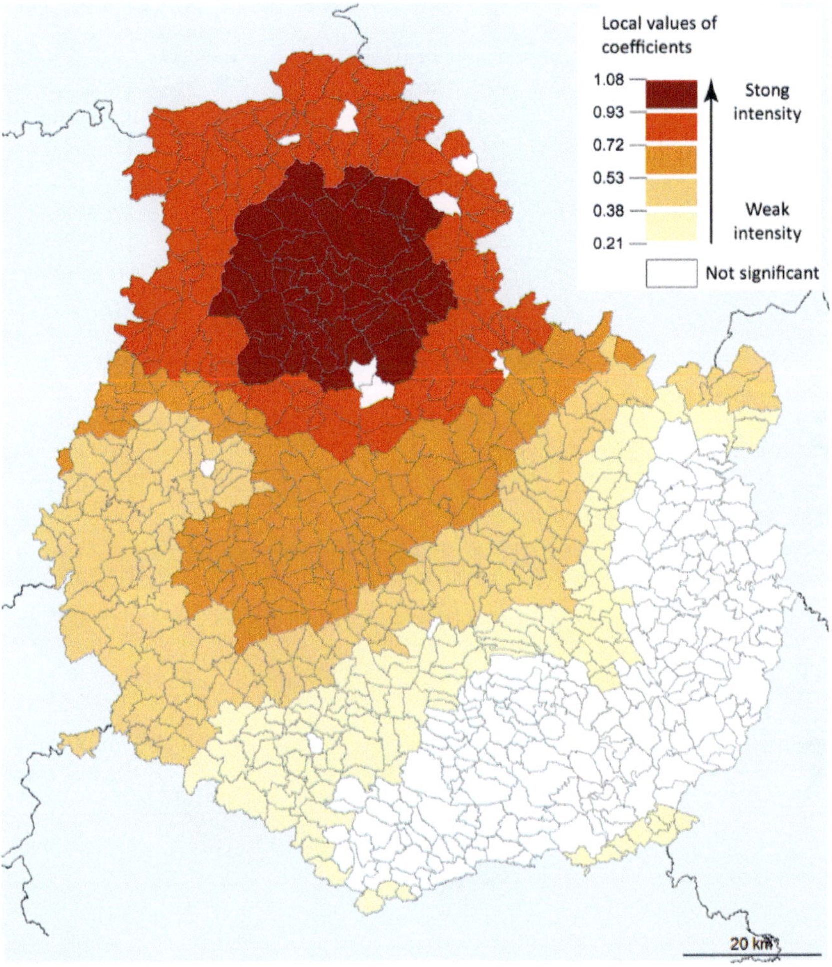


**Figure 5.17A.** *Maps showing spatial variation of local coefficients using the GWR model highlight the intensity of the relationship between house price and attributes. Intensity of the relationship between house price and surface area (sources: DVF and IGN)*

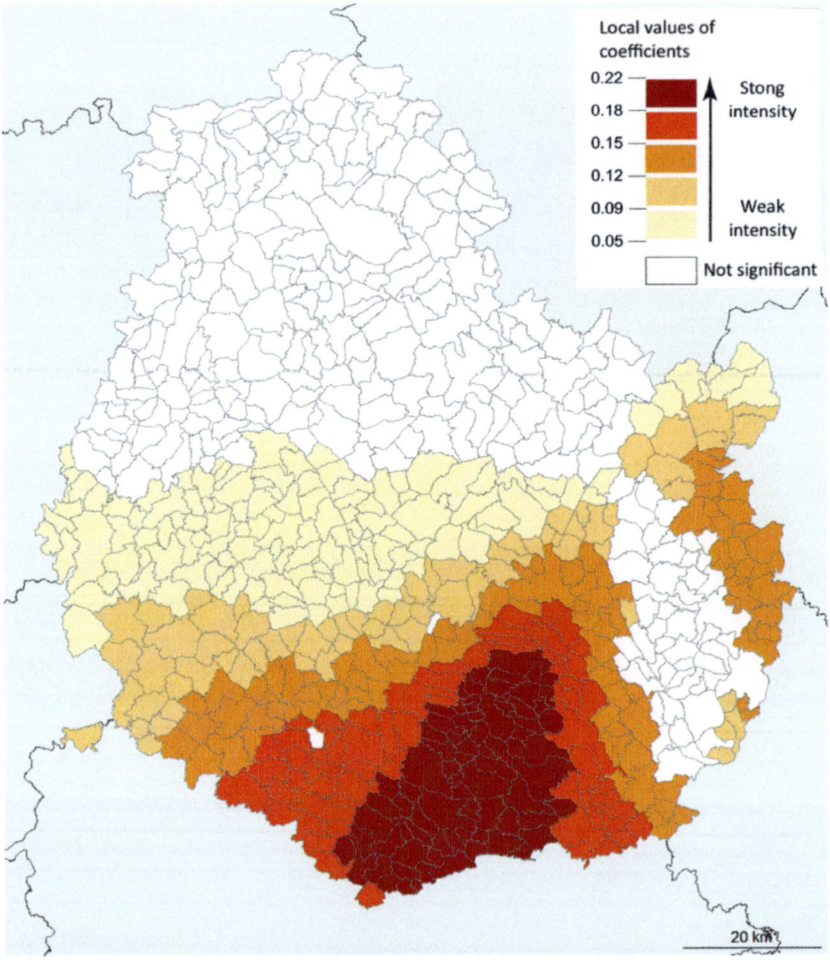


**Figure 5.17B.** *Maps showing spatial variation of local coefficients using the GWR model highlight the intensity of the relationship between house price and attributes. Intensity of the price–number of rooms relationship (sources: DVF and IGN) (continued)*

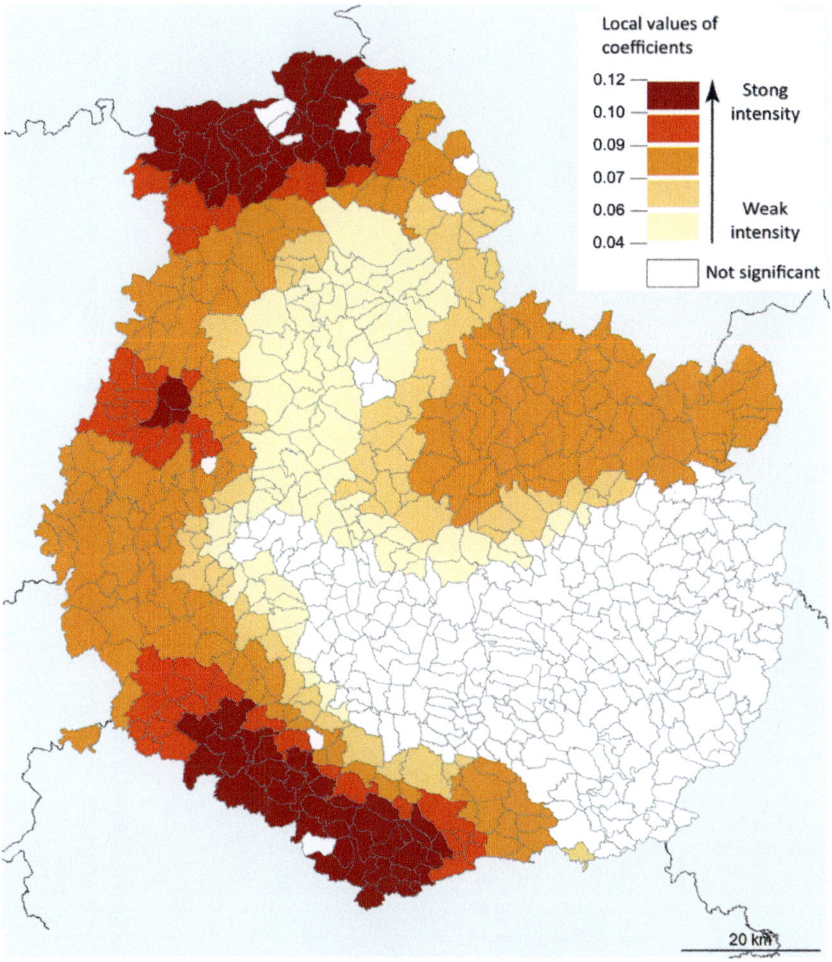


**Figure 5.17C.** *Maps showing spatial variation of local coefficients using the GWR model highlight the intensity of the relationship between house price and attributes. Intensity of the price–lot size relationship (sources: DVF and IGN) (continued)*

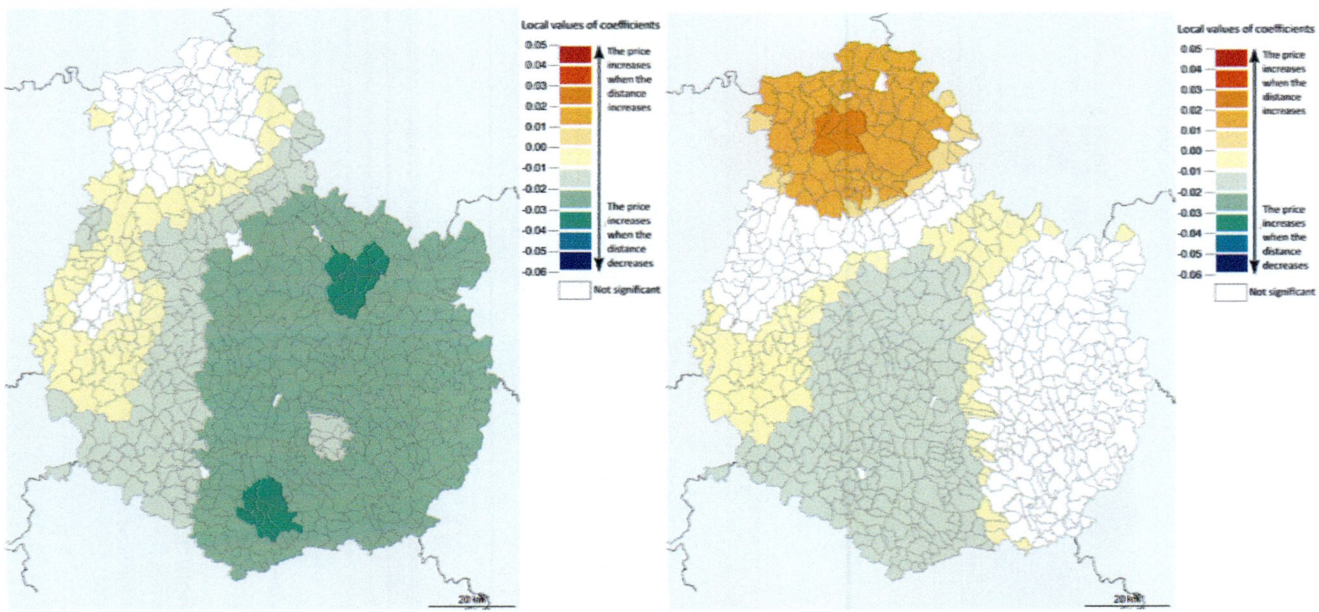


a) b)

**Figure 5.18A.** *Coefficients associated with the spatial attributes of the model estimated using GWR (source: DVF and IGN). (a) Intensity of price–distance relationship in Dijon; (b) intensity of price–distance relationship in urban centers*

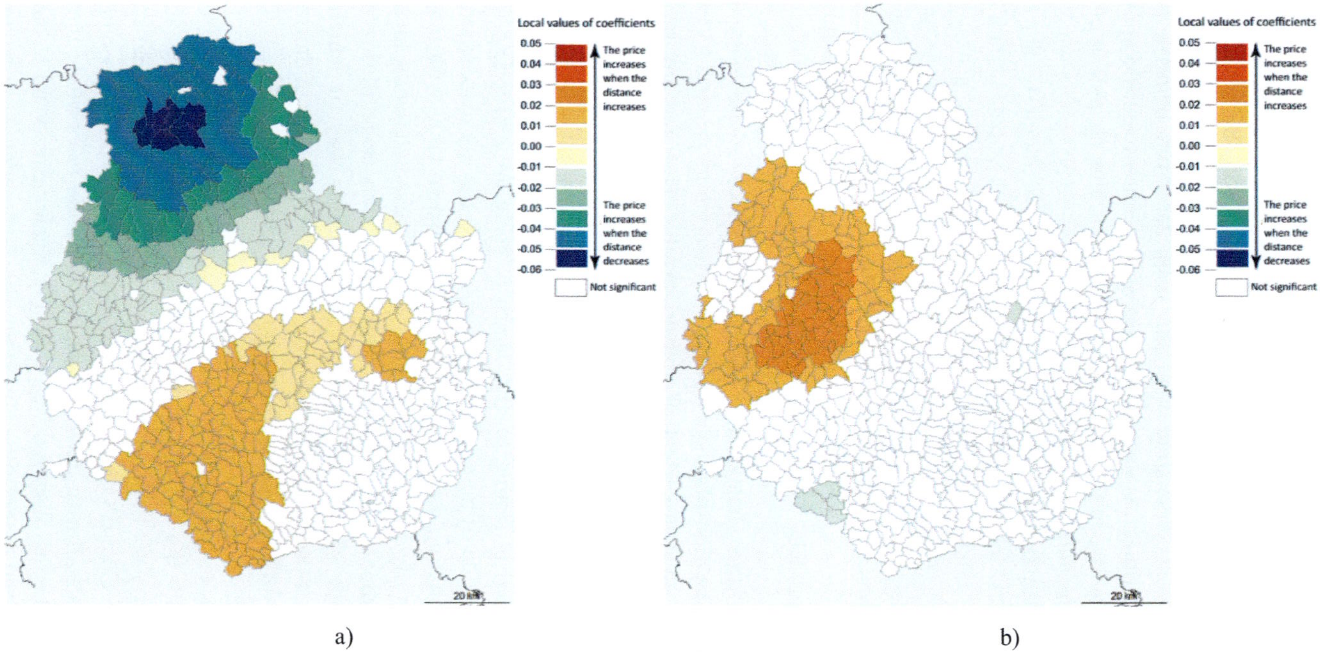


**Figure 5.18B.** *Coefficients associated with the spatial attributes of the model estimated using GWR (source: DVF and IGN). (a) Intensity of price–distance relationship to intermediate centers; (b) intensity of price–distance relationship to local centers (continued)*

## 5.5. Conclusion

This chapter deals with the statistical and spatial modeling of a spatially heterogeneous phenomenon, in this case the real estate market, by systematically combining data, statistical processing and maps in all the proposed phases of analysis. In this way, we demonstrate that the map, far from being a purely descriptive or illustrative object, is an integral part of the entire scientific approach and a key vector in modeling. In the illustration proposed in this chapter, it enables us to interrogate and understand the logic underlying an economic phenomenon with a strong geographical component, such as the formation of single-family home prices in Côte-d'Or, a department that is both urban and rural. The data and methods used are chosen for their operational value. By relying on public data, accessible to all, and statistical and cartographic methods appropriate to the objectives set, we obtain original, verifiable and reproducible results, inviting readers to develop their own empirical applications. The use of R statistical programming software and the availability of the code for all processing should greatly facilitate the appropriation and adaptation of the approach.

The three sections of this chapter each correspond to a key phase in statistical and spatial modeling. The first phase is descriptive. It aims to provide a detailed picture of reality based on data collection and processing. The land and property data used in this chapter come from the DVF database. Derived from notarial deeds and enriched by the DGFiP, these data are recorded in the tabular form, with specific organization and nomenclatures, enabling all legal and administrative objects linked to the sale of a property to be traced and archived. This first section describes the preliminary steps involved in setting up the database: using an appropriate formal model, including the creation of nomenclatures designed to classify dispositions and property types, then extracting data in the form of a statistical table with land and real estate transactions in rows and property types and their characteristics, including their geographical coordinates, in columns. This preparation stage illustrates the need to clarify the scope of the data and to document the way in which it is compiled. It also makes it possible to map the data. These operations authorize and facilitate the use of data for statistical and cartographic processing.

The second section corresponds to an exploratory phase, allowing us to familiarize ourselves with the data by drawing up maps and descriptive statistics. For the sake of simplicity, we have worked on data aggregated to the municipality level, retaining only the Côte-d'Or perimeter. Statistical processing uses relatively simple methods. The maps, drawn up in compliance with the rules of cartography, aim to image and describe the observed data within an analogical spatial framework (Bonin 2014). In a second stage, other methods are implemented to capture the spatial dimension of variable distributions. These treatments are based on a preliminary enrichment: mobilization of geographical reference systems (municipalities, EPCI), definition and construction of a neighborhood matrix. Appropriate spatial statistical analysis methods are used to model the spatial variability of the main variables in the database and highlight the presence of global and local spatial autocorrelation. Maps play a fundamental role in this process. They allow for data enrichment and restores visual exploration, enabling us to formulate questions, generate hypotheses and new treatment ideas.

In the third section, we move into an explanatory phase, with the aim of understanding the determinants and spatial organization of house prices. To achieve this, we use an econometric model based on hedonic regression. An initial regression gives an idea of the overall relationship between price and goods characteristics. However, this relationship is not spatially stable due to the existence of spatial heterogeneity. The use of regression with spatially variable coefficients enables the identification of perimeters that capture the joint spatial effects of housing supply, demographic dynamics and access to employment or services. Maps also play a fundamental role in this phase. Maps accompany and facilitate statistical processing. They contribute to the development and enrichment of statistical modeling, support or refute the explanations put forward, and help determine which of several explanations is the best.

Cartographic representation, in addition to its inventory function (location of places, facts or phenomena) or mediation function (communication of a message), is a key vector for modeling the spatial dynamics of house prices. Combined with statistical processing, it can be used to describe, explore, explain and model spatial phenomena and processes. It is an integral part of the scientific reasoning and approach used in this chapter.

## 5.6. References

Anselin, L. (1980). Estimation methods for spatial autoregressive structures: A study in spatial econometrics. Program in Urban and Regional Studies, Cornell University Ithaca, New York.

Anselin, L. (1995). Local Indicators of Spatial Association – LISA. *Geographical Analysis*, 27(2), 93115.

Badariotti, D. (2004). De la comparaison des cartes électorales. *Cybergeo: European Joural of Geography*, 267 [Online]. Available at: http://journals.openedition.org/cybergeo/3944.

Baltagi, B.H., Fingleton, B., Pirotte, A. (2014). Spatial lag models with nested random effects: An instrumental variable procedure with an application to English house prices. *Journal of Urban Economics*, 80, 76–86. doi: 10.1016/j.jue.2013.10.006.

Baumont, C. and Legros, D. (2013). Nature et impacts des effets spatiaux sur les valeurs immobilières. *Revue économique*, 64(5), 911–948.

de Bellefon, M.-P. and Floch, J. (2018). Régression géographiquement pondérée. In *Manuel d'analyse spatiale. Théorie et mise en oeuvre pratique avec R*, Loonis, V. and de Bellefon, M.-P. (eds). Institut national de la statistique et des études économiques, Montrouge.

Bonin, O. (2014). La carte, modèle analogique et calcul spatial. *Communication & langages*, 180(2), 47–61.

Boulay, G., Blanke, D., Casanova Enault, L., Granié, A. (2021). Moving from market opacity to methodological opacity: Are web data good enough for French property market monitoring? *Professional Geographer*, 73(1), 115–130. doi: 10.1080/00330124.2020.1824678.

Casanova Enault, L., Boulay, G., Coulon, M. (2019). Une aubaine pour les géographes ? Intérêts des fichiers open DVF sur les transactions foncières et immobilières et précautions d'usage. *Cybergeo: European Journal of Geography* [Online]. Available at: http://journals.openedition.org/cybergeo/33602 [Accessed 17 June 2022].

Cerema (2022). De DVF à DV3F [Online]. Available at: http://doc-data-foncier.cerema.fr/dv3f/tuto/objectif_tutoriel [Accessed 24 June 2022].

Chasco, C. and Le Gallo, J. (2015). Heterogeneity in perceptions of noise and air pollution: A spatial quantile approach on the city of Madrid. *Spatial Economic Analysis*, 10(3), 317–343.

Cliff, A.D. and Ord, J.K. (1973). *Spatial Autocorrelation*. Pion, London.

Çöltekin, A., Janetzko, H., Fabrikant, S. (2018). *Geovisualization. The Geographic Information Science & Technology Body of Knowledge*, 2nd Quarter 2018 Edition [Online]. Available at: https://gistbok.ucgis.org/bok-topics/geovisualization.

Comber, A., Brunsdon, C., Charlton, M., Dong, G., Harris, R., Lu, B., Yihe, L., Murakami, D., Nakaya, T., Wang, Y., et al. (2023). A route map for successful applications of geographically weighted regression. *Geographical Analysis*, 55(1), 155–178.

Courville, S. (1995). *Introduction à la géographie historique*. Presses Université de Laval, Quebec.

Dambon, J.A., Fahrländer, S.S., Karlen, S., Lehner, M., Schlesinger, J., Sigrist, F., Zimmermann, A. (2022). Examining the vintage effect in hedonic pricing using spatially varying coefficients models: A case study of single-family houses in the Canton of Zurich. *Swiss Journal of Economics and Statistics*, 158(1). doi: 10.1186/s41937-021-00080-2.

Fotheringham, A.S., Charlton, M.E., Brunsdon, C. (1998). Geographically weighted regression: A natural evolution of the expansion method for spatial data analysis. *Environment and Planning A*, 30(11), 1905–1927.

Le Gallo, J. (2002). Économétrie spatiale : l'Autocorrélation spatiale dans les modèles de régression linéaire. *Économie & prévision*, 155(4), 139–157.

Le Gallo, J. (2004). Hétérogénéité spatiale : principes et méthodes. *Économie et prévision*, 162(1), 151–172.

Giraud, T. and Lambert, N. (2017). Reproducible cartography. In *Advances in Cartography and GIScience. ICACI 2017. Lecture Notes in Geoinformation and Cartography*, Peterson, M.-P. (ed.). Springer, Cham [Online]. Available at: http://link.springer.com/10.1007/978-3-319-57336-6_13.

Giraud, T., Hernangomez, D., Pecout, H., Ysebaert, R., Fellows, I., Lemon, J., Zenoni, F. (2022). R package mapsf: Thematic cartography [Online]. Available at: https://github.com/riatelab/mapsf/.

Guérois, M. and Le Goix, R. (2009). La dynamique spatio-temporelle des prix immobiliers à différentes échelles : le cas des appartements anciens à Paris (1990-2003). *Cybergeo*, 1–25 [Online]. Available at: http://journals.openedition.org/cybergeo/22644.

Hilal, M. and Piguet, V. (forthcoming). Hiérarchie et centralités d'équipements et de services en appui au programme "petites villes de demain". In *Centralités et hiérarchies des réseaux et des territoires*, Fen-Chong, J. and Tannier, B. (eds). ISTE Editions, London.

Hilal, M., Fugeray-Scarbel, A., Gojard, S. (2018). Odomatrix : mesurer des temps de trajet pour améliorer les conditions de vie. Technical Report No. 1325, INRAE, Paris [Online]. Available at: https://www6.inrae.fr/asirpa/Les-60-cas-etudies/Études-de-cas/Économie-territoires-et-numerique/Odomatrix.

Hilal, M., Moret, D., Piguet, V., Bouscasse, H., Cavailhès, J., Diallo, A., Drut, M., Le Gallo, J., Legras, S. (2020). Centralités : comment les identifier et quels rôles dans les dynamiques locales et intermunicipales ? Report, Agence nationale de la cohésion des territoires. [Online]. Available at: https://hal.archives-ouvertes.fr/hal-03064655.

Lancaster, K.J. (1966). A new approach to consumer theory. *Journal of Political Economy*, 74(2), 132–157.

Lu, B., Charlton, M., Harris, P., Fotheringham, A.S. (2014). Geographically weighted regression with a non-Euclidean distance metric: A case study using hedonic house price data. *International Journal of Geographical Information Science*, 28(4), 660–681. doi: 10.1080/13658816.2013.865739.

Muller, J.-C. (1977). Comparaison visuelle des cartes et groupements spatiaux. *Espace géographique*, 6(1), 59–72 [Online]. Available at: https://www.persee.fr/doc/spgeo_0046-2497_1977_num_6_1_1694.

Oliveau, S. (2010). Autocorrelation spatiale : leçons du changement d'échelle. *Espace geographique*, 39(1), 51–64.

Openshaw, S. and Taylor, P. (1979). A million or so of correlated coefficient: Three experiments on the modifiable areal unit problem. In *Statistical Applications to the Spatial Sciences*, Wrigley, N. and Bennet, R. (eds). Routledge and Kegan Paul, London.

Ord, K. (1975). Estimation methods for models of spatial interaction. *Journal of the American Statistical Association*, 70(349), 120–126.

Orozco, V., Bontemps, C., Maigné, E., Piguet, V., Hofstetter, A., Lacroix, A., Levert, F., Rousselle, J.M. (2020). How to make a pie: Reproducible research for empirical economics and econometrics. *Journal of Economic Surveys*, 34(5), 1134–1169.

Paelinck, J.H.P. and Klaassen, L.H. (1979). *Spatial Econometrics/Jean H.P. Paelinck, Leo H. Klaassen, with the Assistance of J.-P. Ancot, A.C.P. Verster, Sj. Wagenaar*. Saxon House, Farnborough.

Rosen, S. (1974). Hedonic prices and implicit markets: Product differentiation in pure competition. *Journal of Political Economy*, 82(1), 34–55 [Online]. Available at: https://www.jstor.org/stable/1830899.

Yule, G.U. and Kendall, M. (1950). *An Introduction to the Theory of Statistics*, 14th edition. Charles Griffin & Co., London.

# 6

# Mapping Time

**Claire CUNTY[1] and Hélène MATHIAN[2]**

*[1] EVS, Université Lumière Lyon 2, France*

*[2] EVS, CNRS, Lyon, France*

*It has to be said that time doesn't leave much choice:*
*invisible, it demands to be illustrated concretely,*
*something other than itself,*
*something perceptible.*
Étienne KLEIN

## 6.1. Introduction

Time is an issue that traverses all the sciences. In each, time raises the question of its meaning, its form, the way it is used to account for changes and evolutions to understand the past and to imagine the future. Depending on the case, time can refer to a notion whose nature is stable, a conventional time, as well as physical, social or lived times. Thus, the nature of time can be multiple and varies according to knowledge, period, societies or even individual abilities. While in physics time is a priori indissociable from space, in the social sciences it is often conflated with the temporal phenomenon itself (Klein 2016). In cartography, the temporal phenomena are spatial processes, such as displacement, diffusion and propagation, and concern many current problems such as migration, the spread of epidemics, climate change, mobilities and so on.

Visualizing the dynamics of a temporal geographical phenomenon involves a series of questions, ranging from ontological considerations about the phenomenon

*Handling and Mapping Geographic Information*,
coordinated by Claire CUNTY and Hélène MATHIAN. 

itself and the objects that make it up, their dynamics, the cartographic objects that can capture them, to considerations about the cartographic language used to serve a particular point of view. Mapping "time" thus raises questions in a number of areas. First and foremost, there are the questions of how to observe this phenomenon, and how to model spatiotemporal data for management purposes, which has given rise to numerous developments over the last 30 years in the field of geographical information science (GIS). These issues also fall within the fields of dynamic modeling and spatial analysis, which enable dynamics to be simplified and summarized before being mapped. Finally, the visualization of geographic phenomena over time is a matter for cartographic language and used methods for representing dynamics (Cauvin et al. 2010b; Klein et al. 2017). The heritage in the field of visualizing spatiotemporal phenomena is numerous and has been widely studied[1] (Palsky 1996; Andrew 2017). The majority of publications or scientific papers on the introduction of time into cartography begin with Minard's remarkable achievement illustrating the Russian campaign (Minard 1869), Marey's graphic illustrating train timetables between Paris and Lyon (Marey 1878, p. 22), or Cheysson's (1888) anamorphosis showing "the acceleration of travel in France over the last two hundred years"[2]. An analysis of the heritage (Kaddouri et al. 2014) shows the diversity of possible visualizations, as well as the ways in which, from the 19th century onwards, maps and graphics are interwoven and can facilitate exchange of ideas, providing a clear glimpse of the issues associated with the simultaneous visualization of time and space.

Yet some have questioned the fact that mapping a temporal phenomenon seems to be "the poor cousin of cartography" (Boria 2012). Imbued with the habits of reasoning specific to geography, centered above all on questions around spatial organization, and with a certain attachment to the representation of an Euclidean geographic space, cartography would then leave little room, literally or figuratively, for the temporal component. This is undoubtedly what has slowed down cartographic innovation and favored the generalization of cartographic animations, which simply unfold successive states of a territory, or set moving entities in motion.

How can we account for such a vast field, given the number of possible inputs (phenomena, data and processing, simplification methods, cartographic choices, etc.)? How can we address the diversity of cartographic solutions and their ability to shed light on temporal phenomena in a context of development of *geovisualization*?

1 These productions are now easy to find online. Some sites present them according to different entries, for example, at www.history.infowetrust.com/ (Andrew 2017) and www.map.cnrs.fr/jyb/puca/ (Kaddouri et al. 2014).

2 Authors' translation.

How can we integrate the fact that this diversity also depends on factors such as the objective pursued, the data available, the target audience, the medium available, etc.? In view of this difficulty, we have decided to present a variety of solutions, starting with the shape of the data. We begin by introducing a categorization of spatiotemporal data and the types of operations required to transform geographic data into cartographic data, as well as the basic knowledge of semiology, necessary for its representation (section 6.2). The next three sections are devoted to operationalizing these elements, illustrating the categorization introduced earlier: changes in a territory (section 6.3), movements (section 6.4) and temporalities (section 6.5).

## 6.2. Formalization

The passage from a phenomenon to its cartographic representation involves a number of formal steps: the modeling of geographic data and their transformation into cartographic data. These steps are traditional in cartography and are not specific to spatiotemporal phenomena. Here, we will focus on the elements specific to the temporal component of phenomena. Two conceptual formalizations that are highly inspiring and consensual in the field of spatiotemporal data management and visualization are proposed here.

### 6.2.1. *Spatial objects and their temporal component*

The first formalization concerns spatiotemporal objects. It is based on the fact that our understanding of spatial processes is based on the decomposition of information into three interrelated cognitive subsystems associated with the *what*, *where* and *when* questions. This formalization is fundamental in the field of spatiotemporal data management and has been widely discussed and illustrated by Peuquet (2002). The objects we observe are thus described in terms of three components:

– Space (noted $(X,Y)$) includes the location (latitude, longitude) and the geometry of the geographic object (the shape of the line, polygon, etc.).

– Theme (noted $Z$) corresponds to semantic information that is an attribute of each geographic object (qualification as qualitative or quantitative).

– Time (noted $T$) is added to the other two components that classically define the geographic objects studied; it can be a date, a period, etc.

The basic categories of the temporal component are instants, represented by points, or periods, represented by intervals. Instants and periods are usually expressed in a calendar frame of reference, enabling ordinal (before – after) and topological (starts before but ends during, etc.) relationships to be managed (Langran 1992). As with the spatial component, it is necessary to discuss the granularity of time, both for its acquisition and processing, and for its representation. This granularity depends on the phenomenon and the question being asked, and most often corresponds to a compromise between the acquisition grain and the temporalities of the phenomenon being analyzed. The finer the granularity, the more we speak of continuous-time observation (see Chapter 2). Otherwise, time is discretized, in relation either to the resolution of the phenomenon (the number of visitors to a tourist site per week), or to the frequency of its observation (censuses every 10 years).

Beyond this initial conceptual formalization, the question of how to represent temporal phenomena also goes back to the choices we make when representing time itself.

In most cases, time is considered to be *linear*, passing and flowing at the same speed over the entire period under consideration. In the case of *cyclical phenomena*, such as daily travel linked to the daily and weekly rhythms of work, or tourist travel linked to the seasons and school vacations, we use an operation that summarizes information for an *average* time, based on the cyclical form of the phenomenon: daily travel patterns according to the day of the week, or visits to tourist destinations according to the month of the year.

We can choose to work in *relative time*, that is, without reference to a calendar, which serves as an absolute reference point: for example, we can evaluate migrations by French department in the first week of the vacations, which in absolute terms can differ from one department to another; or measure the departmental hospitalization rate 30 days after the first case of Covid appears in the department.

Finally, the measurement of time is not necessarily expressed in an arithmetical frame of reference, or even with precision. In history or archaeology, for example, dates are often uncertain and *ordinal* time is used (before, after, long, short, etc.). They may also be expressed in terms of intervals, reflecting uncertainty, or correspond to attestations (date of the first writings citing the entity) rather than to the precise existence of the entity. This characteristic requires specific processing that we cannot go into here. In practice, either dates are reduced to a discrete time scale without regard to precision, or graphical means are used to translate uncertainty associated to them (Mccuaig et al. 2005).

The question of how to represent temporal phenomena thus goes back to the choices we make when it comes to representing time itself.

### 6.2.2. *From spatiotemporal objects to spatiotemporal data*

The second formalization relates to a major issue in terms of data management, that of the *action* of time on either one of the two components, spatial or thematic, of the objects under consideration. If the spatial component is affected, this induces *movement* (change of location) or *deformation* (change of shape) of the object. If the thematic component is affected, this will involve a change in value (quantitative) or state (qualitative). Time can intervene simultaneously on both components, with evolutions going so far as to modify the very identity of the object, which brings us back to the notion of the *lifespan* of objects. These changes pose complex questions from a data management point of view and have given rise to numerous developments and formalizations over many years, integrating issues of object definition, identity tracking and entity genealogy (Cheylan et al. 1994; Claramunt and Thériault 1995; Peuquet 1999; Galton 2004; Worboys 2005; Plumejeaud 2011; Siabiato et al. 2018). In this chapter, we position ourselves downstream of these issues, considering that questions of object identity evolution over time have already been resolved.

We propose here three elementary types of spatiotemporal data to describe these three types of change. Figure 6.1 shows an illustration for each case and a schematic drawing of the associated data.

– $(Z)_T$ refers to the sequence of values or states for one or more thematic attributes over time ($T$), which can be described as a thematic trajectory. This is the case where time changes only the thematic component. Generally speaking, the data concern a set of units (or spatial units) covering a territory (the departments of France, the plots of a neighborhood, the farms of a region). These are data associated with territorial monitoring, such as the observation of an epidemic on a departmental scale, which is done by choosing an indicator (e.g. the Covid test positivity rate or land use) that is regularly surveyed. The associated questions concern what we will generically call trends.

– $(X,Y)_T$ designates the sequence of geometries (locations or shapes) over time ($T$), or the spatial or geometric trajectory. This is the case where time varies the spatial component. This is the case with observation data on the positions of moving objects over time (Box 6.1), the deformation of the spatial envelope of entities (tracking the extension of cities), or a combination of the two (tracking the movement and spatial extent of a flock of sheep). Of course, each geometry may also be associated with thematic information (aircraft altitude in real-time aircraft tracking, for example), but here the focus is only on the change in the geometric component. These types of data are typical of object movements, captured at the object level.

– *f*(*T*) refers to the type of data that specifically describes the temporalities of a phenomenon through measurements expressed in time units. This is the case when questions concern temporalities associated with the life or existence of the entity itself (date of appearance, date of disappearance, duration of existence), or concern events associated with the existence of the entity (sequence of dates for the eruption of a volcano, or the refoundation of a monastery). The term temporality here refers to the way in which data can express time.

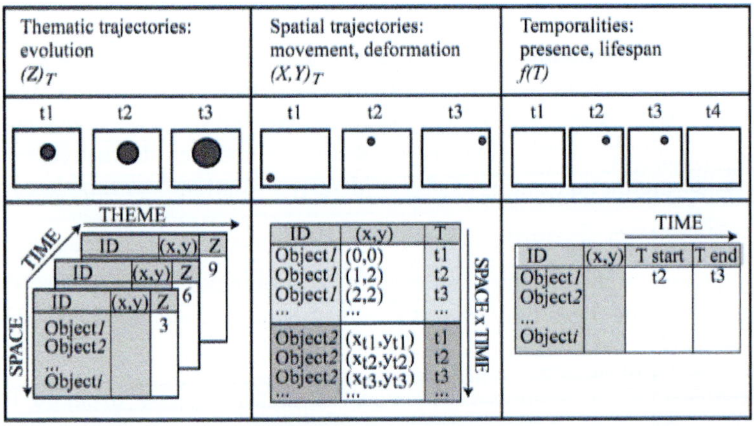


**Figure 6.1.** *Generic forms of phenomena and associated spatiotemporal data (after Klein 2014)*

We postulate that these three types of data constitute elementary building blocks. Any processing or modeling of spatiotemporal data can result in one of these types, and any change studied can be expressed as an elementary type or the combination of several elementary types of change (e.g. a spatial and thematic trajectory such as the variation in the number of soldiers at different stages of the Russian campaign).

**Movement data: spatial trajectories**

Over the last 20 years, the development of data acquisition tools has enabled us to move from observing movements by counting or surveying to tracking devices that record all the points of a movement with a certain frequency. Here, we summarize the various forms of data and associated references:

– a list of dated points describing the geography of an individual journey (hurricane trajectory, cyclist trajectory, animal trajectory, etc.). It can be simply a trace (a list of dated crossing points), a spatial trajectory (differentiating between crossing points and stops) and even a semantic trajectory (an augmented trajectory giving the motives and modes of travel between two stopping points, Kerouanton 2020);

– a list of places frequented with a date that allows them to be ordered, but no trace of the movement between these places. This is the case for movements for which we only know the origin and destination (commuting data, BlaBlaCar data, self-service bicycle data, etc.) or those linked to the activity of their producer (telephony data, participatory fauna or flora inventory data, Flickr data, X (formerly Twitter), etc.).

**Box 6.1.** *Types of movement data*

### 6.2.3. *From geographic data to cartographic data*

| Operation types | Initial types of spatiotemporal data | | | Resulting type after operation |
|---|---|---|---|---|
| | $(Z)_T$ | $(X,Y)_T$ | $f(T)$ | |
| **Time reduction** | – Type of evolution (qualitative var.);<br>– rate of change (quantitative var.) | – Arrival point;<br>– center of gravity;<br>– path | – | ► atemporal Data |
| **Time calculation** | – Event date;<br>– duration of a change;<br>– frequency | – Length of trajectory;<br>– frequency of stops;<br>– speed | Duration of lifespan | ► $f(T)$<br>Date, duration, period, time rhythm |
| **Spatial or spatiotemporal aggregation** | Change of geographical scale | Starting or arrival points of trajectories | Presences on a time scale | ► $(Z)_T$<br>Thematic trajectory |

**Table 6.1.** *Operations and their results depending on the initial type of the data*

The pre-mapping stage involves the processing of geographic data to create cartographic data, that is, the data upon which the graphic choices will be based. In particular, as with the construction of any map, there is the question of associating the spatial component $(X,Y)$ with the thematic component $(Z)$. This association may require transformations of one or another component, or both. Indeed, the data we use may not be directly adapted to the point of view we wish to have on the process underlying the data ("the map must show that there is a gentrification front" or "the visualization must reveal the differentiation between regions"). This is the case, for example, of the maps shown in Figure 6.6(b), where, starting from data recorded at the French department level, the maps show continuous spatial organization to reflect the spread of an epidemic. Here, there is a transformation from the $(X,Y)$ and $Z$ recorded to the $(X,Y)$ and $Z$ represented. We will deal with these transformations here in a specific way, that is, only when they concern the temporal dimension. More precisely, we propose to define three *operations* that transform data from one of the three types of data to another of these three types: $(Z)_T$, $(X,Y)_T$, $f(T)$, or to an atemporal form. Table 6.1 summarizes these transformations.

– *Time reduction* operations: this operation only concerns data of the type $(Z)_T$ or $(X,Y)_T$. The result is quantitative or qualitative data that can be described as atemporal insofar as time has been folded (Mathian and Sanders 2014); Bach et al. (2016) refer to this as *time flattening*. This operation is useful when the interest is in variations between different states or in variation profiles concerning the whole period. It can be used to calculate variations (e.g. a rate of change in population) or to categorize types of change between two dates (e.g. changes in land use). If the chronicles are made up of more than two states, a summary can be established either by a typology of *semantic trajectories* (e.g. the type of municipalities that had continuous population growth, the type of municipalities that had successive decline then growth, etc.). Or, in the case of *spatial trajectories*, we can calculate average geometries (centers of gravity of hurricane trajectories over a century on a global scale) or map successive $(X,Y)$ positions.

– *Time calculation* operations: on the other hand, we may be interested in the temporal component itself, in which case we often create a new thematic variable expressed in a unit of time. For example, for quantitative *thematic trajectories*, the date on which a threshold value has been reached (date on which a municipality reaches its demographic peak), or for *geometric trajectories*, the date (and place) on which an event occurs in the course of the movement (time at which the cyclist had his first breakdown). It can also be a calculation of duration (duration of the longest vacancy of a commercial premises, duration of the movement) or frequency (this premises is vacant every 3 years). Here, the operation on the data leads to the creation of a derived data item which falls into the category of *temporalities* $f(T)$ (see section 6.2.1).

– *Aggregation* operations: the last operation frequently used in the mapping of temporal phenomena is that of aggregation. First, spatial, which is often the means of harmonizing the spatial component over time and, at the same time, providing a different perspective on the data. This operation is particularly useful for geographic data of types $(X,Y)_T$ or $f(T)$, which describe objects (geometric trajectories or life) and for which analysis requires an aggregated view at the level of a territorial grid. Aggregation operations make it possible to identify, for example, the places of origin or destination of geometric trajectories, or the presence of people in the area at different times of the day, based on data on how people spend their time. This spatial aggregation can be accompanied by a temporal aggregation (change of spatial and temporal granularity), such as the aggregation of all robbery-type events by neighborhood and day of the week to observe the daily rhythms of delinquency in a city. The result of this operation is data of type $(Z)_T$ at another geographical level.

### 6.2.4. *Visualizing time*

There are many different ways to account for or introduce time into a cartographic visualization. These are listed here briefly and very generically (for a detailed presentation, see Cauvin et al. (2010b, pp. 126–131) and Vasiliev (1997)). The contexts in which these are used are manifold, and their use may or may not be possible, depending on the question, the means, the audience and so on. The following entries illustrate the wealth of cartographic representations made from their implementation and discuss, in context, the use of these processes.

– *A semiotic approach to time*: time can be seen as a quantity, and the simplest way to represent it is to use a visual variable:

- for dates, the visual variable value (variation in intensity from darkest to lightest), as defined by Bertin (1967), is generally used. This visual variable is used to represent the order of dates;

- in the case of duration, the visual variable size can be used to translate shifts in proportional relationships between durations (this period was twice as long as another);

- finally, signs can be replaced by statistical graphs illustrating the evolution of a variable over time for each spatial entity. These maps are called *chart-maps*[3].

---

3 By mistake or misuse of language, diagram maps have been sometimes called *cartograms*.

– *Time as an additional dimension*: "When the two planar dimensions are utilized to represent space, no planar dimension remains available to represent the 'time' component" (Bertin 2011, p. 342). In practice, various visualizations have been devised and are in use.

The solutions that have been most widely used are kinetic, involving the construction of series of images. In English, these series are generally called small multiples, as the multiplication of maps on the surface of the sheet of paper leads to a reduction in the size of each map. The visualization of these *small multiple maps* was applied as early as 1863 by Galton (1863, pp. 51–52) to map meteorological indicators (temperature, wind/rain and pressure) three times a day for every day in December. It was popularized in the field of data visualization by Tufte (1990, p. 67).

Series of images were progressively animated, in line with the development of digital media. Animation (Box 6.2) is a process that matches the time of the phenomenon to the time of visualization: time is represented by time.

Another visualization solution is the use of the third dimension to represent time, that is, a vertical axis perpendicular to the plane, traditionally used for altitude.

– *Time on space or space-time*: the last category includes cartographic representations that transform geographical space, where time is read on space. This is particularly true of certain anamorphosis (or cartographic transformations of position), where the metric of the map background is no longer space, but time. In these maps, space and time are indissociable, as in the conception of physical time; they are called space-time cartographies (L'Hostis 2003; Cauvin et al. 2010a; see Chapter 7).

The solutions listed above are mainly static representations. In addition to animation, current developments are introducing new visualization possibilities through interactivity. These can all be enriched by interactive use, with basic map navigation functions such as zoom and move tools, stop and play for animations, and on-demand queries and visualizations.

The next three sections are devoted to exemplifying the graphical and cartographic choices presented above and are structured around the types of data presented in section 6.2, namely, thematic trajectories and the monitoring of territorial evolutions (section 6.3), spatial trajectories or the visualization of movements (section 6.4) and, finally, the visualization of temporalities (section 6.5). Contextualizing these solutions makes it possible to discuss the choices made and the point of view on the phenomenon that the result proposes, depending on the audience for which it is intended.

**Snapshot animations**

The most elementary animations are based on the principle of snapshots presented successively on the screen (Figure 6.2), which, if the scrolling speed is fast enough, can be perceived continuously due to the physical phenomenon of visual persistence (as in flipbooks or stop-motion cartoons).

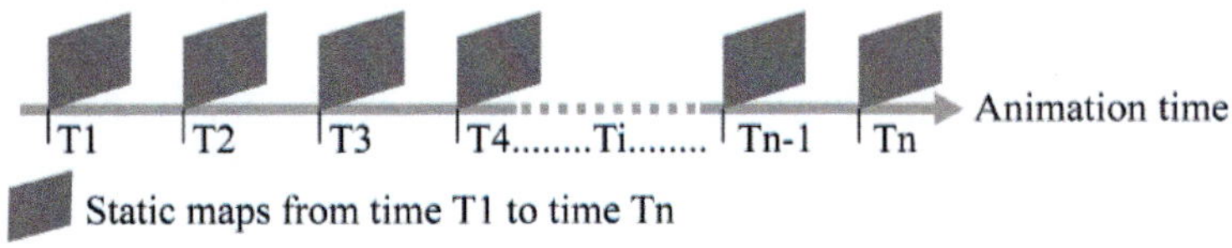


**Figure 6.2.** *Static snapshots appear one after the other*

**Object animations**

Other animations can focus on each of the map's constituent geographical entities separately. The animation can then be continuous, with a graphic effect of interpolation of shape or movement to move from one state to another, creating intermediate states that often do not correspond to any known or measured reality of the phenomena (Figure 6.3). The question of the relevance of this type of object animation also arises for different states corresponding to qualitative data and represented by colors, insofar as the intermediate colors resulting, for example, from a change of situation from green to red are questionable (Cunty et al. 2012).

**Figure 6.3.** *Object animation using shape and/or motion interpolation*

**Animated visual variables**

Animation leads to the definition of new animated visual variables to transcribe time through time (DiBiase et al. 1992; Cauvin et al. 2010b, pp. 53–58). These include order and duration. Other variables have been proposed, such as frequency and velocity, but in our view, these relate more to the characteristics of the phenomenon in time (and space in the case of velocity) than to visualization alone. However, it seems fairly obvious that animation facilitates the perception of frequency and speed.

**Box 6.2.** *Animation processes and techniques*

## 6.3. Monitoring territorial changes

Observing, studying and analyzing the evolution of a territory is a major challenge in today's world, whatever the geographical scale. It can be real-time monitoring or a reconstruction of the past, with a view to decision-making, prevention or simply understanding: what are the population dynamics of a city's or a country's neighborhoods? What are the differences in the economic evolution of a territory (changes in the types of companies present, employment sectors, etc.)? How do pollutants spread within different parts of a watershed? These observations are supported by appropriate visualizations based on data describing a territory according to a certain division (administrative grid, regular tessellation, etc.) and with regard to a theme over time, data we have called thematic trajectories.

### 6.3.1. *Visualizing temporal phenomena*

The Covid epidemic is a phenomenon that has been widely observed and, undeniably, in an unprecedented way. This observation has led to the creation of a pool of data for health monitoring, on an international scale and with a temporal granularity that has probably never been matched on this scale for health monitoring. When the *data visualization* trend is expanding, the global dimension of pandemics and the opening up of data have stimulated the production of graphic and cartographic representations. These visualizations, focused on a single subject, have enabled the testing of their uses on a massive scale (Genevois 2020). In this section, dedicated to the representation of $(Z)_T$ thematic trajectory data, we draw on these achievements, as they are all the more appropriate given that the spatial component is an important explanatory factor in the spread of the epidemic. Clearly, maps can be used to present the health status of a territory, as well as to demonstrate the spatial logics at work in the spread of the phenomenon.

#### 6.3.1.1. *Small multiple maps and animation*

Whether for real-time monitoring or the a posteriori analysis of acquired data, one of the first challenges of visualization is to look at evolution over time, that is, to reconstruct, step by step, the state of a territory. Figure 6.4 shows a number of screenshots associated with key dates and repeated over 2 years, creating a series of small multiple maps. A common legend is chosen for all the maps, revealing the shape of the spatial organization of the phenomenon over the entire period. Other options can be chosen in the legend to allow further comparisons from one map to another (Box 6.3).

With the development of web-based visualizations, the representation mode of small multiple maps has been supplanted by one in which the evolution of a phenomenon is visualized by its unfolding over time: animation (Box 6.2). In this case, we say that time (of the phenomenon) is represented by time (of the visualization).

While Bertin (1967) notes that "the number of images is limited by the reading process: with a long series it is difficult to suggest motion" (p. 342), small multiple maps have the added advantage of giving the reader time to make comparisons between maps, which animation does not. So, while these two types of representation appear to be close, they are also in competition in terms of their cognitive capacity (Fabrikant et al. 2008).

The maps of states over time are displayed simultaneously in small multiple maps, allowing for the construction of reasoning based on comparison. Conversely, animation makes it difficult to memorize the spatial distributions revealed successively, unless they outline a clearly identifiable form of organization. Indeed, animations are particularly effective when dealing with phenomena that are highly organized in space.

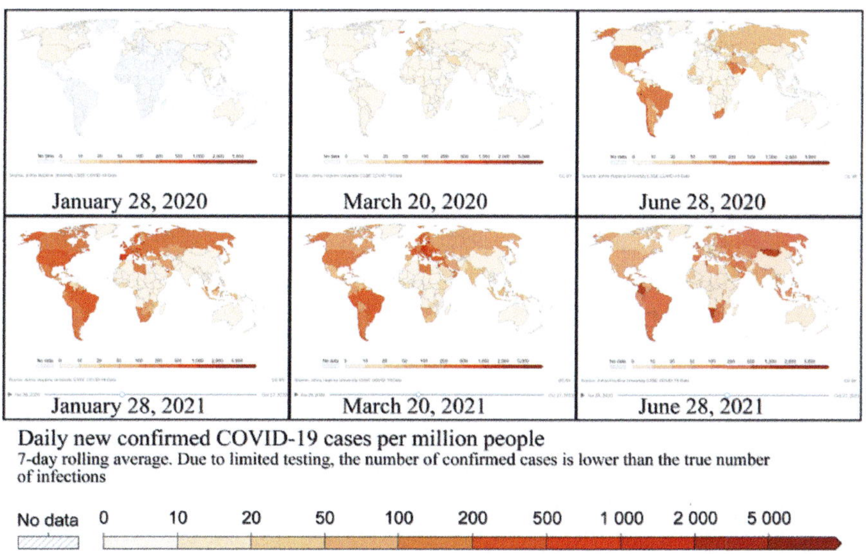


**Figure 6.4.** *From animation to multiple maps: the example of the Covid daily incidence rate (source: Ritchie* et al. *2020)*

**Absolute or relative comparison of data over time**

In the case of quantitative variables (evolution of a territory), whether presented in animated sequences or in the form of small multiple maps, it is necessary to make choices for the circle sizes or discretization used to enable the comparison of one map with another.

– When the legend is common to all time intervals, the comparison is made on an absolute basis (overall, the rate increases or decreases in relation to a maximum over the entire period). In this case, the cartographer seeks to understand the overall geographical footprint of the phenomenon.

– When the legend is specific to each time interval, we choose either a circle size corresponding to the maximum of each date, or a discretization allowing comparison against indicators of central tendency and dispersion (mean and standard deviations, median and quantiles). In this case, the positions of values in the statistical distribution are interpreted in relative terms, that is, we are more interested in the hierarchy of values than in their levels: for example, such and such a country is always in the top quartile of incidence rates (whether the incidence rate is globally high or globally low). In this case, the cartographer wishes to emphasize local spatial heterogeneities.

**Box 6.3.** *Specific or common legend and discretization for animated maps or small multiple maps*

#### 6.3.1.2. *Replaying the phenomenon of interest: animation and interaction*

The constraint of animation is its obvious unsuitability for paper publication. Bertin (1967) rejects this possibility in the introduction of *La sémiologie graphique*, stating that "to analyze graphic representation precisely, it is helpful to distinguish from [...] writings percieved in a linear or temporal sequence and the animated image, gouverned by the laws of cinematographic time" (Bertin 2011, p. 2). This remark does not take into account the democratization of digital media since the 2000s, when cartographic animation will also benefit from the exploration capabilities enabled by interactivity.

Animation is the simplest mode, and certainly the most intuitive, since it lets you unfold time as you experience it: a priori, linear time. The advantage is that you can perceive the rhythms: accelerations, decelerations, stops, cycles, etc., when you replay time, even if it is accelerated.

Insofar as cartographic animations are generally built on raw data (without prior operations as described in section 6.2.3), interaction is necessary to be able to explore the data and conduct reasoning as identified by MacEachren (1994, p. 6). Some digital environments for exploring these spatiotemporal data offer interactive timelines, enriched by a graph showing the statistical distribution of the phenomenon over time.

The user can select a date or period on the timeline and/or graph, and synchronously view the geographical distribution on a map. These elements of interaction are generally essential to contextualize the phenomenon and overcome the fascination of moving images and the ephemeral nature of disappearing images.

Figure 6.5 is a screenshot of a dynamic scene: the environment from which it is taken makes it possible to animate time and discover the construction of the curve as time passes, a way of reliving the phenomenon. The time axis is interactive, allowing you to select and zoom in on certain periods, and move backwards and forwards. Time can be explored, stopped, accelerated and so on. These interactive processes help you to understand different temporal components.

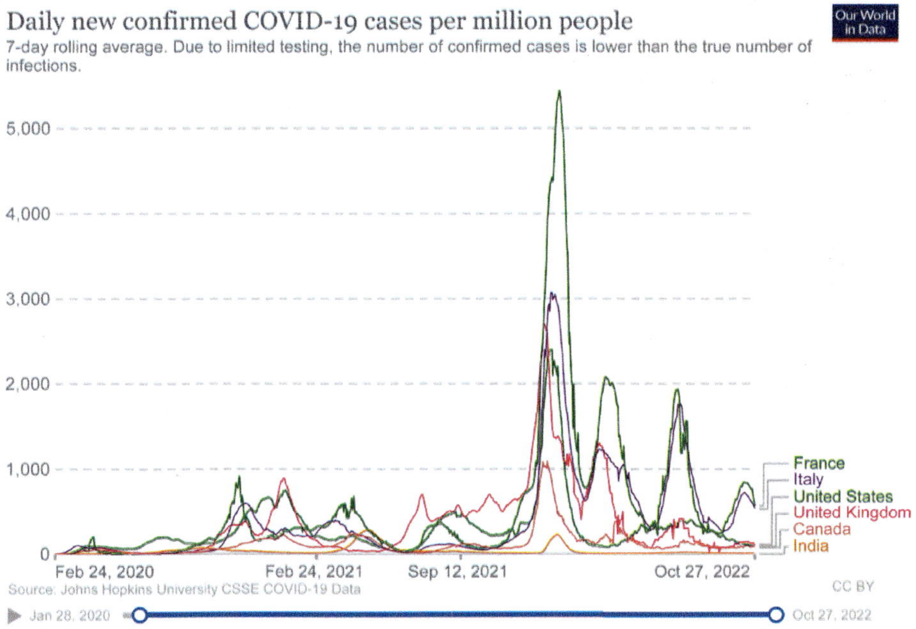


**Figure 6.5.** *Graph monitoring new Covid cases (source: Ritchie et al. 2020)*

### 6.3.1.3. *Telling the story: animation and scriptwriting*

Animation also offers the cartographer the opportunity to play on different forms of representation to support a message and to build a narrative. By way of example, here we take a look at the maps produced by Lambert (2021) on France, using semiological elements to support a point of view (Figure 6.6). The data used show, by department, the number of people declared ill and the number of deaths. The cartographic processes employed show either hospital overcrowding (counting points, Figure 6.6(a)), or a wave of diffusion (smoothing, Figure 6.6(b)). In this way, the animation invites the cartographer to set the scene in order to translate their intentions.

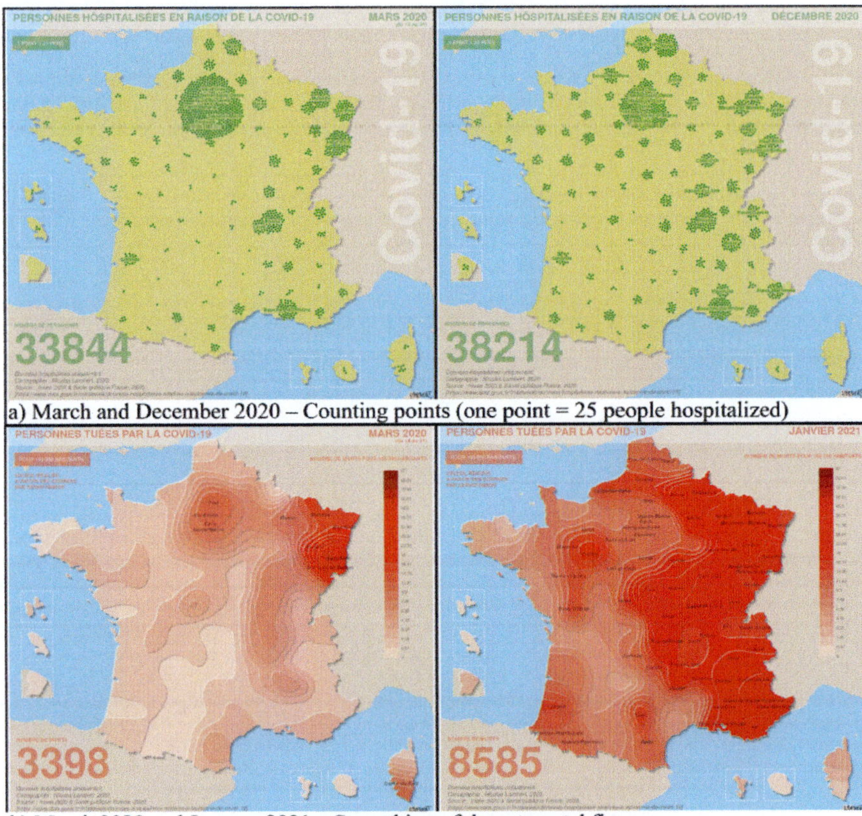


a) March and December 2020 – Counting points (one point = 25 people hospitalized)

b) March 2020 and January 2021 – Smoothing of departmental figures for deaths per 100,000 inhabitants

**Figure 6.6.** *Animating and reporting a point of view: cartographic representations of the Covid epidemic in France (Lambert n.d.)*

### 6.3.1.4. *Carto-graphics*

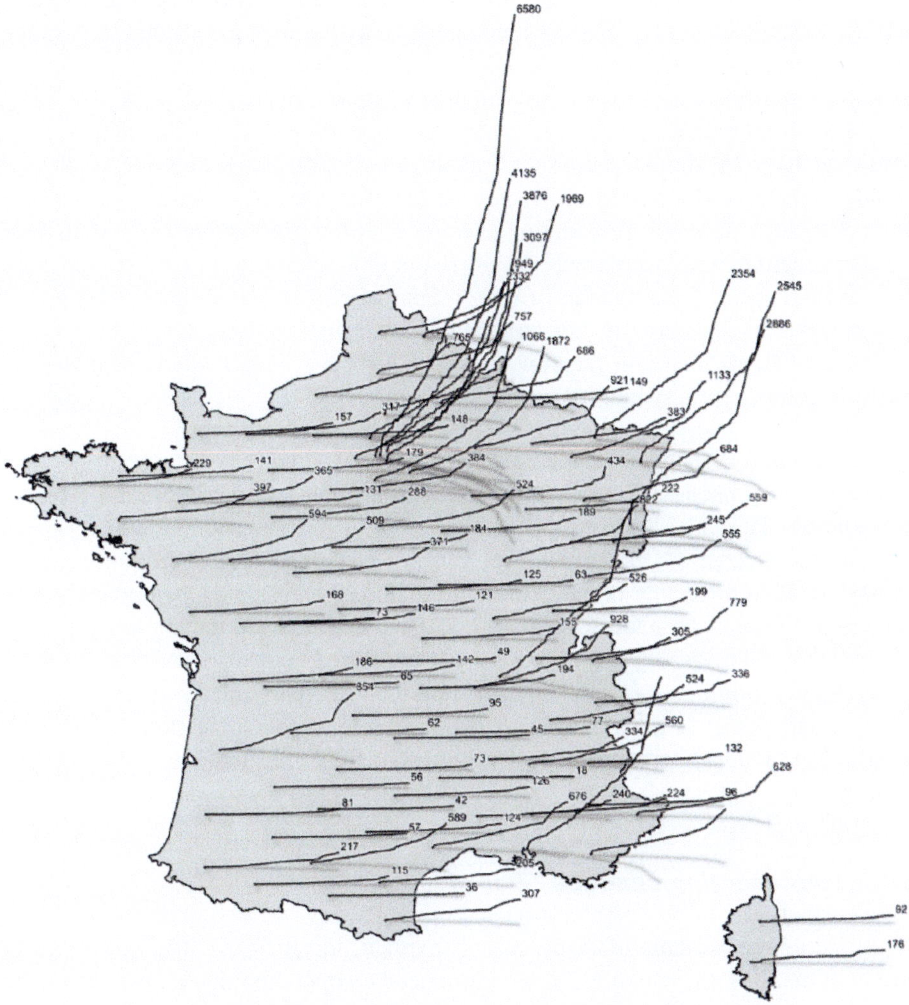

**Figure 6.7.** *Example of a chart-map: COVID incidence rate by department in France (Rajerison 2021)*

As we saw earlier, graphs showing changes in quantities or rates make it possible to read and compare the rhythm and intensity of change (Figure 6.5). Proposals have emerged in which time is projected on space, that is, each geographical entity on the map is associated with a temporal graph. Maps produced in this way are sometimes referred to as chart-maps. The advantage of these representations is that they reflect the diversity of epidemic trajectories in each department. Local trends are given priority over global trends. The graphs describe trends by department and, rather than the overall dynamic, it is the heterogeneity of the trajectories that is targeted (Figure 6.7). As the phenomenon is linked to geographic diffusion, there is a strong spatial autocorrelation, and the spatial organization of the metropolis is clearly visible, with the north-east of the country standing out, as well as the departments bordering Italy. A similar proposal (Baruch et al. 2020), visible online on the *Le Monde* website and updated daily, reveals another spatial organization, focused more on time, with graphs illustrating changes over the past month. Local dynamics are more legible, with each graph not overlapping neighboring graphs. In this map, the original spatial component *(X,Y)* (departments) is modified. Position of departments on the map are not strictly on the right locations. Only the relative positions of departments in relation to each other have been retained to make the thematic component's spatial transfer legible.

These types of maps have been around for a long time (e.g. Cheysson's helical diagrams, 1884, used on Covid data by Field, 2021), but present issues of graphical overlapping. To get away from this, some authors (Forlines and Wittenburg 2010; Thakur and Hanson 2010; Gautier et al. 2021) have proposed that temporal graphs be positioned at the center of entities using the third dimension as the time axis. The authors of these proposals discuss their cognitive abilities and remain fairly critical of reading possibilities when phenomena are organized neither in space nor in time. Other highly original forms have been put forward, particularly in the field of *DataViz* and design (e.g. Cruz et al. 2019). All these cartographic experiments help to visualize the local heterogeneity of the phenomenon and stimulate spatiotemporal writing.

### 6.3.2. *Representing changes*

In the above example of maps, time is explicit, linear or cyclical, visualized or lived (animation). In section 6.2.3, we introduced an operation that allows time to be summarized (this can be referred to as *folded time*). When the focus is on quantitative or qualitative variations between states or the evolutionary profile over a given period, we reduce the temporal component by creating an atemporal variable (Figure 6.8). If we are calculating a rate of change, we need to represent both positive and negative values on the map. In this case, the visual variable value is used, with a divergent progression of hue intensity on either side of zero. Warm hues generally represent positive variations, and cold hues represent negative ones (Figure 6.8(a)). If we have more than two dates, we can classify thematic trajectories (quantitative or qualitative) to obtain a typology describing evolutions (qualitative variable). This is the example shown in Figure 6.8(b), which presents demographic trajectory types by community. The diversity of trajectories is summarized by four classes derived from a hierarchical ascending classification, each associated with a specific evolution. Chapter 3 presents an illustration of these two types of approach applied to the issue of afforestation.

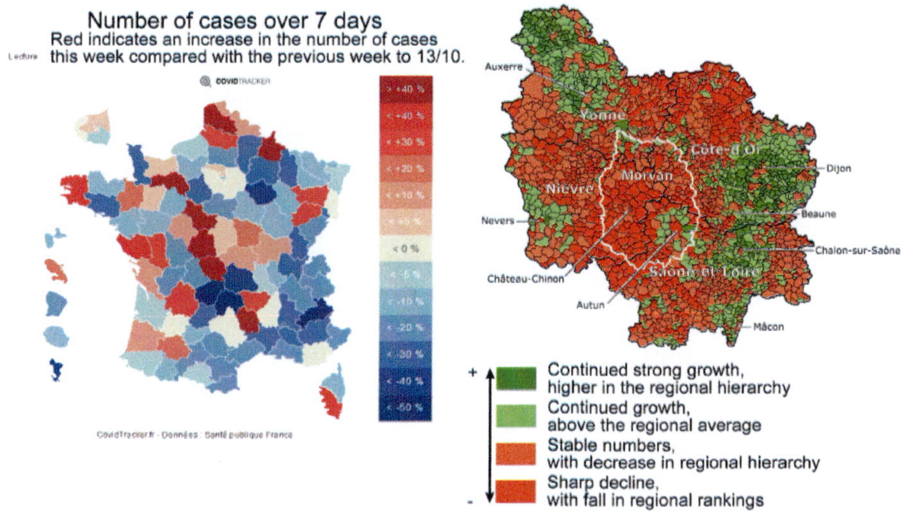


a) Cloropleth rate-of-change map for COVID incidence rates (Rozier 2020)

b) Typology of communal demographic trends in the Morvan between 1968 and 2017 (Cossart 2021)

**Figure 6.8.** *Mapping change*

## 6.4. Representing phenomena associated with movement

A second class of cartographic issues is about the representation of phenomena involving changes in position (*displacement, movement*) or form (*deformation*), that is, changes impacting the spatial component $((X,Y)_T)$. These phenomena, which are spatiotemporal by definition, are frequent and are at the root of the transformations observed in the territory.

We begin by presenting the cartographic issues associated with travel data, namely spatial trajectories (section 6.4.1). In section 6.4.2, we demonstrate how individual journeys can be aggregated to form lived territories. Finally, we shall study the visualization of more specific deformation phenomena (section 6.4.3).

### 6.4.1. *Representing movements: from points to trajectories*

The simplest form of individual trajectory data is points (Box 6.1). From a visual point of view, it is quite easy to go from a series of points (trace) to a path, as the eye easily infers the passage from a sequence of points to a line or arrow materializing this path. In the case of punctual implementation, the variation in color intensity visually translates the succession of positions (Figure 6.9(a)). However, displacement can mobilize new graphic elements to translate "the 'trace' and direction of a moving body" (Bertin 2011, p. 342). It is commonly assumed that there is a continuous evolution between positions, which is graphically translated by the trace of movement, which can take the form of one or more arrows (Figure 6.9(b)).

These solutions are the most widely used and are easy to read when movement is simple and unidirectional. When movements in space are more complex, such as in battles where forces advance and retreat one after the other, or in individual mobilities, arrow representations quickly prove difficult to read on a static representation. In this case, it is not only the spatial inscription that can be difficult to read (intersecting, overlapping arrows), but also the temporal succession of the various movements (did the advance take place before the retreat? etc.). The increasingly frequent use of animation helps to solve the problem of understanding temporal succession and to apprehend the speed and rhythm of movements.

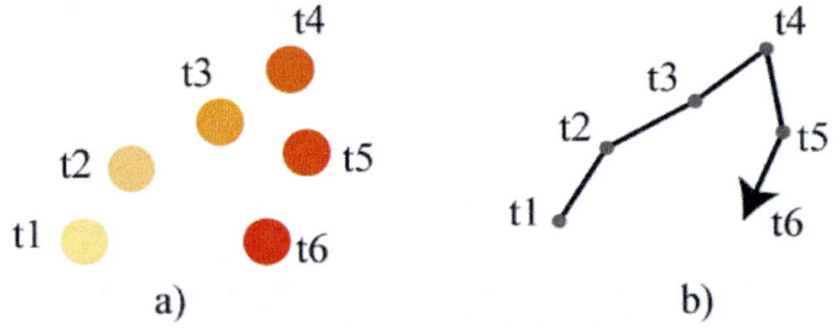


**Figure 6.9.** *Representing a trajectory (from Bertin 1967; Klein 2007)*

Another visualization solution is the use of the third dimension to represent time. This representation has been used extensively to illustrate the conceptual framework of *Time Geography*, proposed by geographer Hägerstrand in the 1960s–1970s for the analysis of daily activities and travel. This type of visualization is generally referred to as a “spatiotemporal cube”, a famous reworking of which is the Minard map by Kraak (2003). This transposition is powerful because the trajectory of a single object is visualized, while the troops are modeled as a moving object. The complexity of the representation increases drastically when we try to represent several trajectories.

The various methods described above are only operational if all the movements have a certain coherence or similarity. This is the case for daily trips, which are relatively concentrated, both spatially (employment hubs) and temporally (working hours).

These strong constraints organize displacements and, in this case, representations make it possible to read the more global form of the movement (area of attraction of employment poles) while reintegrating a fine granularity, both at the level of entities and spatial and temporal components.

The characteristics that a displacement phenomenon must have for its representation to be operational, particularly when the data are massive, are widely discussed in Andrienko and Andrienko (2013).

In any case, the alternatives to 2D representation are very powerful in this case. The third dimension then provides a better perspective and enables us to grasp the convergence or otherwise of trajectories and their spatiotemporal organizations.

The *Time Geography* framework has been very inspiring, and numerous experiments have been carried out using the spatiotemporal cube (Kwan 2000; Thévenin et al. 2007; Figure 6.10(a)). Representation in an environment where you can enter the scene enables you to explore the different dimensions of the spatiotemporal cube, with interactive exploration allowing selections, particularly if the data includes semantic attributes such as transport mode or reason for travel. Recent experiments have focused on immersive exploration (using a virtual reality headset) of these visualizations (Wagner et al. 2019).

Other environments make it easier to reduce the complexity of the data to be represented, through transformations or data modeling, such as geometric trajectory classification, which summarizes the diversity of trajectories (Figure 6.10(b)).

a) Spatio-temporal cube: Spatio-temporal trajectories of around 400 individuals on their way to the Bonal stadium (Sochaux-Montbéliard) for a soccer match.

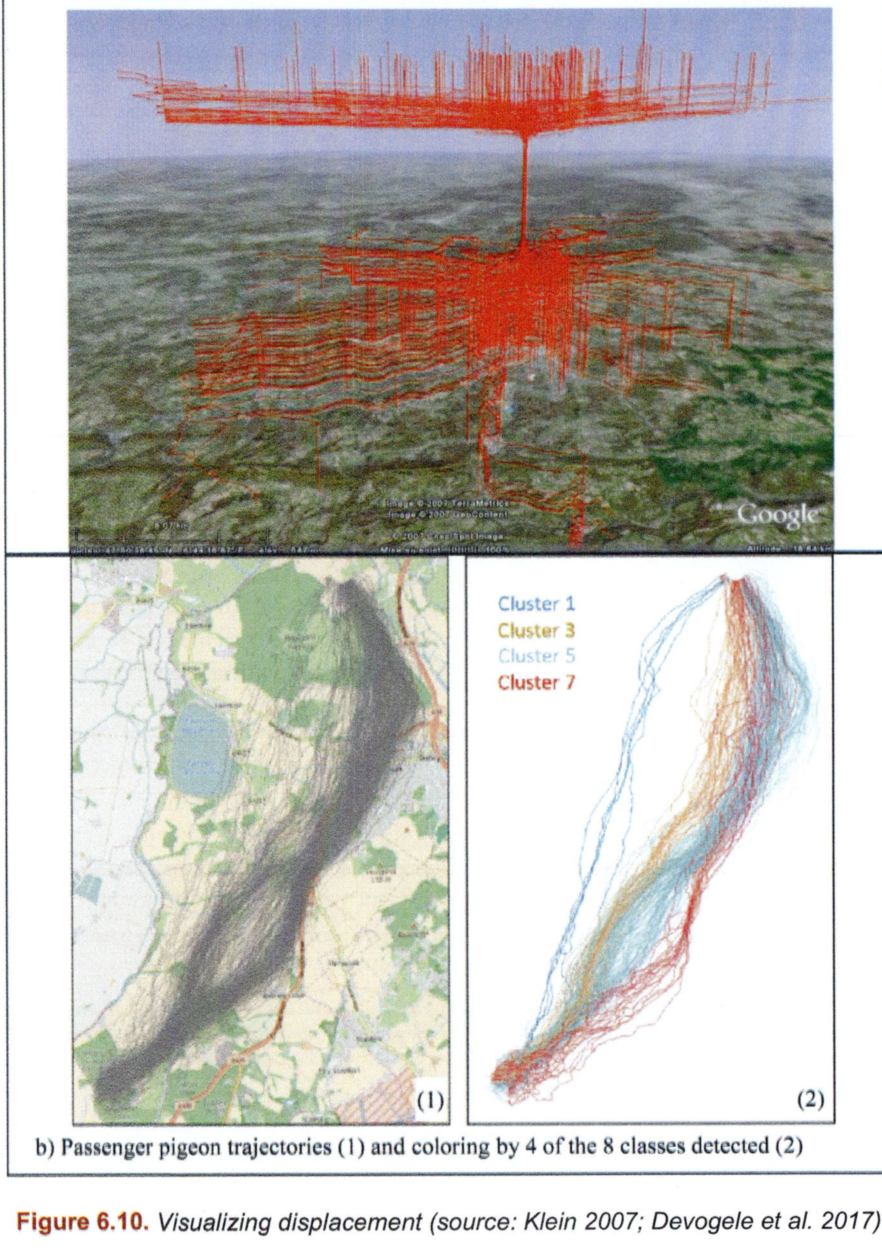


b) Passenger pigeon trajectories (1) and coloring by 4 of the 8 classes detected (2)

**Figure 6.10.** *Visualizing displacement (source: Klein 2007; Devogele et al. 2017)*

### 6.4.2. *From trips to their registrations on the territory*

Movements can also be related to the territory. This may correspond to the data collected (Box 6.1) or to a transformation of individual trajectory data. The operation used in this case is *spatial and temporal aggregation*. It involves *projecting* the trajectories onto a grid of the territory to account for the passages on each of the cells, adopting a certain temporal granularity. At the level of each grid and for each time interval, the number of trajectories is counted. We then obtain data of the type $(Z)_T$: each territorial grid is described by the number of trajectories that have passed through it over a certain number of periods.

This process has been applied to daily commutes (Banos and Thévenin 2005; Klein 2007; Hurez et al. 2012). Given the cyclical nature of commutes, which are spatially polarized around employment centers, animation is particularly well suited to their representation: the sequence of events reflects a marked spatial and temporal pattern that the authors have likened to urban pulsations. Without depicting displacements, it is easy to see how city centers suck in the suburbs in the morning and empty out in the evening. 3D is used to visualize the intensity of presence over a period of time, and animation unrolls the film over the course of a day (Figure 6.11(a)).

We can imagine all kinds of summaries of these journeys across the territory. For example, Figure 6.11(b) shows a calculation of average travel speed per grid. In each cell, the square figures are proportional in size to speed values, allowing us to see the road network and its hierarchy in relation to traffic speeds.

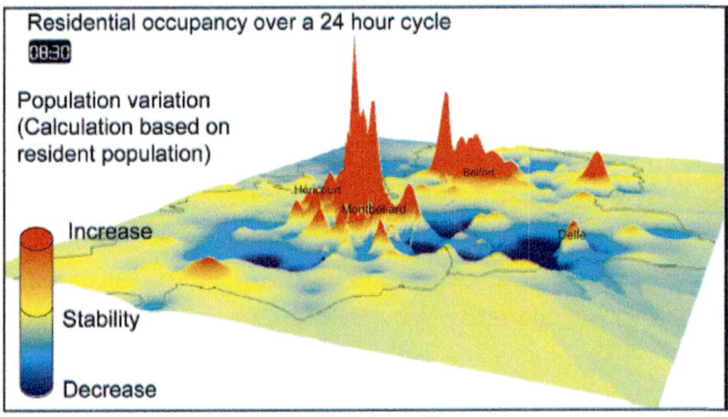


a) City pulses (Klein 2007)

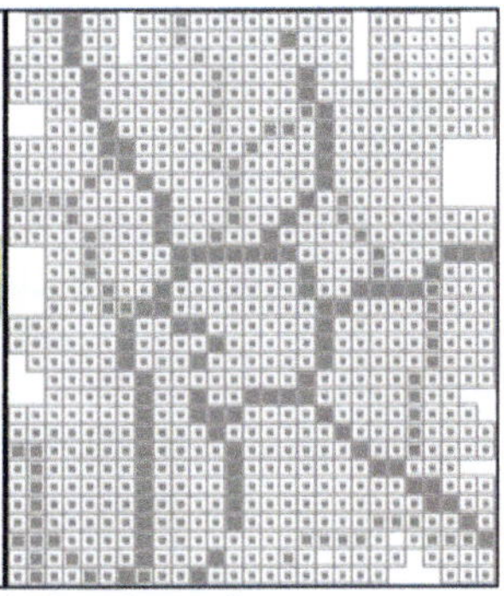

b) Travel speed
(from Andrienko et al. 2007)

**Figure 6.11.** *Projecting trajectories on the territory*

The most frequent data transformations in movement analysis are those involving the simple aggregation of movements by place of origin and place of destination, either because the data are supplied in this form (so-called O-D data, such as the French statistic institute *INSEE*'s commuting data), or because their individual analysis is meaningless (*BlaBlaCar* data, for example, Mericskay 2019). From this spatial aggregation, we derive either data of the form $(Z)_T$, or data of the flow matrix type, characterizing exchanges between locations. In the first case, we find ourselves back in the situation discussed above (section 6.4.2). In the second case, we focus on the logic of exchanges and interrelations between locations. Flow mapping is a field in its own right that we do not address here, but which has been the subject of an entire chapter to which we refer (see Bahoken 2024).

### 6.4.3. *Shape changes*

Motion data $(X,Y)_T$ are positional data of a moving object, the representations of which we have just described in detail, but they can also be based on the geometric deformation of entities.

The focus may be on the movement of objects with geometries more complex than a point. This can be either an envelope created around moving elementary objects, the whole of which forms a coherent higher-level entity[4] (a herd, a procession of demonstrators, etc.), or a form modeled to account for the evolution of an organization. An example of this is a pioneer front, which is shaped by the successive positions of new settlements, or a gentrification front, which can be approximated by an indicator such as the evolution of the types of shops in a neighborhood. This front will give rise to the visualization of a line whose shape and position over time can be translated by the visual variable value. Figure 6.19 illustrates this.

However, beyond objects whose shape is the result of an aggregation of singular objects, there may also be objects whose shape changes over time. For example, a watercourse or a coastline can have successive courses (Figure 6.12). In this case, it is the geometry $(X,Y)$ of the objects that changes. We can then associate a hue with the time and vary the intensity in relation to the ordinal character of the time. In Figure 6.12, the choice is made to associate darker hues with older dates and lighter ones with more recent dates. We could imagine the opposite (as in Figure 6.9(a)), as

4 A characteristic of perception established by Gestalt theory, known as the *law of common fate,* states that objects forming a set moving at the same speed in the same direction are perceived and mentally interpreted as a single entity.

a metaphor for the attenuation of visibility of older river courses in the field compared to more recent ones.

As shown in Figure 6.12, the main limitation of static representations of shape changes is that their effectiveness is low when deformations are not unidirectional. How can we translate deformations that see a succession of dilatation and retraction of the shape without using small multiple maps or animation? In this case, it seems necessary to evaluate the change, compare the shapes and construct global measurements that accompany the image with graphs, or to model a shape synthesizing the whole, which is more legible, but in which local information is lost.

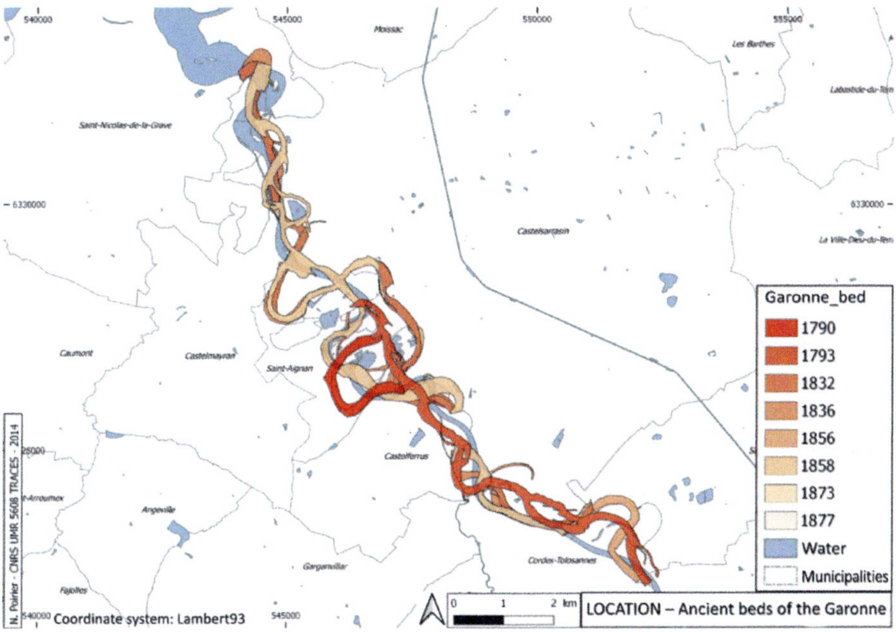


**Figure 6.12.** *Divagations of the Garonne riverbed from Napoleonic cadastral plans (Poirier 2021)*

## 6.5. Representing temporality

Following the formalization proposed in section 6.2, this last part discusses the representation in space of temporal data (date, duration, frequency) noted $f(T)$. These relate to the very existence of objects, or to a temporal thematic variable resulting from a *time calculation* operation (section 6.2.3): a date (e.g. the date on which the maximum value was reached), a duration, etc.

We begin by looking at representations of events, which constitute remarkable units of date, and then generalize these approaches to representations of lives, which add a depth to time – the duration. Lastly, we associate "space-time" representations of the temporalities associated with travel.

### 6.5.1. *Tracking the spatial organization of events*

The notion of event is fundamental to the question of temporal representations, and an event can be defined in two ways:

– the objects processed are themselves events, that is, entities with an instantaneous lifespan: either because, on the physical time scale of our experience, their duration is almost nil (lightning strikes), or because the choice of temporal granularity leads us to consider their duration as almost nil on the time scale taken into account (volcanic eruptions studied over a century);

– it is a construct based on data. The event is the result of an operation that associates a date or duration with each entity studied: date of population doubling for communities, duration of heat peak by neighborhood, time of return to departure point for travels, date of entry into the European Union for European countries, etc.

As already mentioned for movements, it is first possible to map these dates of occurrence using a visual variable (continuous or discrete scale) that translates the order of occurrence of each event over time (Figure 6.13). This procedure also works for visualizing the dates of an event on surface entities (e.g. the date on which a country joined the EU, or the date on which a municipality reached its demographic peak).

This type of visualization is very effective from a cognitive point of view if the entities are few in number (the case of European countries) or if the phenomenon has a simple spatial structure (lightning strikes). When spatial structures are more complex, or when events occur recurrently in identical locations, they become less legible, as information is superimposed. Consequently, to observe the dynamics of territories, it may be necessary to make temporal aggregations to produce map series of events or object appearances over a given period (e.g. the number of shops that opened between 2000 and 2005 on a given street section). In the latter case, the interest lies in the dynamics of the territory and not in tracking a specific entity (a particular shop).

Alternatively, an animation (Box 6.2) can be produced that unfolds an accelerated version of time, but depending on the spatial organization of the phenomenon, the result may be a flashing display of appearances (this is the case, for example, with lightning impact maps). Such visualizations enable us to clearly perceive the rhythms of appearing, but any concentrations in space are more difficult to discern, which is why we often keep the display of entities. We can work on semiotics to improve retinal persistence and use the imprint effect proposed by the *CartoMouv* group (Cunty et al. 2012). For example, real-time lightning strikes (Blitzortung.org, n.d.) remain displayed by default for 1 h, during which they gradually turn from yellow to brown (it is also possible to request the display of strikes over the last 24 h, as in Figure 6.13). The occurrence of an event is signaled by an accentuation effect: each new impact is visualized by a circle that diminishes in size and disappears after 60 s.

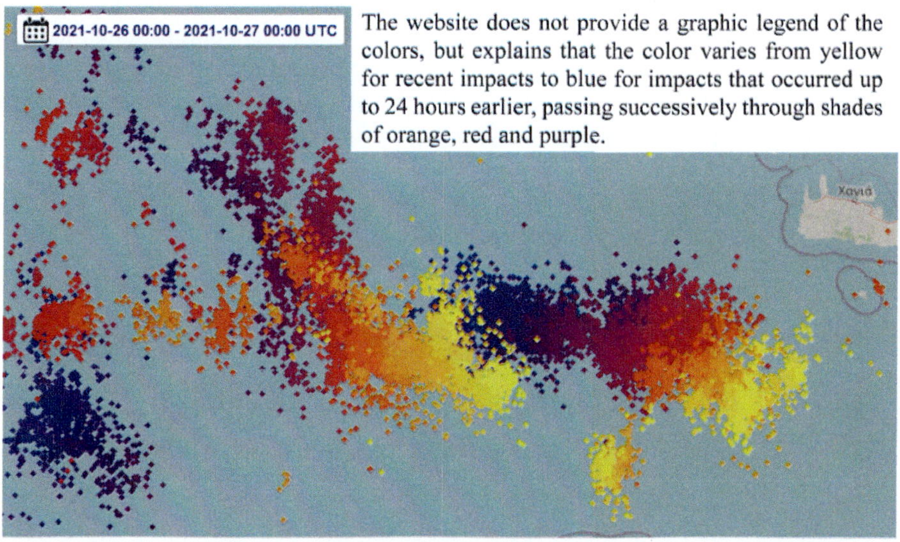


**Figure 6.13.** *Lightning strikes recorded between October 26 and 27, 2021 in the Mediterranean Sea southwest of Greece (source: blitzortung.org)*

Some authors have experimented with visualizing events such as earthquakes, in interactive exploration environments that unfold time as a third dimension, like *Time Geography*, enabling exploration to detect spatiotemporal clusters (Gatalsky et al. 2004).

Other static mapping proposals have been put forward to represent space and time simultaneously. Figure 6.14 presents an original proposal to translate the “spatiotemporal localization of Covid cases in Paris between March 19 and May 16, 2020” (Gautier et al. 2021). Covid cases are spatially aggregated, the number of cases is translated by the size of the circle and the dates of appearance are presented by a double progression of colors (in red the oldest dates and in blue the most recent dates). This makes it possible to identify areas where the epidemic started early and stopped (red dots), areas where the epidemic was present throughout the period (red to blue circles) and areas where it started later (blue dots). This type of representation, known as the Growth Ring Map, was proposed by Bak et al. (2009) and adopted by Andrienko et al. (2011). The latter used this process to combine linear and cyclical representations of the locations of photographic shots published on *Flickr*. The number of photos is aggregated by shooting location, expressed by the size of the circle, and dates discretized by season are marked by the color associated with a value variation (photos taken in winter are all of a blue hue with light blues for 2005 and dark blues for 2009). In the various uses of this process, the choice has always been made not to transform the initial data and to retain all the diversity of the profiles over time. These representations, often of an experimental nature, reflect the influence of "dataviz", which tests direct graphic representations of data, without applying any prior transformation for exploratory purposes.

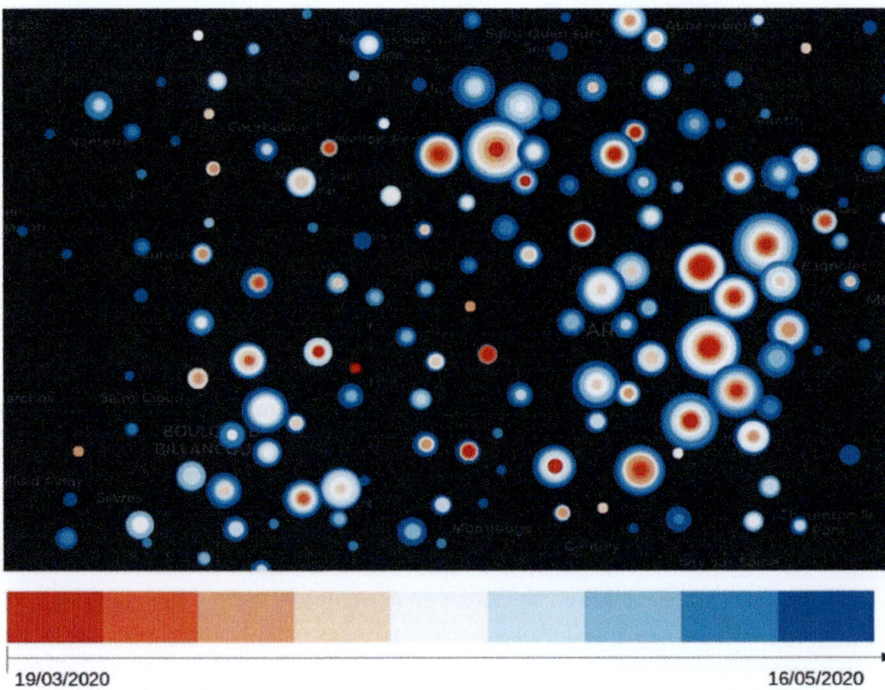


**Figure 6.14.** *An extract from the Growth Ring map showing the spatiotemporal distribution of Covid cases in Paris between March 19 and May 16, 2020*

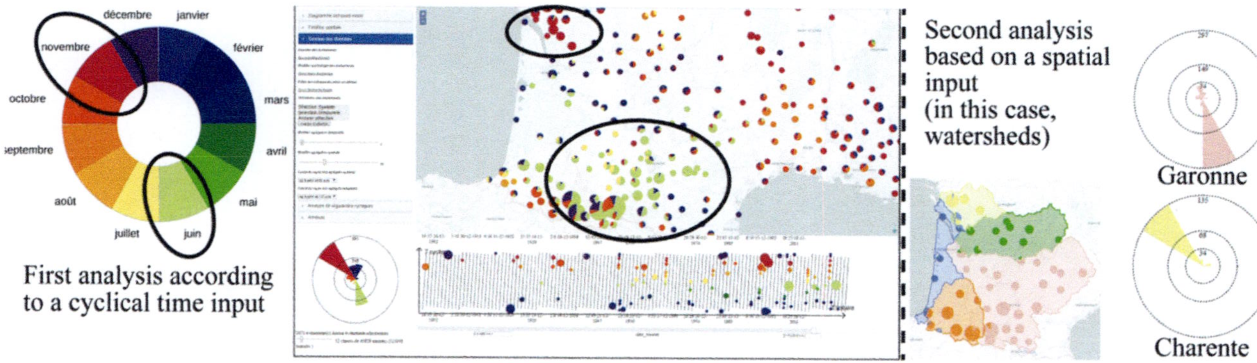


**Figure 6.15.** *The GrapHist interface for simultaneous analysis of the linear and cyclical characteristics of flooding events in southwest France between 1900 and 2015 (Gautier et al. 2019)*

Interesting proposals along these lines have been integrated into geovisualization interfaces such as GrapHist (Gautier 2018), which explores and visualizes events in their both linear and cyclical temporal components (Figure 6.15). By mapping the number of floods in south-west France according to their month of occurrence, we visually identify two zones that clearly correspond to two different cycles: zone 1 tends to recur in June and zone 2 in November. The interface also offers the spatial aggregation of measurement points by zone, in this case by river basin, thus highlighting two flood regimes: in June in the Garonne basin and in November in the Charente basin. The interface can be used to explore different cycles and thus identify spatiotemporal regularities.

### 6.5.2. *Representing lifespan, duration and change*

In cases where the duration of existence of the entities studied is significant (trades, types of culture, religious establishments), we work with entities characterized by two dates, corresponding respectively to their appearance and disappearance. A first solution, if the entities are immobile, is to revert to data of the form $(Z)_T$: we define a temporal reference frame by a series of dates or periods, generally regular, and each entity is described by a binary value, 1/0, indicating its existence or non-existence at each time interval. Next, we can choose from the visualization solutions already described above, either animation or 3D. We find ourselves in a situation similar to that of event representation (section 6.5.1), and the perception that remains will depend on the existence of forms of organization (densification, diffusion, extension, etc.).

Geovisualization environments, by synchronizing cartographic windows with a graphic timeline, whatever its form (continuous line, curve, histogram), enable us to experiment with temporal granularity (selection of a period on the timeline) and to unfold the spatiality of the phenomenon. Figure 6.16 illustrates this type of visualization interface for the presence of religious establishments in medieval France according to their religious order (Mathian et al. 2020). The shaded selection on the temporal graph filters out settlements existing at that period. The map is updated as the user moves this period on the timeline, and settlements that are not present at the selected period appear with a lower opacity. This semiological effect compensates for the visual loss associated with animation. In this way, we keep the reference to all the establishments studied, whatever the period.

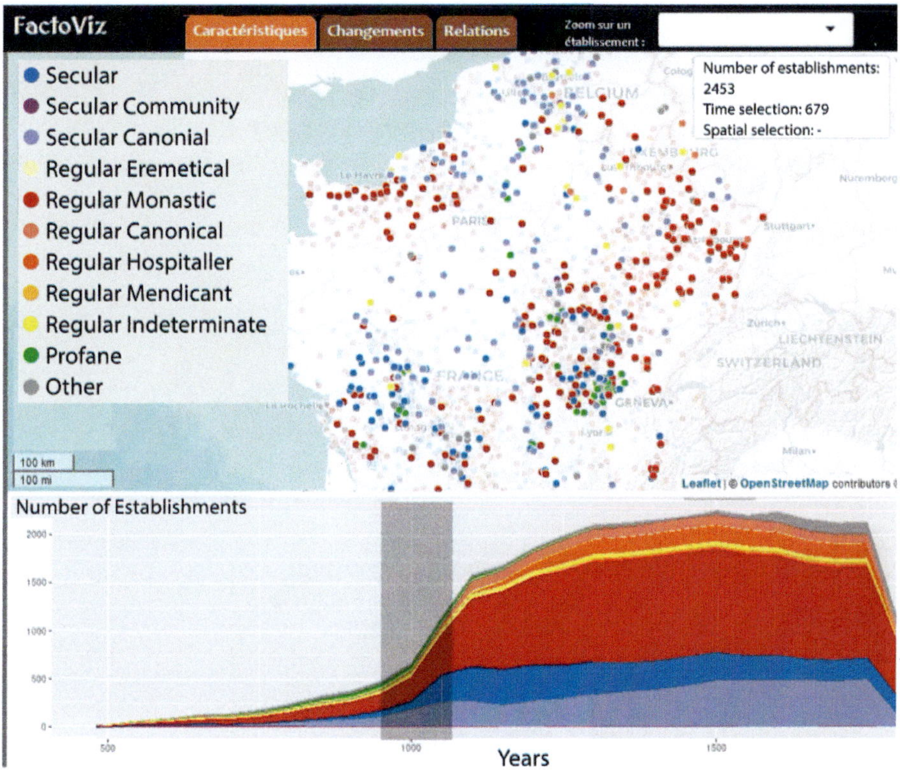


**Figure 6.16.** *Displaying presences over time: example of the FactoViz user interface (Mathian et al. 2020)*

A second solution is to work on other forms of temporality, such as durations ($f(T)$), which can be represented by proportional signs, expressing an index of durability. Or, if biographical information is known, such as the status of the establishments in the previous example (e.g. abbey, collegiate church, etc.), we can represent constructed indicators, such as the frequency of changes over the period, called the turbulence index in the example in Figure 6.17.

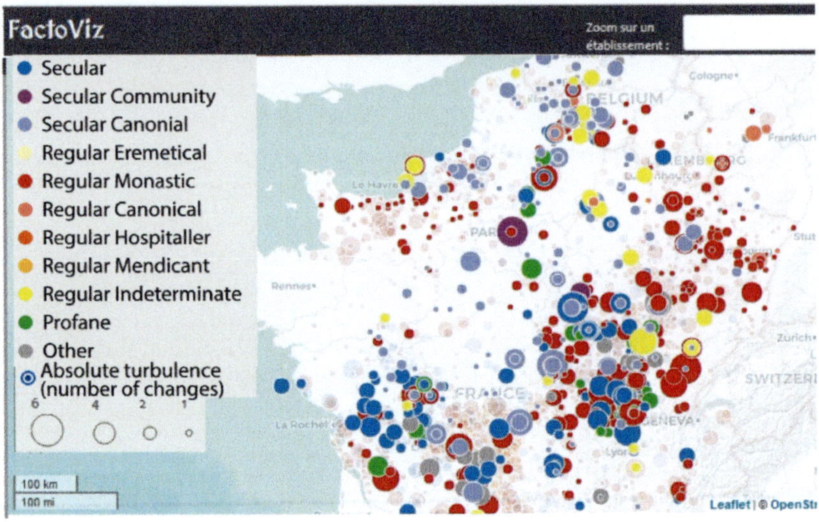


**Figure 6.17.** *Processing change indicators: the example of FactoViz (Mathian et al. 2020)*

### 6.5.3. *Representing space-time*

In all the above examples, the representations preserve geographical space (i.e. they do not affect $(X,Y)$): time does not transform space, and is either represented as a variable, or supported by representations as a timeline, or it constitutes a third dimension. However, as mentioned above in section 6.4, in movement and displacement, space and time appear as two inseparable notions: space-time. Movement takes time, and time is a constraint on movement, as Hägerstrand's aquarium (1970) perfectly illustrates. What is more, we live in this space-time, and our language incorporates it: "We live fifteen minutes from our work" or "it has been raining across 100 km"[5] (Cheylan 2007; L'Hostis 2009). Space-time representations are those in which time is read on space. Essentially driven by questions of movement, this category includes accessibility maps (e.g. isochrone maps) and all maps in which Euclidean space is transformed to reflect the time it takes to travel through it.

5 Author's translation.

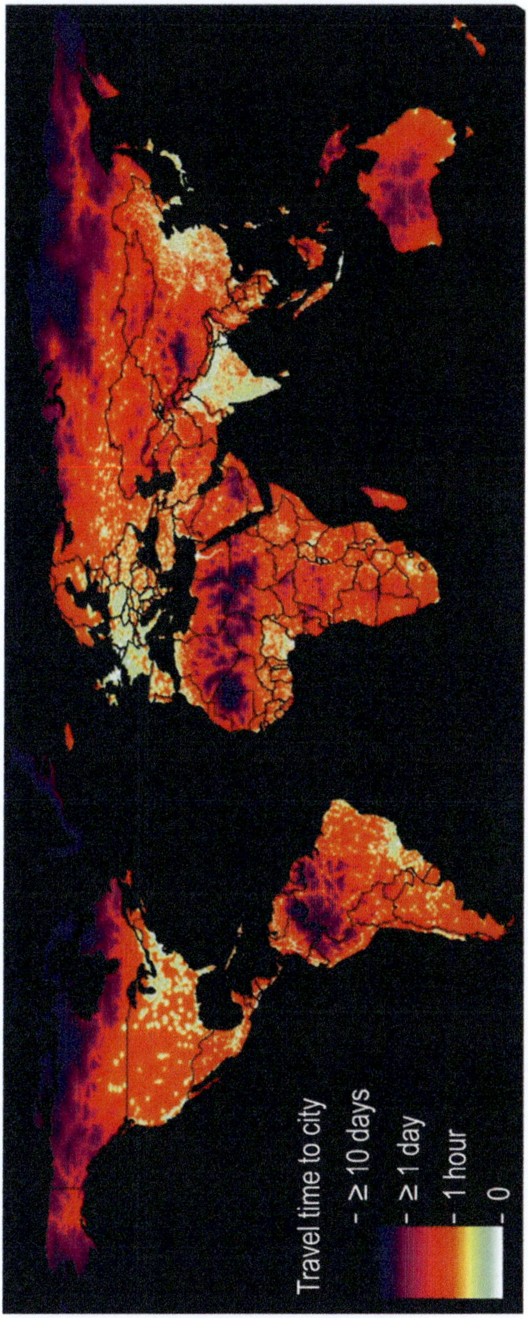


**Figure 6.18.** *A global map of travel time to cities (Weiss et al. 2018)*

Accessibility maps are the most widely used: the simplest allow you to read off a measure of access time to one or more resources at each point in space, such as the accessibility of cities worldwide (Figure 6.18). The data are therefore made up of a matrix of distances between each 1 km$^2$ cell and a set of cities. One way of summarizing this matrix of distances for each portion of land is to extract the smallest distance and map it, illustrating a logic of access to the nearest resource, in this case the nearest city. These maps are a very effective way of expressing spatial inequalities in accessibility, showing gaps and fills. The calculations associated with this type of mapping are generally very complex, as they have to integrate different modes of transport, hence multiple, combined networks and even timetables (Richer and Palmier 2012).

Another family of emblematic representations of space-time are anamorphic maps. As these cartographic transformations are the subject of an entire chapter in this book, we shall not deal with them here, but instead refer the reader to Chapter 7. Such maps are undoubtedly the most accomplished and original of space-time representations.

## 6.6. Conclusion

Throughout this chapter, we have gone through a series of examples and experiments of cartographic treatments for spatiotemporal organizations. We have structured our discussion by exploring different visualization options based on simple spatiotemporal data forms. We have focused on the temporal component of these data for didactic purposes, and all the solutions described can of course be enriched with cartographic and semiological treatments of the other components. By focusing on relatively simple data shapes, we were able to determine which visualization solutions, or combinations of solutions, could improve our understanding of certain aspects of temporal phenomena, and conversely, which aspects of the phenomena lend themselves better to certain methods.

Table 6.2 summarizes correspondences between characteristics of the targeted phenomenon we are seeking to analyze and visualization method. The main correspondences encountered during our inventory are thus listed. In this table, we refer only to those processes that bring the temporal dimension into play through specific visualization, as discussed in this chapter: the third dimension, graphics, time (animation) and interactivity. We have deliberately not included solutions that fold back time (a temporal data) in order to focus on those that, on the contrary, make time visible.

| Characteristic of the targeted phenomenon | Processes* |
|---|---|
| (1) Spatiotemporal organization | |
| General form of the process | Animation, 3D |
| Spatial changes from near to far | Multiple maps |
| Local spatial heterogeneity | Chart-map |
| Overall trend of evolution | Legend covering the entire period |
| Changes to the organization of relative positions in the statistical distribution | Legend specific to each time interval |
| (2) Temporality of the phenomenon | |
| Rhythm, frequency | Animation (1)<br>Chart-map (2)<br>Interactivity (3)<br>Animation + Visual variable value |
| Speed | Animation<br>Time calculation |
| Duration | Visual variable size<br>(possibly coupled with animation) |
| Cyclicity | Chart-map |
| Lifespan | Animation (1)<br>3D (2)<br>Timeline interactivity (3) |
| (3) Time and relations between places | |
| Accessibility | Isodistance<br>Isochronous |
| Space-time | Anamorphic |

*When there are several, the methods are ranked by efficiency.

**Table 6.2.** *Characteristics of the targeted phenomenon and cartographic methods presented in this chapter*

In any case, the various examples have shown that solutions are not unique and that choices depend primarily on the phenomenon and the more or less complex form of the dynamics of its spatial footprint. The effectiveness of the chosen method depends to a large extent on the way in which the phenomenon is organized, both spatially and temporally. Therefore, the preliminary stage, which consists of treating spatiotemporal data upstream to target an essential aspect of the dynamics, remains primordial. It is classically articulated with the combined choice of data processing and semiological solution to support a point of view or even a demonstration. A perfect illustration of this creative process is the map of gentrification in Paris between 1970 and 2000, which expresses the synthesis of a territorial dynamic, derived from numerous statistical treatments and promoted by advanced schematization (Figure 6.19). This map tells the story of a "gentrification front" in Paris.

Figure 6.19 re-examines the combination of methods linked to new digital environments for data creation and visualization and graphic choices. Without a doubt, environments integrate animation and interactivity, or immersive 3D exploration, thus allowing for original visualizations of spatiotemporal data. However, their effectiveness must be discussed in relation to the objectives of the map, as categorized by MacEachren (1994), namely, data exploration, analysis, synthesis and storytelling (Kaddouri et al. 2014). Here, socio-spatial dynamics in Paris are represented statically and lie between synthesis and narrative. An animation process would reinforce the temporal dimension of the narrative, adding to the synthesis work that semiotically integrates time (visual variable value that translates different periods). At present, animation or other method (interactivity, 3D) is often used to represent spatiotemporal phenomena without prior data treatment. These visualizations have often fascinated users with their originality, but their effectiveness in helping to understand the phenomenon has yet to be proven: we look at the data rather than the phenomenon. The complexity of implementing these methods has sometimes led to the representation of raw data in the belief that they would speak for themselves, as part of an exploratory approach. Still, static representation of a dynamic phenomenon calls for even more original graphic inventiveness. This inventiveness is fueled by the reflections generated by the treatment of the data to be represented and the analysis of the phenomenon, both of which are essential prerequisites for its representation. There is a lot of originality to be found in productions from the fields of *dataviz* and *datajournalism*. We can assume that this originality stems from a double freedom: that linked to computer skills, which enable us to free ourselves from standard data modeling, and that linked to the objective, which is precisely that of showing a data set that enables us to free ourselves from the phenomenon and thus from heritages and habits, in particular with regard to the representation of the temporal dimension. This highly stimulating production should also give rise to more formal developments, leading to graphical innovations in more conventional visualization environments.

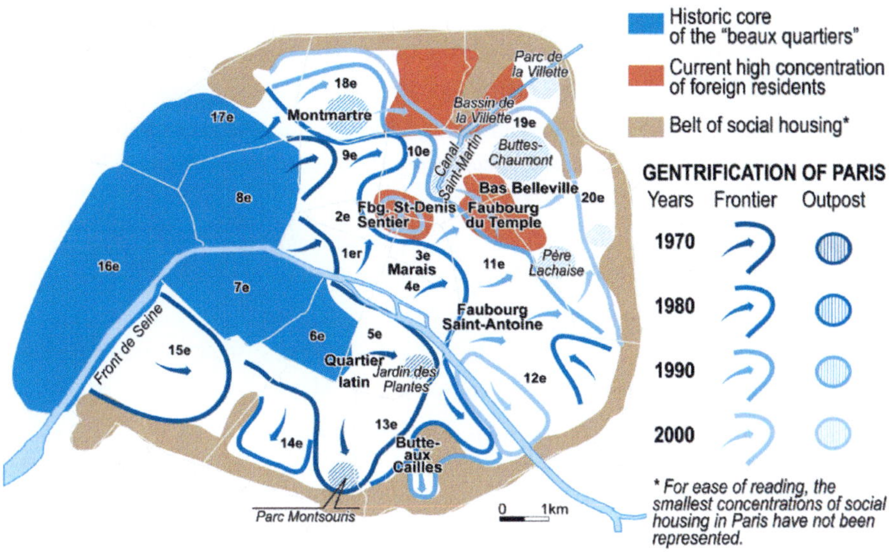


**Figure 6.19.** *Socio-spatial dynamics in Paris (Clerval 2010): a story of data processing and graphic choices (sources: APUR base map; Atlas des Parisiens, 1984; INSEE field surveys, RGP 1982, 1990,1999 – CS household population and housing characteristics)*

It is a safe bet to say, that currently, the massive arrival of data from the digital humanities, and with them diverse natures of time such as historical time and individual lived time, will stimulate new experiments in representation.

## 6.7. References


Andrew, R.J. (2017). Map of firsts: An interactive timeline of the most iconic infographics [Online]. Available at: https://history.infowetrust.com/ [Accessed 26 January 2023].

Andrienko, N. and Andrienko, G. (2007). Designing visual analytics methods for massive collections of movement data. *Cartographica*, 42(2), 117–138.

Andrienko, N. and Andrienko, G. (2013). Visual analytics of movement: An overview of methods, tools and procedures. *Information Visualization*, 12(1), 3–24.

Andrienko, G., Andrienko, N., Bak, P., Keim, D., Kisilevich, S., Wrobel, S. (2011). A conceptual framework and taxonomy of techniques for analyzing movement. *Journal of Visual Languages & Computing*, 22(3), 213–232.

Bach, B., Dragicevic, P., Archambault, D., Hurter, C., Carpendale, S. (2016). A descriptive framework for temporal data visualizations based on generalized space-time cubes. *Computer Graphics Forum*, 36(6), 36–61.

Bahoken, F. (2024). Mapping flows and movements. In *Cartographic Communication: Graphic Semiology, Semiotics and Geovisualization*, Mericskay, B. (ed.). ISTE Ltd, London, and John Wiley & Sons, New York.

Bak, P., Mansmann, F., Janetzko, H., Keim, D. (2009). Spatiotemporal analysis of sensor logs using growth ring maps. *IEEE Transactions on Visualization and Computer Graphics*, 15(6), 913–920.

Banos, A. and Thevenin, T. (2005). Révéler les rythmes urbains quotidiens par la carte animée. *Revue internationale de géomatique*, 15(1), 11–31.

Baruch, J., Aubert, R., Ferrer, M., Sanchez, L., Dagorn, G., Breteau, P. (2020). Covid-19 : le tableau de bord de l'épidémie. *Le Monde* [Online]. Available at: https://www.lemonde.fr/les-decodeurs/article/2020/05/05/coronavirus-age-mortalite-de-partements-pays-suivez-l-evolution-de-l-epidemie-en-cartes-et-graphiques_6038751_4355770.html [Accessed 26 January 2023].

Bertin, J. (2011) *Semiology of Graphics: Diagrams, Networks, Maps*. Translated by W.J. Berg (University of Wisconsin, 1983). ESRI Press, Redlands, CA.

Blitzortung.org. (n.d.). Real time lightning map [Online]. Available at: https://www.lightningmaps.org [Accessed 26 January 2023].

Boria, E. (2012). Mapping time [Online]. Available at: https://www.espa-cestemps.net/articles/mapping-time/ [Accessed 26 January 2023].

Cauvin, C., Escobar, F., Serradj, A. (2010a). *Thematic Cartography 2. Cartography and the Impact of the Quantitative Revolution*. ISTE Ltd, London and John Wiley & Sons, New York.

Cauvin, C., Escobar, F., Serradj, A. (2010b). *Thematic Cartography 3. New Approaches in Thematic Cartography*. ISTE Ltd, London and John Wiley & Sons, New York.

Cheylan, J.-P. (2007). Les processus spatio-temporels : quelques notions et concepts préalables à leur représentation. *M@ppemonde*, 87(3) [Online]. Available at: http://mappemonde-archive.mgm.fr/num15/articles/art07303.html [Accessed 26 January 2023].

Cheylan, J.-P., Lardon, S., Mathian, H., Sanders, L. (1994). Les problématiques liées au temps dans les SIG. *Revue internationale de géomatique*, 4(3/4), 287–305.

Cheysson, E. (1884). Mouvement quinquennal de la population Française par département depuis 1861 jusqu'en 1881. *Album de statistique graphique*, plate 25.

Cheysson, E. (1888). Accélération des voyages en France depuis 200 ans. *Album de statistique graphique*, plate 8.

Claramunt, C. and Thériault, M. (1995). Managing time in GIS: An event-oriented approach. In *Recent Advances in Temporal Databases*, Clifford, J. and Tuzhilin, A. (eds). Springer, London.

Clerval, A. (2010). Les dynamiques spatiales de la gentrification à Paris. *Cybergeo: European Journal of Geography* [Online]. Available at: http://journals.openedition.org/cybergeo/23231 [Accessed 26 January 2023].

Cossart, É. (2021). "Le Morvan est-il maudit ?", analyse géographique d'un territoire hyper-rural. *Territoire en mouvement, Revue de géographie et aménagement*, 50 [Online]. Available at: http://journals.openedition.org/tem/7276 [Accessed 26 January 2023].

Cruz, P., Wihbey, J., Ghael, A., Shibuya, F., Costa, S. (2019). Dendrochronology of U.S. immigration. *Information Design Journal*, 25, 6–20.

Cunty, C., Ribardière, A., Mas, A., Lizzi, L., Lambert, N., Ben Rebah, M. (2012). Premiers pas vers une proposition de sémiologie graphique animée. *Bulletin du comité français de cartographie*, 211, 21–31.

Devogele, T., Étienne, L., Esnault, M., Lardy, F. (2017). Optimized discrete Fréchet distance between trajectories. In *BigSpatial'17: Proceedings of the 6th ACM SIGSPATIAL Workshop on Analytics for Big Geospatial Data*. Association for Computing Machinery, New York.

DiBiase, D., MacEachren, A.M., Krygier, J.B., Reeves, C. (1992). Animation and the role of map design in scientific visualization. *Cartography and Geographic Information Systems*, 19(4), 201–214.

Fabrikant, S., Rebich Hespanha, S., Andrienko, N., Andrienko, G., Montello, D. (2008). Novel method to measure inference affordance in static small-multiple map displays representing dynamic processes. *The Cartographic Journal*, 45(3), 201–215.

Field, K. (2021). Corona Coxombs [Online]. Available at: https://arcg.is/1qbaum0 [Accessed 26 January 2023].

Forlines, C. and Wittenburg, K. (2010). Wakame: Sense making of multi-dimensional spatial-temporal data. In *AVI'10: Proceedings of the International Conference on Advanced Visual Interfaces*, Rome. Association for Computing Machinery, New York.

Galton, F. (1863). *Meteorographica. Methods of Mapping the Weather*. Macmillan, London.

Galton, A. (2004). Fields and objects in space, time, and space-time. *Spatial Cognition & Computation*, 4(1), 39–68.

Gatalsky, P., Andrienko, N., Andrienko, G. (2004). Interactive analysis of event data using space-time cube. In *Proceedings of 8th International Conference on Information Visualization*. IEEE Computer Society.

Gautier, J. (2018). GrAPHiST. Une approche d'analyse exploratoire pour l'identification des dynamiques des phénomènes spatio-temporels. PhD Thesis, Université Grenoble Alpes, Grenoble.

Gautier, J., Davoine, P.-A., Cunty, C. (2019). GrAPHiST : un environnement de visualisation pour l'analyse exploratoire de données spatio-temporelles. In *SAGEO Conférence*, Clermont-Ferrand.

Gautier, J., Lobo, M.-J., Fau, B., Drugeon, A., Christophe, S., Touya, G. (2021). Covid-19 geoviz for spatio-temporal structures detection. *Proceedings of the International Cartographic Association*, 4(37), 8.

Genevois, S. (2020). La cartographie des épidémies entre peur de la contagion et efforts de prévention. Exemple à travers la diffusion du coronavirus [Online]. Available at: https://cartonumerique.blogspot.com/2020/01/cartographie-epidemies.html [Accessed 26 January 2023].

Hägerstrand, T. (1970). What about people in regional science? *Papers in Regional Science*, 24(1), 7–24.

Hurez, C., Jardinier, L., Mermoud, F., Mathieu, L., Baroudi, G., Grassot, L., Cunty, C. (2012). *Les pulsations urbaines, un outil pour suivre la répartition géographique et temporelle des habitants d'une agglomération sur une journée*. CERTU, Paris.

Kaddouri, L., Blaise, J.-Y., Davoine, P.-A., Mathian, H., Saint-Marc, C. (2014). État des lieux des représentations dynamiques des temporalités des territoires. Report, ESPACE/MAP-GAMSAU/Laboratoire d'informatique de Grenoble [Online]. Available at: http://www.map.cnrs.fr/jyb/puca/ [Accessed 26 January 2023].

Kerouanton, C. (2020). Analyse spatio-temporelle des mobilités de randonneurs dans le PNR du Massif des Bauges. PhD Thesis, Université de Savoie, Chambéry.

Klein, O. (2007). Modélisation et représentations spatio-temporelles des déplacements quotidiens urbains : application à l'aire urbaine Belfort-Montbéliard. PhD Thesis, Université Louis Pasteur de Strasbourg, Strasbourg.

Klein, O. (2014). Géovisualisation des espace-temps quotidiens. Propositions d'outils d'aide à la réflexion. *URBIA. Les Cahiers du développement urbain durable*, 16, 129–146.

Klein, É. (2016). *Le facteur temps ne sonne jamais deux fois*. Flammarion, Paris.

Klein, O., Drevon, G., Gwiazdzinski, L. (2017). Représenter les temps et les rythmes urbains. In *Chronotopies. Lecture et écriture des mondes en mouvement*, Drevon, G., Gwiazdzinski, L., Klein, O. (eds). Élya Éditions, Seyssinet-Pariset.

Kraak, M.J. (2003). The space-time-cube revisited from a geovisualization perspective. In *Proceedings of the 21st International Cartographic Conference*. International Cartographic Association, Durban.

Kwan, M.-P. (2000). Interactive geovisualization of activity-travel patterns using three-dimensional geographical information systems: A methodological exploration with a large data set. *Transportation Research C: Emerging Technologies*, 8(1/6), 185–203.

Lambert, N. (n.d.). Vague après vague. *Carnet (neo)cartographique* [Online]. Available at: https://neocarto.hypotheses.org/12035 [Accessed 26 January 2023].

Lambert, N. (2021). Cartographier la Covid-19 : quelles narrations ? *Revue francophone sur la santé et les territoires* [Online]. Available at: http://journals.openedition.org/rfst/853 [Accessed 26 January 2023].

Langran, G. (1992). Time in geographic information systems. *Geocarto International*, 7(2), 40.

L'Hostis, A. (2003). De l'espace contracté à l'espace chiffonné. Apports de l'animation à la cartographie en relief des distances-temps modifiées par les réseaux de transport rapides. *Revue internationale de géomatique*, 13(1), 69–80.

L'Hostis, A. (2009). The shrivelled USA: Representing time-space in the context of metropolitanization and the development of high-speed transport. *Journal of Transport Geography*, 17(6), 433–439.

MacEachren, A.M. (1994). Visualization in modern cartography: Setting the Agenda. In *Visualization in Modern Cartography*, MacEachren, A.M. and Taylor, D.R.F. (eds). Elsevier, Oxford.

Marey, E.-J. (1878). *La méthode graphique dans les sciences expérimentales et principalement en physiologie et en médecine*. Masson, Paris.

Mathian, H. and Sanders, L. (2014). *Spatio-temporal Approaches: Geographic Objects and Change Process*. ISTE Ltd, London, and John Wiley & Sons, New York.

Mathian, H., Cunty, C., Joliveau, T., Janin, O., Leroy, O. (2020). FactoViz : analyser les données de Coll&Mon, Collégiales et Monastères [Online]. Available at: https://analytics.huma-num.fr/Helene.Mathian/FactoViz/FactoViz_Stable/ [Accessed 26 January 2023].

Mccuaig, J., Hetzler, B., MacEachren, A., Gahegan, M., Pavel, M. (2005). A typology for visualizing uncertainty. In *Proceedings of SPIE, The International Society for Optical Engineering*. Electronic Imagining 2005, San Jose.

Mericskay, B. (2019). Potentiels et limites des traces (géo)numériques dans l'analyse des mobilités : l'exemple des données de la plateforme de covoiturage BlaBlaCar. *Cybergeo: European Journal of Geography* [Online]. Available at: http://journals.openedition.org/cybergeo/31990 [Accessed 26 January 2023].

Minard, C.-J. (1869). *Carte figurative des pertes successives en hommes de l'armée française dans la Campagne de Russie 1812-13 (comparées à celle d'Hannibal durant la 2ème Guerre Punique)*. Régnier & Dourdet, Paris [Online]. Available at: https://gallica.bnf.fr/ark:/12148/btv1b52504201x [Accessed 26 January 2023].

Palsky, G. (1996). *Des chiffres et des cartes : naissance et développement de la cartographie quantitative française au XIXe siècle*. Comité des travaux historiques et scientifiques, Paris.

Peuquet, D.J. (1999). Time in GIS and geographical databases. In *Geographical Information Systems: Principles, Techniques, Management and Applications*, Longley, P.A, Goodchild, M.F, Maguire, D.J., Rhind, D.W. (eds). Wiley, New York.

Peuquet, D.J. (2002). *Representations of Space and Time*. Guilford Press, New York.

Plumejeaud, C. (2011). Modèles et méthodes pour l'information spatio-temporelle évolutive. PhD Thesis, Université de Grenoble, Grenoble.

Poirier, N. (2021). Les défis d'une approche multisource et diachronique de l'occupation du sol : l'exemple de la vallée de la Garonne. *Humanités numériques*, 3 [Online]. Available at: http://journals.openedition.org/revuehn/1228 [Accessed 26 January 2023].

Rajerison, M. (2021). Geosparklines [Online]. Available at: https://github.com/datagistips/geosparklines [Accessed 26 January 2023].

Richer, C. and Palmier, P. (2012). Mesurer l'accessibilité territoriale par les transports collectifs : proposition méthodologique appliquée aux pôles d'excellence de Lille Métropole. *Cahiers de géographie du Québec*, 56, 427–461. doi: 10.7202/1014554ar [Accessed 26 January 2023].

Ritchie, H., Mathieu, E., Rodés-Guirao, L., Appel, C., Gavrilov, D., Giattino, C., Hasell, J., Macdonald, B., Dattani, S., Beltekian, D. et al. (2020). Coronavirus pandemic (Covid-19) [Online]. Available at: https://ourworldindata.org/coronavirus [Accessed 30 August 2022].

Rozier, G. (2020). COVIDTRACKER: Dashboard Départements [Online]. Available at: https://covidtracker.fr/dashboard-departements/ [Accessed 26 Jaunary 2023].

Thakur, S. and Hanson, A. (2010). A 3D visualization of multiple time series on maps. In *Proceedings of the 14th International Conference on Information Visualisation*. IEEE Computer Society.

Thévenin, T., Chardonnel, S., Cochey, É. (2007). Explorer les temporalités urbaines de l'agglomération de Dijon. *Espace populations sociétés*, 2/3, 179–190.

Vasiliev, I. (1997). Mapping time. *Cartographica*, 34(2), 1–51.

Wagner, J., Stuerzlinger, W., Nedel, L. (2019). Evaluating an immersive space-time cube geovisualization for intuitive trajectory data exploration. *IEEE Transactions on Visualization and Computer Graphics*, 26(1), 514–524.

Weiss, D. J., Nelson, A., Gibson, H. S., Temperley, W., Peedell, S., Lieber, A., Hancher, M., Poyart, E., Belchior, S. et al. (2018). A global map of travel time to cities to assess inequalities in accessibility in 2015. *Nature*, 553, 333–336.

Worboys, M. (2005). Event-oriented approaches to geographic phenomena. *International Journal of Geographical Information Science*, 19(1), 1–28.

# 7

# Cartograms, Anamorphic Maps: Transformed Territories

**Anne-Christine BRONNER**
*SAGE, CNRS, Université de Strasbourg, France*

## 7.1. Introduction

Anamorphic maps hold a special place in the cartographic processing of spatial data; deciphering them leads us to question the shapes of geographic space, the represented territory and the base map. The transition from the Earth's geoid to a "flat", two-dimensional representation creates different shapes, and de facto, every map is an anamorphic map[1].

The conception of the base map is related to the dataset's type, to mathematical, statistical, graphic and cartographic processing, as well as to the objectives of the final map. The positions and shapes of spatial entities, whether represented by points, lines or surfaces, are determined by various operations such as geolocation or georeferencing, generalization or simplification, change of spatial base, etc. This spatial dimension is referenced in a XY geographic coordinate system. The anamorphic map differs from other cartographic treatments in one specific way: in an anamorphic map, the "geographic" space, the planimetric space, fades away and

1 Note to English-speaking readers: We decided to use "anamorphic map" rather than "anamorphosis" directly translated from the word *anamorphose*. Anamorphic map identifies the type of cartographic representations and methods identified in the anamorphosis system of correspondence and transformations between spatial and thematic dimensions. In English or French, what we call a cartogram (*cartogramme*) is more inclusive, including some statistical maps.

*Handling and Mapping Geographic Information*,
coordinated by Claire CUNTY and Hélène MATHIAN. 

is modified by the theme represented, revealing phenomena that cannot be observed directly. On a planisphere, where population determines the size of each country, China will appear almost 10 times larger than Russia. For a European businessperson, distances between European and African metropolises would be reduced to a few hours. The same places would be positioned differently to map the space-time of migrants from sub-Saharan Africa.

In the anamorphosis system[2], the position of locations is defined by the attribute data Z: $X = f(Z)$, $Y = g(Z)$. This leads us to focus on the reality we are mapping, and on the specific link between locations and attribute data. This link is present in many of the works of American cartographer Waldo Tobler: locative and substantive geographic data are combined in one (or a succession of) cartographic transformations to meet the uses, production contexts and objectives of the map. The spatial base is created and "calculated" depending on one's field of interest: angles preserved for plotting a course for navigation, accurate land surface areas for the surveyor, computing a map centered on one location, depicting a territory in proportion to the population to analyze statistical results or taking into account distances between locations to plan a metropolitan area, etc. (Tobler 1979).

In this chapter, we will be talking about the *anamorphic map* or *cartogram* (see Box 7.1), two terms that today, in the French-speaking cartography field, seem to have become generic and interchangeable, even if this equivalence is highly discussed, and refer to the two most common types of anamorphic map[3]:

– those where the positions and shapes of areal spatial entities are determined by the value of a variable, for example, population, which impacts the size and contours of administrative units;

– those that identify transformations that reshape the territory by shifting positions according to distances between locations, such as access times.

Understanding and approaching the methods used to construct anamorphic maps is one of the keys to avoiding confusion between them and to improving their interpretation. In this chapter, we present an overview dedicated mainly to the

2 See the cartographic systems of correspondence and transformations between the spatial and thematic components XY and Z (Cauvin et al. 2010).

3 In English, various terms or expressions coexist: cartogram, area cartogram, continuous or contiguous cartogram, equal density map, area-by-value map, equal area map, isodemographic map, rubber map, anamorphic map, etc. In French: *anamorphose*, *carte par anamorphose* or *en anamorphose*, *représentation anamorphique*, *cartogramme* or *anamorphose scalaire*, *de surface*, *surfacique*, *vectoriel*, *de distance*, *cartogramme continu*, *contigu*, etc. It should be underlined that, depending on the context, these names refer to different methods and types of maps.

different types of cartograms that resolve the question of mapping count data by transforming the surface area of entities (section 7.2), distinguishing them from space-time anamorphic maps (section 7.3). Only after presenting this set of cartographic processes do we return to cross-cutting aspects, highlighting common methodological features and offering a few details on the presentation and reading difficulties of these maps (section 7.4).

Among French-speaking cartographers, the term *cartogram* is associated with statistical data's graphic treatment and has been, and still can be, used to identify any type of visualization of statistical geographic information, such as diagram maps. While the immediate meaning tends to refer to the continuous representation of count data by a proportional variation of surface area (restricted meaning), it extends de facto to a wide range of (carto)graphic processes applied to spatial data, sometimes specified in an epithet: Dorling cartogram, Olson non-contiguous area cartogram, distance cartogram, grid cartogram, etc.

The word anamorphosis comes from the Greek *anamorphoûn*, meaning "to transform", which in Latin refers to *transformare*, from *trans*, and *formare* for change of form. The term emerged in cartography (among French-speaking cartographers) to identify maps whose geographic forms are reshaped. Like anamorphosis in art or the graphic arts, these anamorphic maps, whether or not they are the result of mathematical processes, intrigue the viewer. We might mention perspective or optical effects (such as *anamorphic illusions*), where the image varies according to the viewing position (with the well-known reference to the skull in Hans Holbein's painting *The Ambassadors*).

**Box 7.1.** *About the terms cartogram and anamorphosis*

This chapter concentrates on the principles of the methods, without going into mathematical or algorithmic demonstrations, which are otherwise detailed in the source papers. Two databases illustrate this point: an electoral dataset of 2017 for the Grand Est region (France), to test the distortion effects of an unfamiliar territory, and rail travel time in 1980 on a French scale, for the more complex question of space-time representation. The reference list at the end of the chapter cites the main software used.

## 7.2. Cartograms to represent count data associated with areal units

The basic principle underlying the "area" cartogram is to calculate, for each entity, a surface area proportional to the value of the variable. As an output of this basic principle, the cartogram reads like a treemap. If a spatial unit occupies a quarter of the surface area of the map, it contains a quarter of the total account; it is possible to perform a rough count by the visual grouping of several entities, or to assess the share of this group in relation to the global. The countable quantities–size variation association is a common and well-recognized tacit rule in graphical representation, with the histogram, the pie chart, as well as the treemap or mosaic plot and the bubble chart, current in dataviz. The spatial component imposes an additional constraint: the configuration of the geographic space.

To exemplify the methods, we chose the Grand Est region of France, the shape of which is not very well known, allowing us to better focus on methodological aspects and cartographic results. Processing was carried out using R software and various packages (Jeworutzki et al. 2020; Giraud et al. 2021), ScapeToad (Andrieu et al. 2008) and QGIS.

### 7.2.1. *Why use cartograms?*

Cartograms, the term used here in the restricted sense of an "area-by-value" cartogram or map, encounter a twofold challenge: to solve the problem of size variation of areal units, and to provide an appropriate base map, particularly for mapping sociodemographic variables.

#### 7.2.1.1. *Solving the problem of size variation of areal units*

In his reference work, *Semiology of Graphics*, Bertin (2011, 1st edition, 1983) identifies an obstacle to the use of the visual variable size: "In the area representation, the area cannot vary in size. [...] Here, the extent of variation is limited by the size of the area involved"[4]. In this case, the proportional symbol map and the dot map are used.

At the same time, Tobler came up with the idea of transforming the base map, formulating it as follows:

> Imagine that one could stretch a geographical map so that areas with many people would appear large, and areas with few people would appear small. If such a map could be made one would expect all voting districts to be the same size since they should contain equal numbers of people. (Tobler 1973)

4 Translated by W.J. Berg.

### 7.2.1.2. *Transforming the choropleth base map*

On a choropleth map, the visibility of a variable attached to its geographic entity is linked to its surface area. This is not a problem if the map's purpose is to depict the "visible" world, such as land use rates (urbanization, deforestation, etc.), in other words when the denominator of the rate corresponds to the surface area of the entity. On the other hand, when this so-called "classic" base map is used to display a population rate (typically when working with sociodemographic, electoral or epidemiological data), the map visually subtracts what is happening in small administrative localities and overexposes the phenomenon in large spatial units. The bias is introduced by two facts: (1) small territorial entities, generally census tracks, are not the least populated and vice versa; and (2) a rate is sensitive to the value of the denominator (the value of a rate can vary all the more strongly the smaller the value of the denominator).

This imbalance can be seen from the concentration curve in Figure 7.1: in the Grand Est region, a quarter of the land area displays just 3% of the population; half of the municipalities of the region have fewer than 400 inhabitants, totaling less than 10% of the population. The 168 populated areas with more than 5,000 inhabitants, that is, half the region's population, are concentrated in 5% of the territory.

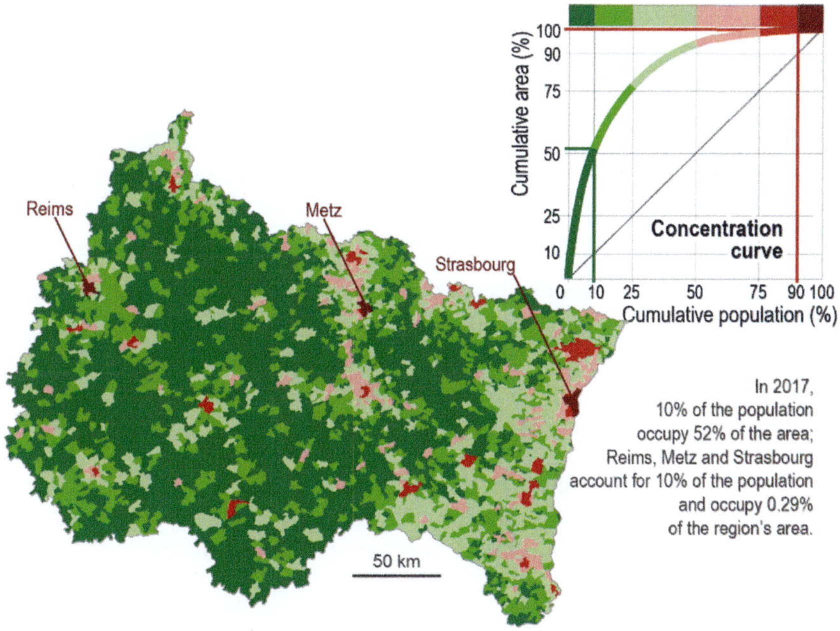


**Figure 7.1.** *Choropleth map and visibility of mapped information. Grand Est region, municipality level (sources: IGN ADMIN EXPRESS, INSEE)*

The usefulness of cartograms cannot be reduced to the improvement of the visibility of small administrative units – inset maps for static representations, and zoom/pan functionality for dynamic displays enables zooming in on each part of the map – or to their "communicational" potential. Although their interest or reading problems remain a matter of debate, and their use is limited, there is nevertheless a regular renewal of algorithms to calculate them. They are implemented in tools developed on an ad hoc basis (e.g. *ScapeToad* or the online application *go-cart.io*), in some mapping software (e.g. the online software magrit.cnrs.fr), in plugins, as in GIS, or as scripts in various programming languages (e.g. the *cartogram* or *cartogramR* packages in R).

Please note that some cartographic treatments similar to cartograms will not be discussed, as they do not, stricto sensu, take place in the anamorphic map's family:

– Olson's non-contiguous cartogram (1976). In criticism to Tobler's rubber map (1973), which produced a continuous representation of a discrete variable and an unrecognizable territory, Olson's proposal appears as a simple variation of the proportional symbols, retaining the shape and geographic anchorage of the spatial entity. This cartographic solution is very interesting for carrying the metaphor of a fragmented and unequal territory, like illustrating socio-political discussion on a continental or global scale;

– tile grid maps (or any other combination of the terms tile, grid, map, cartogram), accepted in the sense of dataviz productions, where each spatial unit is represented by a geometric shape of identical size (square, hexagon), whatever the surface area or the value of the variable. These unitary components, distributed to recall the overall shape of the territory, generally display qualitative or quantitative data by colors, values, textures or any combinations of visual variables, or even graphs such as histograms, curves, etc. (see Figure 6.7; Ribecca 2018);

– a new process which combines the dot density map solution (one dot equals a fixed quantity) and a non-overlapping function, generating visual variations of the surface area drawn by the dots (Lambert 2020). The dynamic animation of the map collection visualizes the evolution of hospitalizations during the Covid crisis (see Figure 6.6(a)).

As with any cartographic process, faced with the variety of methods, the research context, the objectives of the analysis and the characteristics of the source dataset all need to be taken into account in order to make an appropriate choice.

### 7.2.2. *Links between methods and data*

It is necessary to describe the data's proprieties for which the cartogram can be used:

– The data to be mapped is a countable quantity, a population, a count data, a stock, etc., meaning a discrete quantitative variable (ratio level), which can be totalized and which, by convention, is displayed by a size variation in graphic and cartographic treatments. If the variable is a rate, we proceed in two steps: first, we calculate an area cartogram based on the denominator's values, and then we map the rate on this new base map.

– Population is generally the basic attribute to compare different variables (electoral results, medical diagnoses, social datasets – wealth, level of education, etc. – or demographics). In specific situations, the denominator variable creates the transformation: the votes cast for the result of an election, the registered voters for the abstention rate, the labor force for the unemployment rate, the global waste for the plastic waste rate, etc.

– The basic principle of cartograms – calculating the size area as a function of a count data – excludes a priori datasets with zero or missing data, which identify units that are not concerned, especially with the continuous cartogram methods.

### 7.2.3. *Anatomy of the cartogram: methods*

There are several methodological solutions to the problem of size variation of areal units: the so-called continuous or contiguous cartogram[5], which reshapes territorial and sub-territorial boundaries, and a series of methods, which can be seen as graphic treatments of geographic data, taking into account the spatial dimension. While pointing out that some works propose to identify, classify and evaluate methods (Bortins et al. 2002; Tobler 2004; Cauvin et al. 2010; Nusrat and Kobourov 2016; Nusrat et al. 2018; Markowska 2019), we here simply outline the principles and illustrate the different types. We will look successively at the continuous, Dorling, Demers, rectangular and mosaic cartograms.

#### 7.2.3.1. *The continuous cartogram*

The continuous cartogram is the result of an iterative process that redistributes the surface areas of the entities, reshaping the boundaries until the "thematic" surface area is reached. The share of the total account determines the size of the

5 There are two qualifiers: continuous, referring to the continuous transformation of space, and contiguous, referring to the preservation of contiguity (or even propagation by contiguity).

areal unit. Two terms are more explicit: area-by-value map and density-equalizing map projection[6].

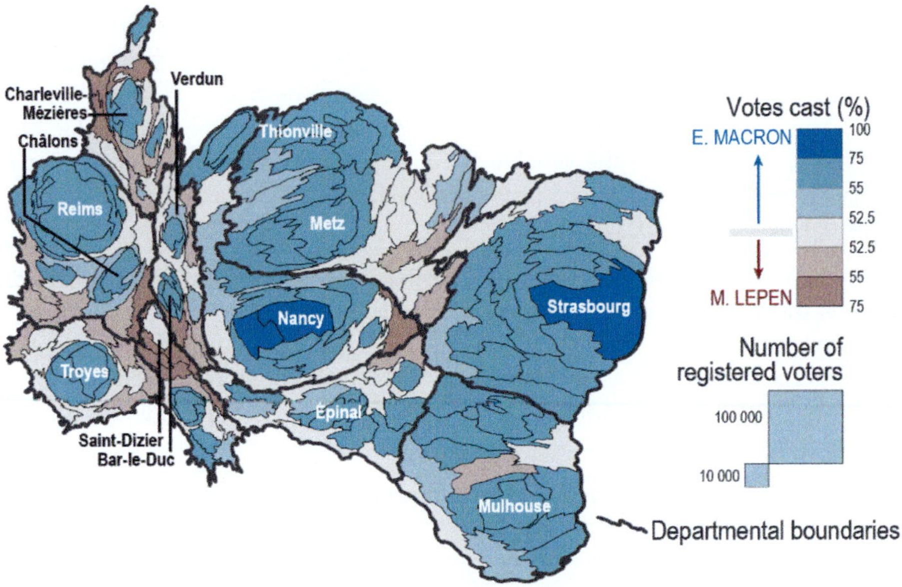


**Figure 7.2.** *A continuous cartogram showing the second round of the 2017 French presidential election. Grand Est region, canton level (sources: IGN ADMIN EXPRESS, data.gouv.fr)*

Popularized in particular by *Worldmapper, the world as you've never seen it before*, the continuous cartogram is proving appropriate for mapping socio-spatial inequalities, as in *The Social Atlas of Europe* (Ballas et al. 2014), or for election maps to point out contrasts between urban, suburban and rural areas or between regions. In Figure 7.2, the second round of the 2017 presidential elections in the Grand Est region shows that urbanized cantons behave differently than suburbs, and more generally, that the two east departments, which are more metropolitanized, stand out from the rest of the territory.

#### 7.2.3.1.1. Multiple algorithms

It is difficult to list all the methodological solutions which, moreover, are regularly renewed (Gastner et al. 2018; Henriques et al. 2009; Sagar 2014; Sun

6 A dot map of the base map before and after the transformation helps us to understand the qualifier *density-equalizing map projection*.

2020). Area error, shape and topology preservation, equations that are numerically easier to solve, computational complexity, open code… make it possible to evaluate their efficiency. Let us mention, without wishing to be exhaustive, the metaphors or principles to be found behind the methods: the "rubber sheet" for Tobler (1973), Dougenik et al. (1985) or Sun (2013), cellular automata for Dorling (1996), springs for House and Kocmoud (1998), Fourier transform for Keim et al. (2004), Gastner and Newman (2004), the density and diffusion process for Gastner et al. (2004, 2018) or the reference to projections for Tobler (1986), Kadmon and Shlomi (1978), Selvin et al. (1984), Langlois (2003), Gastner et al. (2004, 2018).

The most usual algorithms found in geoprocessing tools are those of Dougenik et al. (1985) and Gastner and Newman (2004) or Gastner et al. (2018); the first authors apply the idea of a rubber sheet that is locally stretched: the border coordinates move according to a distance-weighted force. The latter refer to the idea of the diffusion of small particles across a layer of water: the aim of the model applied is to achieve a constant probability of finding a particle (density-equalizing map projection). The two algorithmic solutions of Gastner et al. (2018) produce better results, both in terms of area accuracy and shape preservation.

#### 7.2.3.1.2. Measurement error

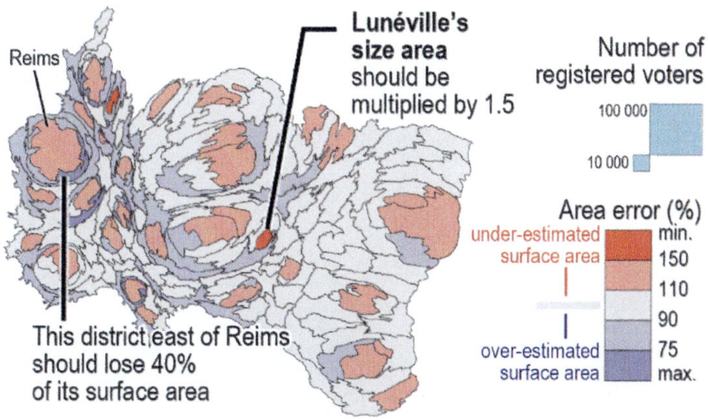


**Figure 7.3.** *Continuous cartogram and area error. Grand Est region, canton level (sources: IGN ADMIN EXPRESS, data.gouv.fr)*

Whatever the method, a measurement error compares the theoretical "thematic" area with the area calculated by the algorithm. It identifies over- or under-estimated surface areas (to put it briefly, large initial districts that have not "shrunk" enough, and small geographic units that have not "expanded" enough). To analyze and validate the accuracy of the cartogram, we refer to central tendency and dispersion measures.

The area error map (see Figure 7.3) is used to localize under-estimated urbanized units and over-estimated countryside cantons, where the number of inhabitants is low, especially if the geographic units concerned are large. Reiterating the process after an initial transformation could help to increase the accuracy. However, the error will remain high in the case of strongly heterogeneous distributions, with extreme values, whether surface area or variable. When area errors cannot be reduced, we recommend applying another cartographic method.

When the drawing of spatial unit boundaries has no direct link with the variable, or when spatial boundaries are blurred or unknown, we propose a hybrid of a tessellation and a continuous cartogram algorithm (see Box 7.2).

Freely inspired by an original idea from Bunge (1966), which introduces the weight of population in a partitioned territory (Christaller model), we propose to couple a change of spatial base with the continuous cartogram. Thiessen polygons display a partition of the territory based on discrete points (chief towns, castles, waste disposal sites, etc.). These can be applied when boundaries are unknown (theoretical example: a partition based on the location of medieval castles to approach old administrative limits), or to avoid the complex shapes of administrative districts while preserving topology. On the map in Figure 7.4, only the area errors generated by the continuous cartogram algorithm will degrade the accuracy of the mapped variable.

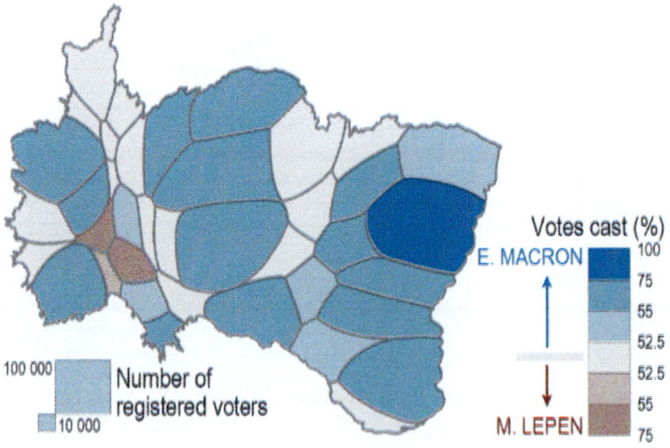


**Figure 7.4.** *Hybrid process: tessellation and continuous cartogram. Second round of the 2017 French presidential election. Grand Est region, arrondissement level (sources: IGN ADMIN EXPRESS, data.gouv.fr)*

**Box 7.2.** *Continuous cartogram on a Voronoi diagram (Thiessen polygons)*

Without further ado, let us consider the impact of different administrative divisions on the analysis of the mapped variable.

#### 7.2.3.1.3. Continuous cartogram and administrative divisions

Whether using a planimetric or transformed base map, it is interesting to observe the variations in the spatial distribution at different administrative divisions, as in the example of the Grand Est region (see Figure 7.5): one region, ten departments, 39 arrondissements, 198 cantons, 169 living areas, 149 EPCIs (inter-municipal cooperations), 31 employment districts, 5,136 municipalities.

Fine granularity inevitably generates a graphic density problem that blurs the information readability: it concerns mostly small spatial units on the cartogram. The arrondissement level hides the heterogeneity between cities and the suburbs, a gradient that appears at smaller administrative levels.

Note that similar territorial divisions in terms of number of entities show different results: the electoral level of the 198 cantons highlights the city of Nancy and underlines the difference between the town of Saint-Dizier and the surrounding territories; the living area level (defined according to access to facilities) differs from the inter-municipal cooperation level (EPCI), the latter showing more homogeneous administrative units.

The continuous cartogram is not the only cartographic solution that attempts to resolve the issue of size variation of areal units.

#### *7.2.3.2. The many shapes of the cartogram*

Alongside the continuous cartogram, various cartographic methods, listed in Figure 7.6, resolve variations in variable values by transforming surface areas, starting with Dorling's (1996) solution, which reduces the spatial unit to a circle, a principle declined in a square by Demers. The latter should not be confused with the rectangular cartogram or the mosaic cartogram.

Non-continuous cartograms, with surface areas strictly proportional to the values of a variable, are appropriate for display datasets with zero or missing data (unconcerned spatial units vanish). However, they generate more or less significant topological errors.

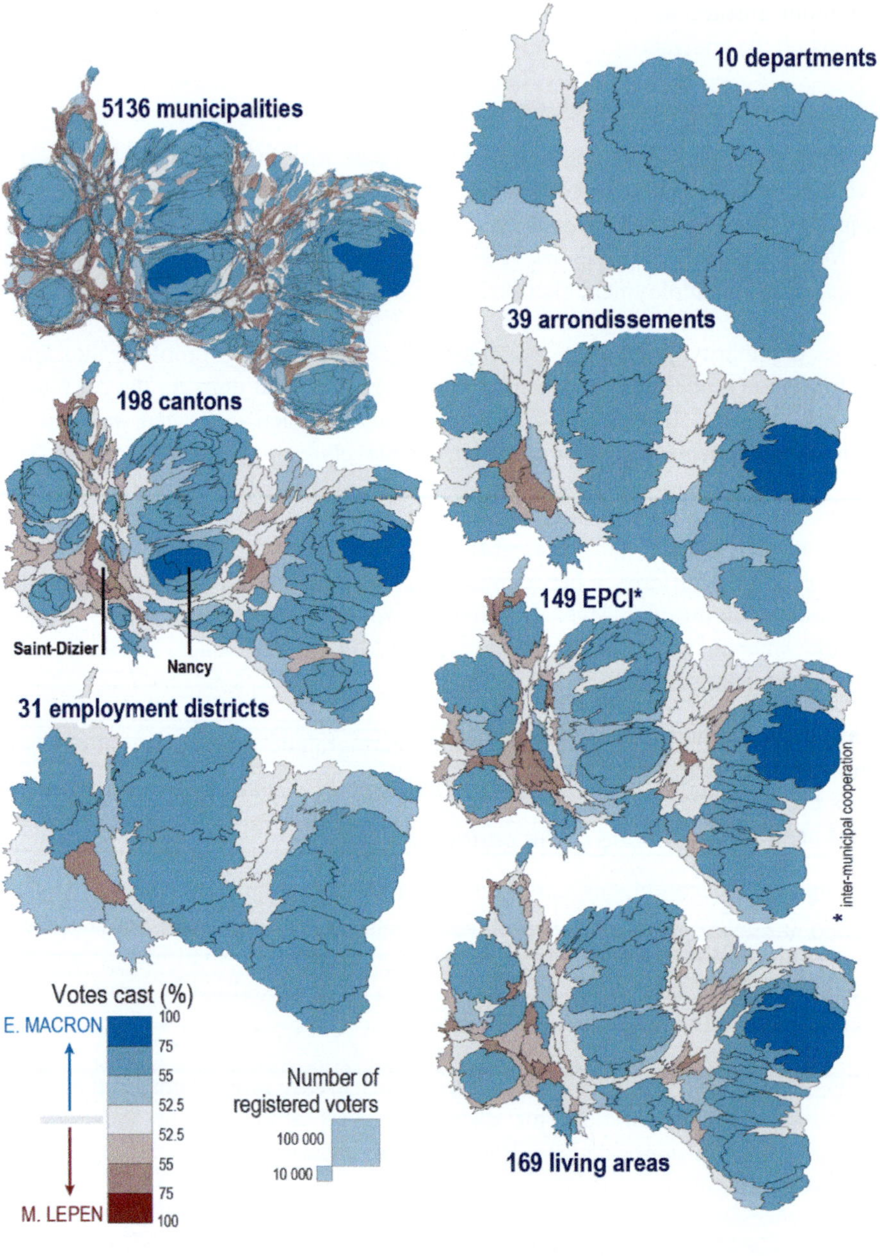


**Figure 7.5.** *Cartogram and administrative divisions. Second round of the 2017 French presidential election. Grand Est region (sources: IGN ADMIN EXPRESS, data.gouv.fr)*

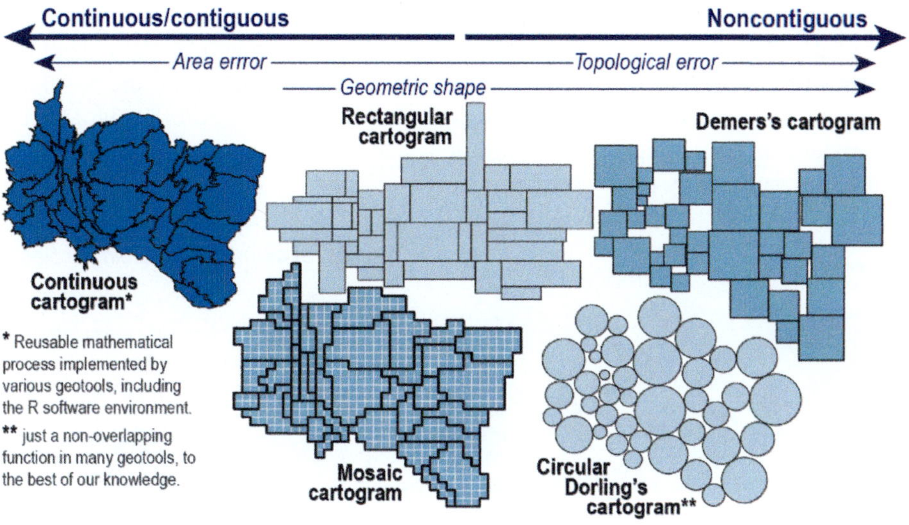


**Figure 7.6.** *Cartogram types (proportional size variation – countable quantities – areal units; sources: IGN ADMIN EXPRESS, data.gouv.fr)*

### 7.2.3.2.1. Dorling's circular cartogram...

In his thesis (Dorling 1991) and the book *Area Cartograms: Their Use and Creation* (Dorling 1996), Dorling presents, among other innovative cartographic treatments, two processes for varying the surface area of an entity in proportion to a countable quantity, one based on cellular automata, the other he calls a *circular cartogram*.

This second process, identified as Dorling's cartogram, is described as very simple and attractive (Bortins et al. 2002; Nusrat et al. 2018): the circle appears as a familiar shape, due to its wide use in maps with proportional symbols. Spatial units are represented by circles with their area in proportion to their data value, which will organize themselves by combining *repelling and attractive displacements*. The advantage of this method is that it produces accurate surface areas, but at the expense of geographic shapes[7] and topology. Dorling's cartogram offers a generally relevant and legible image for many regions, countries or continents, notably Great Britain, on which Dorling has extensively experimented his circular cartogram, the Europe region or the world. For a better analysis of the map, geographic names can be added and displayed with a "mouse over" action.

---

7 "If, for instance, it is desirable that areas on a map have boundaries which are as simple as possible, why not draw the areas as simple shapes in the first place? Here an algorithm is described which does just that, using circles as the simplest of all shapes" (Dorling 1996, p. 32).

In Figure 7.7, the – partial[8] – application of this method to the arrondissement level of the Grand Est region displays a cluster of circles where spatial coherence appears with the arrangement of Thionville, Metz, Nancy, Épinal and the positions of Charleville-Mézières, Reims and Troyes or Strasbourg. Whatever the algorithm's performance or the settings, relative locations and contiguities are upset: Briey and Saverne are particularly poorly located.

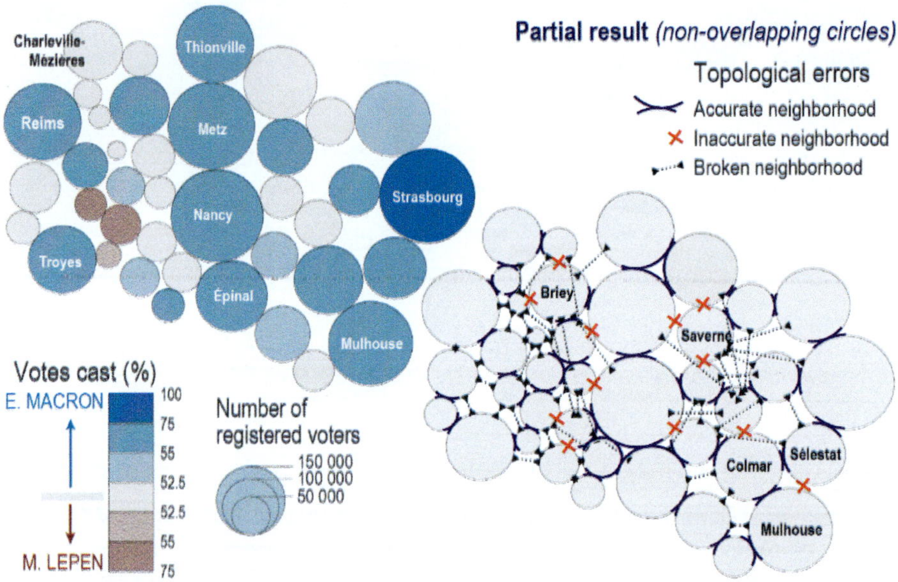


**Figure 7.7.** *Circular cartogram alias Dorling's cartogram. Grand Est region, arrondissement level. Second round of the 2017 French presidential election (sources : IGN ADMIN EXPRESS, data.gouv.fr)*

### 7.2.3.2.2. … and Demers's Squares

Demers's cartogram may appear as a variant of Dorling's circular cartogram, which can be associated with Pierre Émile Levasseur's historical squared statistic map of Europe, cited notably by Tobler (2004). This cartographic method is of little

8 In many geotools (Rpackages, packcircles or cartograms, www.blog.datawrapper.de/cartograms/, www.observablehq.com/collection/@joewdavies/dorling-cartograms), there seems to be confusion between the non-overlapping function and the Dorling cartogram. Voids (sometimes very large ones) between the circles appear, due to the lack of an attractive force function.

interest to us: automating the calculation is complex (Nickel et al. 2019) and the square shape causes multiple topological errors and numerous voids. To illustrate this (see Figure 7.8), we opted for a handmade creation after having calculated proportional squares. Verdun cannot maintain neighborhood relationships with the six arrondissements that surround it.

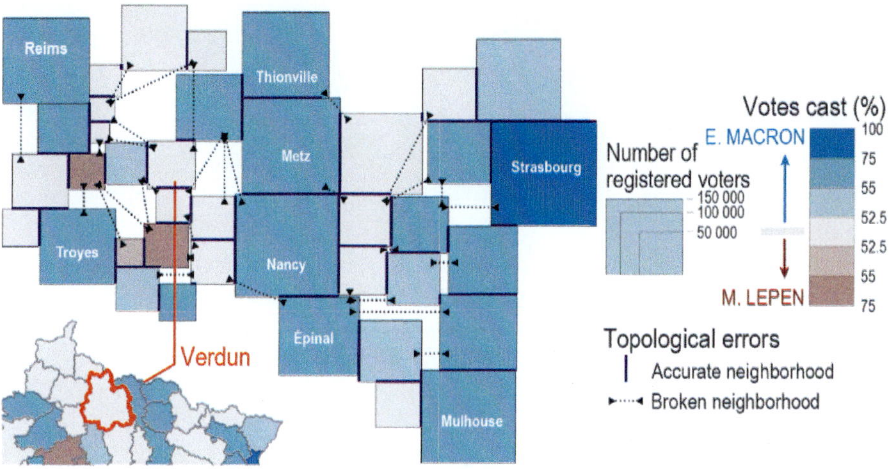


**Figure 7.8.** *Demers's cartogram. Second round of the 2017 French presidential election. Grand Est region, arrondissement level (sources: IGN ADMIN EXPRESS, data.gouv.fr)*

The slightly more flexible rectangular shape can be another option, particularly when geographic entities have rectilinear boundaries, such as the states of the USA or certain urban divisions.

#### 7.2.3.2.3. The rectangular cartogram

The rectangle (see Figure 7.9) recalls the appearance of handmade cartograms on graph paper, or the classic historical reference of Raisz's population map of the United States (1934). Computing it is a challenge: the multiple possibilities of height-width adjustments while respecting surface areas enable the majority of topological relationships to be maintained (Heilmann et al. 2004; van Kreveld and Speckmann 2007).

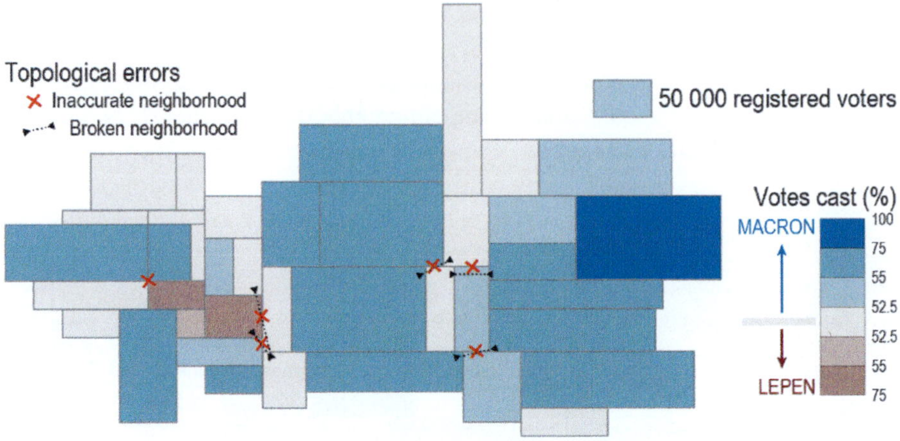


**Figure 7.9.** *Rectangular cartogram. Second round of the 2017 French presidential election. Grand Est region, arrondissement level (sources: IGN ADMIN EXPRESS, data.gouv.fr)*

### 7.2.3.2.4. The mosaic cartogram

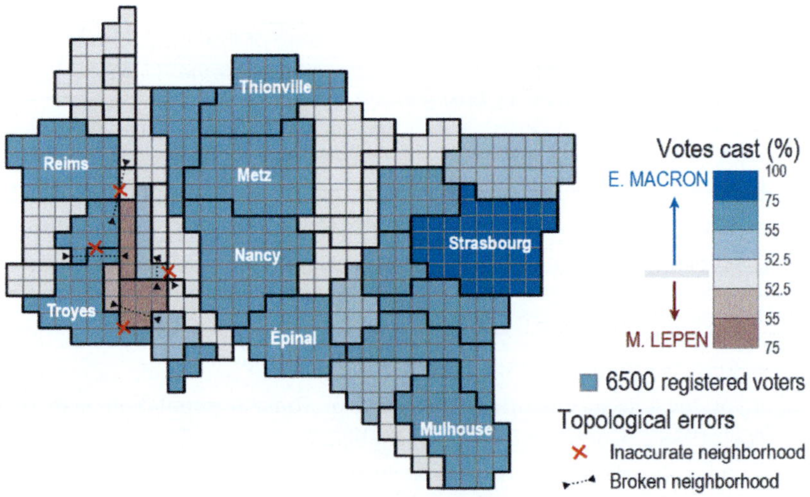


**Figure 7.10.** *Mosaic cartogram. Second round of the 2017 French presidential election. Grand Est region, arrondissement level (sources: IGN ADMIN EXPRESS, data.gouv.fr)*

We suggest using the term mosaic cartogram, after Cano et al. (2015), for the cartogram type based on the repeated symbol principle. A tessellation of unitary squares or hexagons associated with a reference value reshape each spatial unit by assembling the required number of geometric pieces (see Figure 7.10). This process should not be confused with the geoprocessing tools which create rectangular, square or hexagonal grid. The mosaic cartogram can minimize errors in topology, surface and spatial unit shape. It enables us to get a fine approximation of the value of the variable and is quite attractive, as this world population cartogram by Roser (2018), where each tessera represents 1 million people, or the online tool *Tilegrams* (Pitch Interactive), which propose mosaic cartograms for a few territories and variables.

While all the processes described above are designed to represent count quantities associated with areal units, there is a cartographic solution that transforms space using data such as rates and ratios: the piezopleth map method.

### 7.2.4. The piezopleth map to represent rates associated with areal units

The piezopleth map is the result of a cartographic transformation conceived in 1979 by Schneider (Cauvin and Schneider 1989), calculated by *Anaplaste* software (2011). This transfer of a civil engineering model likens the mapped territory to a thin slab of concrete, on which forces (the values of the variable) will exert themselves, causing the material to stretch or shrink. Note, unlike the continuous cartogram (the surface area is proportional to the count data), the transformation reflects deviations from a central tendency parameter (mean or median). The greater the deviation, the more the surface dilates or shrinks. The advantage of the piezopleth map lies in its ability to transform a rate into a cartographic shape. On a given territory, instead of comparing a series of value-varying maps, we look for resemblances between shapes, as in Figure 7.11 between the share of managers in the participation of the labor force and the votes cast for the center-right candidate Emmanuel Macron in 2017.

So, just as projection systems cannot preserve angle, shape and area when flattening the Earth's ellipsoid, a cartogram cannot preserve surface areas, topology and shapes. These methods, which must be carefully selected according to datasets and objectives, offer several ways of reading a territory. Basically, they enable a simple display of the variable distribution, then can analyze territorial inequalities by comparing shapes at different levels of aggregation or themes: for example, a population cartogram versus a wealth cartogram. Finally, there is the possibility of constructing a base map adapted to the phenomenon displayed; this is the choice of commuting the planimetric ground into a thematic reference base map, especially for mapping social, environmental, health datasets.

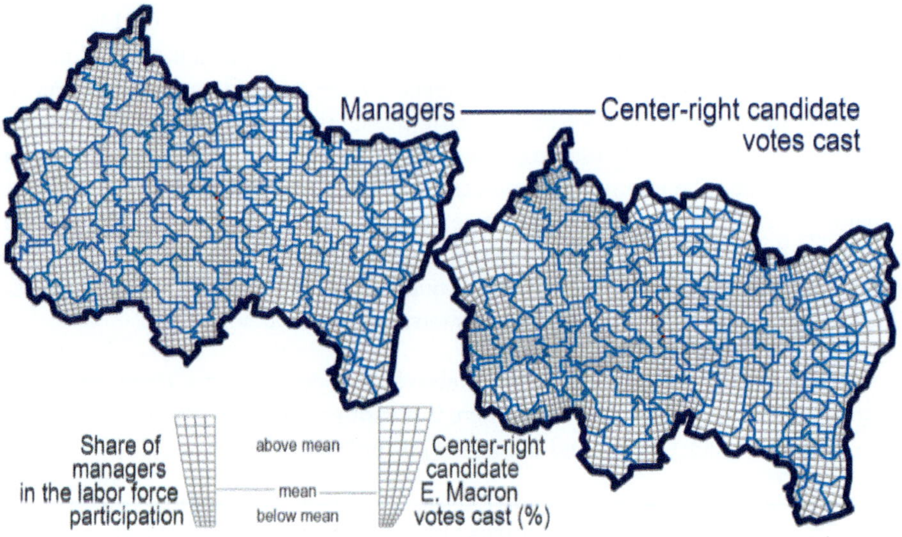


**Figure 7.11.** *Piezopleth map. Comparison of the share of the managers in the participation of the labor force and the votes cast for the center-right candidate Macron in the 2017 French presidential election. Grand Est region, EPCI level (inter-municipal cooperation) (sources: IGN ADMIN EXPRESS, data.gouv.fr, INSEE)*

A second series of cartographic transformations focuses on displaying the distances between places. We shall be focusing in particular on the cartographic representation of time-distances, a small contribution to the complexity of processing and mapping space-time phenomena in geography (for further reading see Chapter 6).

## 7.3. Anamorphic map for the representation of space-time

The world in which we live is an extremely mobile world, where women and men, goods, animals, plants, viruses, etc., get around. If places are generally anchored spatially, their positions can also be defined within these circulations in terms of proximity and distance, by the relationships which link them, and by their junctions. The anamorphic map is one possible response among others for the visualization of these connections, of this relational, functional, practiced, experienced, imagined space. The most common anamorphic map is that of space-time, a configuration where "geographic" distances vanish to display travel times between places[9]. These maps emerged in the 1990s, notably with the work of Cauvin and his Strasbourg team[10] (Cauvin 2005; Cauvin and Reymond 1987) and Spiekermann and Wegener (1993).

These cartographic maps find their place in transportation or mobility studies. Moving de facto includes a temporal dimension (navigation systems often aim to minimize travel time). Representing space-time makes it possible to characterize, analyze and compare existing or simulated networks, spatial practices and representations, to reveal territorial inequalities, to improve the understanding of movements and relationships between places.

We voluntary omit the theoretical framework and research hypotheses underlying the type of space-time studied[11] and everything about the measurement of distance, as well as the various descriptive statistical analyses and maps that can be made from a distance matrix. As with any dataset, the sources must make visible the series of choices that lead us to spatially and temporally reliable and thematically consistent distance measurements (gathered or available datasets). Informing on data collection choices and a minimum understanding of the methodological process are necessary for analyzing, interpreting and communicating results.

It should be noted that other types of thematic distances can be analyzed (cost, social, etc.): as long as this measure is a numerical quantity, the matrix or origin-destination vector can be processed by the methodological proposals presented below. Unlike the cartogram methods seen in the first part of this chapter, which are a relatively direct graphic response (the size visual variable expresses a count data), space-time anamorphic maps articulate several methods that vary according to whether we are interested in:

– on the one hand, what we call multipolar accessibility, which takes into account the complete matrix, all the nodes being treated simultaneously;

– on the other hand, the unipolar accessibility, based on specific nodes, collected specifically or extracted from a distance matrix (row or column).

9 Other names such as tempographic map, time map or chronotopes may appear, without clearly identifying the type of representation they designate.

10 Research by or under the direction of Colette Cauvin, mainly commissioned for public institutions (scientific reports in most of case) and, occasionally, original maps produced on request, e.g., for atlases, the press, school books, etc.

11 A "functional" space where travel time characterizes the transport service? Simulated time-distances for planning proposals? A practiced space, measured, for example, by our digital traces? A "lived", socio-cognitive space, which could be collected by estimated distances between places or mental maps? And so on.

In both cases, the process joins several methods (see Figure 7.12):

– The first aims to calculate coordinate pairs based on distance value, either by nodes from an origin according to the distance (Isodist method, section 7.3.1.2), or by reducing a distance matrix in a two-dimensional space (MDS and related methods, section 7.3.2.1).

– The second, in two steps, (a) compares access time locations with original positions (usually geographic), displayed by displacement vectors; then (b) generalizes these local displacements across the whole studied territory to create an anamorphic map (bidimensional regression, see Box 7.3).

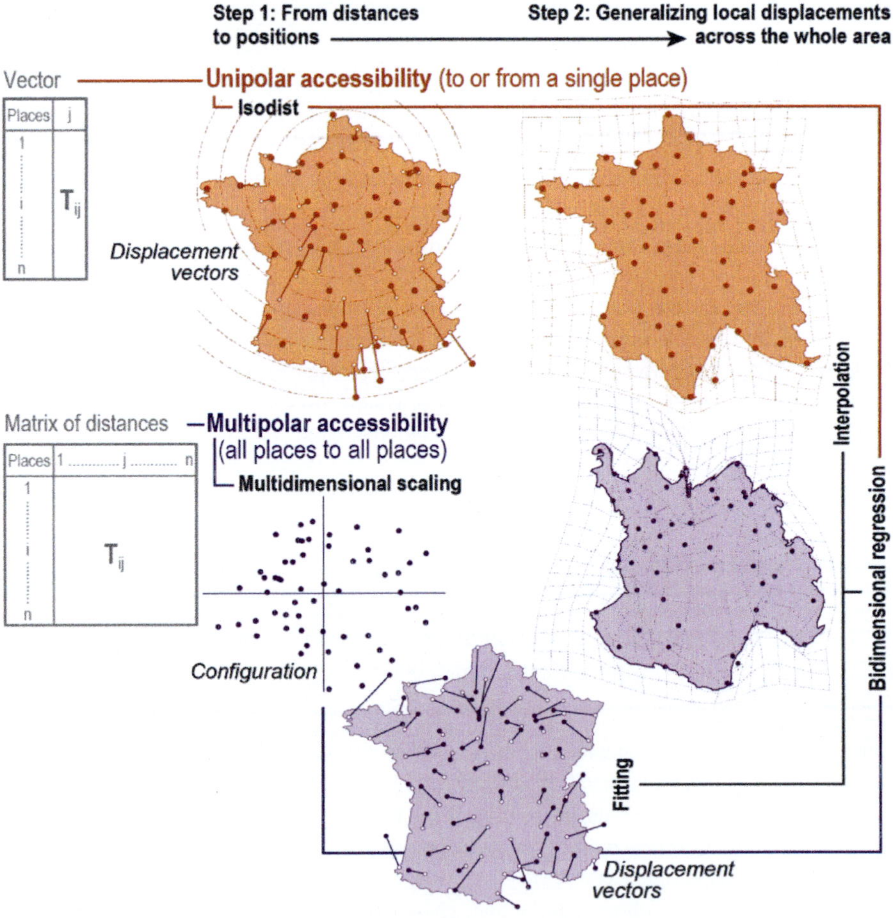


**Figure 7.12.** *From distance values (access time) to anamorphic map: methods (sources: IGN ADMIN EXPRESS, SNCF official timetable 1980)*

We use the latest version of *Darcy* software based on the work of Tobler (1977), developed by Vuidel in Java in 2009. This issue (Version 2.4.2, 2021) incorporates a multidimensional scaling technique and Cauvin's Isodist method (extracted from the *IsoDistAngle* application (Vuidel 2013)); it produces various statistical indices (including measures of error) and (carto)graphical results, to be articulated with other analyses.

Bidimensional regression is a method developed in the 1970s by Tobler (1977), implemented in a program called *Darcy*, in reference to the works of D'Arcy Thompson (1917). Its aim is to compare two shapes, described by homologous point coordinates, and consists of a fit and an interpolation. The cartographic and statistical results characterize the transformations between the original and transformed coordinates: where are the changes located? How significant are they? In which directions is space being transformed? Small differences reveal a probable similarity in the underlying structures behind these shapes.

The author applies his method on several examples: the Gough Map, the growth of a tobacco leaf, a perceived space, two skulls, two drawings of a face. Cauvin (1998, 2002, 2010) extends the applications of the method, in particular for the analysis of functional and cognitive spaces. We use it to study distances[12], but also to analyze sketch maps, a collection of cognitive spatial representations. Cauvin uses the term "differential cartographic transformations of position" in her classification of CTPs, and speaks of model and template. In various application fields, it requires us to redefine the links between geographic hypotheses and to reinterpret the input and output of the entire process of transformations data to treatments and then results.

**Box 7.3.** *Bidimensional regression*

The simplest situation in terms of cartographic processing is to work on a single vector (one origin to several destinations or several origins to one destination).

### 7.3.1. *Accessibility to a place (unipolar accessibility)*

Unipolar accessibility is easy to grasp. It involves answering questions such as: What places can I access from where I am? What is the service area of this resource place? Answers will of course vary in time and space, and according to the networks taken into account.

12 In this research context, bidimensional regression compares two series of homologous points describing: geographical space versus space-time, space-time at time *t* versus space-time at time *t-1* or *t+1*, space-time measured on a network versus space-time derived from individual or collective practices versus space-time estimated by individuals or groups.

#### 7.3.1.1. *Isoline and anamorphic maps: two complementary maps*

The variable is a set of values attached to pairs of points: measured, simulated, practiced access times, etc., expressed in minutes, hours, days, etc. The most efficient visualization for communicating information is an isoline map of time-distances (via an interpolation technique), which requires measurement points covering the whole studied territory (see Figure 7.13).

If the transport network is homogeneous and isotropic, the further you go, the longer the access time, and the isoline pattern is more or less circular: this is the case for medium and small-scale road and pedestrian time-distances. Less regular shapes can be interpreted just as easily: indentations highlight corridors of greater accessibility, while curvatures mark peripheral and landlocked areas (e.g. the Rhône corridor versus mountainous areas in France).

France's railway network is centered on Paris, with very few transverse connections: the 1980 map shows territorial imbalances that will become even more marked as the high-speed train network is becoming important. If outlying towns (Nice, Brest, etc.) are necessarily far from other places, what about Bourges, which is nevertheless close to Paris?

Time maps, travel time maps and other isochronic maps[13] are not new. F. Galton's *Isochronic Passage Chart for Travellers*, designed in 1881, is often cited. As for the anamorphic map, we propose two references: Cheysson's 1889 map of the acceleration of travel in France, with a rather metaphorical point of view; Bunge's 1960 isometric map of Seattle referenced in Tobler's (1961) thesis in a more analytical approach, which also enables us to understand the passage from planimetric space to space-time. This passage is illustrated in Figure 7.13, with the cartographic transformation of irregular isolines (top maps) into circular isochrones (bottom maps).

#### 7.3.1.2. *Unipolar anamorphic map: Isodist and bidimensional regression interpolation*

A first mathematical method, Isodist, developed in 1991 by Cauvin (2010), moves geographic locations using a vector calculation that takes into account the value of the link (in our case, travel time) to or from an origin place.

---

13 Various software, plugins or web applications implement computational algorithms to display isodistances or isochrones based on a network, e.g., the ISO4APP application (K-SOL).

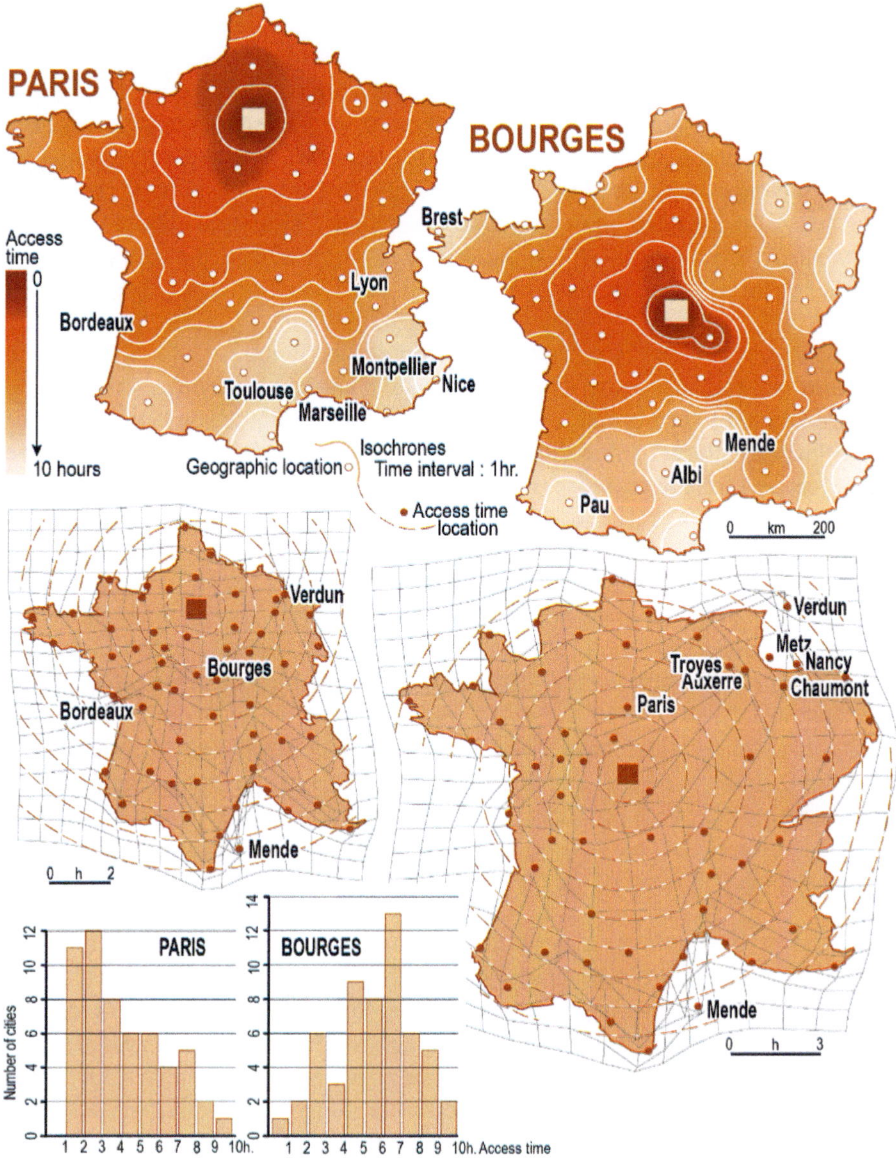


**Figure 7.13.** *Two representations of unipolar accessibility by train in 1980: isopleth map and anamorphic map (sources: IGN ADMIN EXPRESS, SNCF official timetable 1980)*

An equivalence coefficient (see Figure 7.14, left), which can be a calculated[14] or fixed average speed (in a comparative analysis requiring the same space-time), is used to bring access times T (expressed in minutes, hours, days, etc.) and distances (metric, kilometric… unit) into the same reference frame. Subtracting distance DT from distance D computes the length of the displacement vector (the norm) and its direction (the sign of the norm). The angle formed with the position of the origin defines the direction of the displacement. This calculation transforms the "geographical" location into an access time location as it moves closer or farther away from the origin.

Isodist produces various results, as shown in Figure 7.14: access time coordinates; several maps displaying isochrones (visualizing the map's new scale: time); displacement vectors and access time positions, distinguishing between places that are "moving closer" (time-distance values lower than those obtained with the equivalence coefficient) and places that are "moving away" (time-distance values higher than those obtained with the equivalence coefficient). The contour line and initial locations are the traces of the data input.

The interpolation[15], second step of the bidimensional regression, generalizes local displacements to each point in the territory, providing a continuous representation of the phenomenon. The vector field and map of deformation forces at the bottom of Figure 7.15 are additional results to the anamorphic map, and illustrate this interpolation phase. They visualize displacements (vector orientations and lengths), on the one hand, and highlight movements of convergence (surfaces in green tones, where areas contract on the anamorphic map) or movements of repulsion (surfaces in red tones, corresponding to less accessible mountain territories, which will expand), on the other.

The transformed grid is a decorative feature to improve the analysis of the transformation, and, more particularly, to identify inversions in geographical space: areas that are close in planimetric terms moving "behind" more distant places. For example, Mende, not very accessible by train, jumps into the Mediterranean Rim, on the south of the Mediterranean coastal towns. If we return to Figure 7.13, which compares Paris and Bourges, the latter, less accessible overall, presents a dilated space-time and appears poorly connected to its eastern neighbors.

---

14 Calculated on a vector or matrix, speed equals the sum of access time values divided by the sum of distances. Distances are calculated from coordinates, or supplied when the planimetric reference frame is not appropriate.

15 "Average" displacements are calculated at each node of a weighted grid, and extended to all map features (map outline, network, etc.).

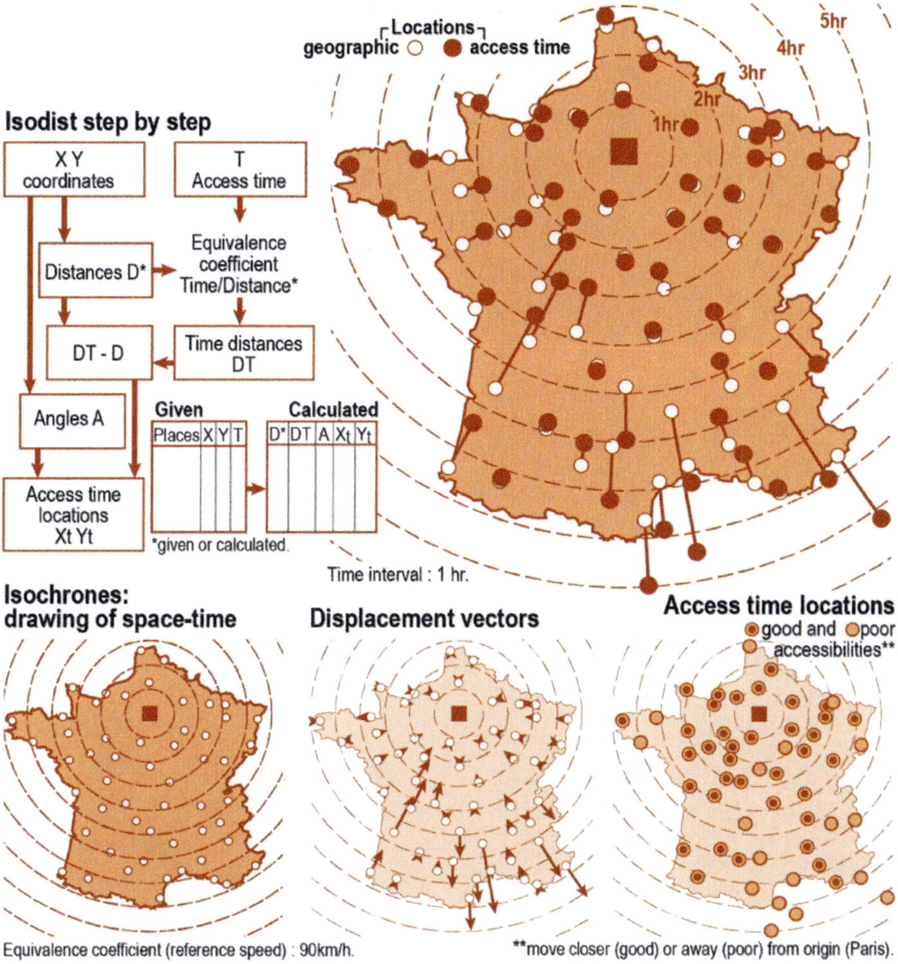


**Figure 7.14.** *From geographical coordinates to access time locations. Accessibility by train to Paris in 1980 (sources: IGN ADMIN EXPRESS, SNCF official timetable 1980)*

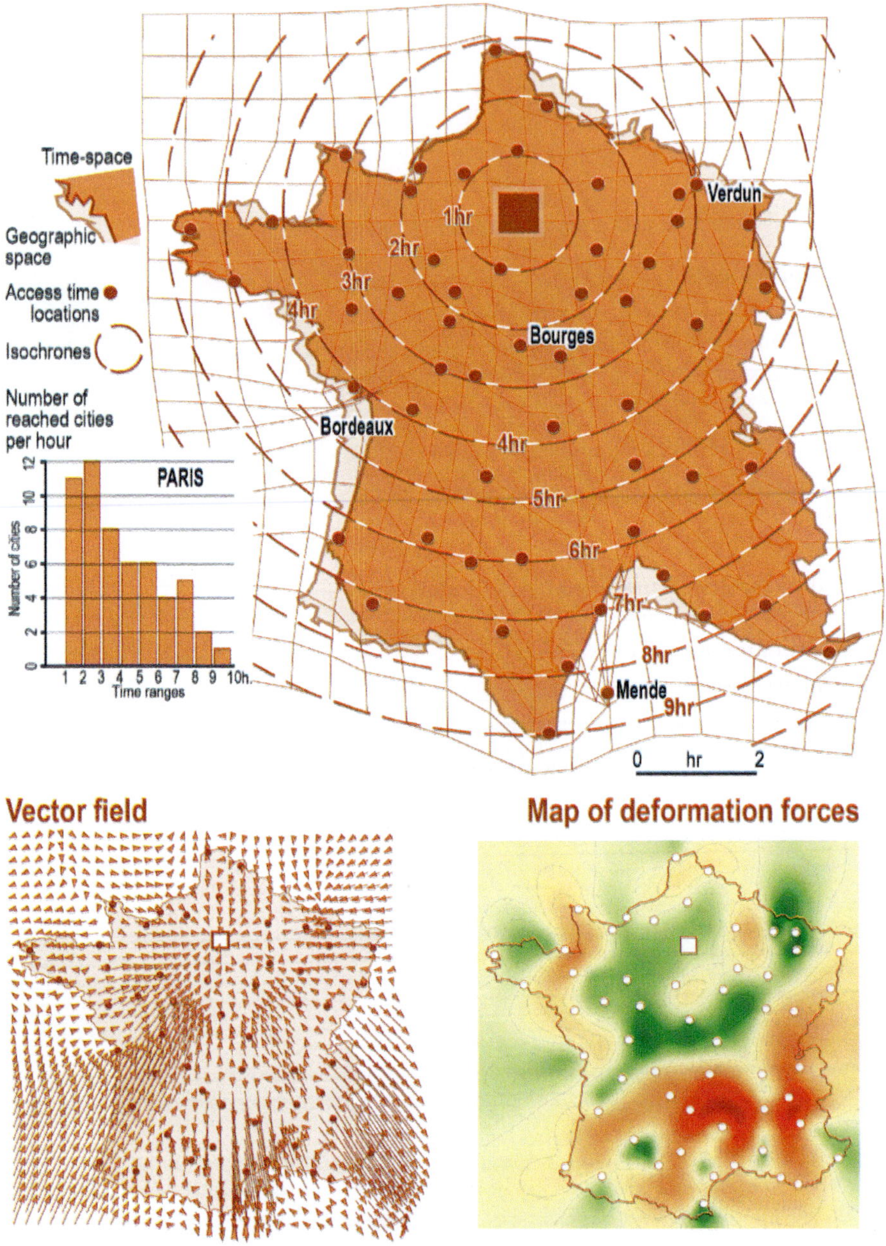


**Figure 7.15.** *Unipolar anamorphic map. Accessibility by train to Paris in 1980 (sources: IGN ADMIN EXPRESS, official SNCF timetable 1980)*

Alongside accessibility targeted at one place, it is interesting to treat time-distances from all places to all places simultaneously to create a global vision.

### 7.3.2. *Accessibility between all places (multipolar accessibility)*

Eminently more complex to read (or even design), the multipolar anamorphic map operates on the complete matrix. The cartographic transformation, based on a relative positioning of places derived from time-distances, brings together places that are better connected to each other and to the whole territory, and moves away, generally on the margins, peripheral or enclosed nodes that are less accessible or poorly connected.

#### 7.3.2.1. *Reducing a distance matrix into two dimensions: MDS*

Multidimensional scaling (MDS) identifies various statistical techniques that analyze distances (Groenen and Borg 2015; Mair et al. 2019). Used in genetics, psychology, sociology and geography, MDS is intended to "reduce" to two (or even three) dimensions an n X n matrix of distances (in our case, the origin-destination matrix of measured time-distances between spatially localized places). The aim is to obtain a configuration of points whose distances correspond to the values in the initial matrix, or at least are as close as possible to these input values. Today, the algorithms available could be sorted by type into classic, metric, and non-metric MDS, to be used in relation to the meaning of the data to be processed.

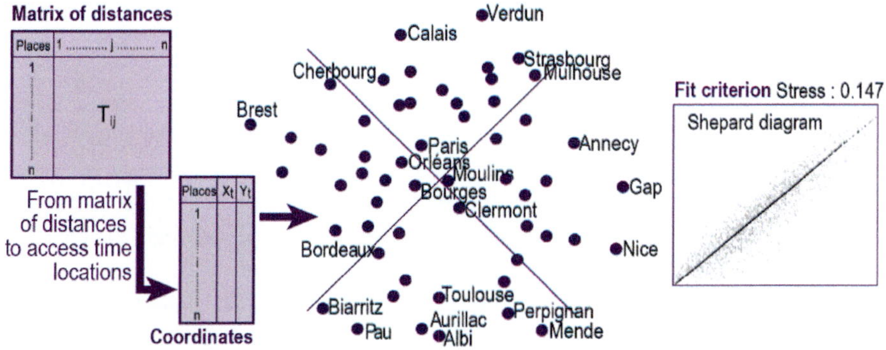


**Figure 7.16.** *Multidimensional scaling (sources: IGN ADMIN EXPRESS, SNCF official timetable 1980)*

Whether using MDS or other techniques intending to do the same issue (Cauvin et al. 2010), the results can be visualized in a configuration associated with, depending on the method, a fit criterion (GoF, stress, R2, etc.). Figure 7.16 shows a configuration derived from the 1980 matrix of time-distance by train, the result of a non-metric MDS (Kruskal 1964; Mair et al. 2019; Shepard 1957), an iterative process that configures locations by minimizing stress, a fit criterion calculated on the basis of the distances (between positions in the configuration) and proximities (input values, i.e. time-distances here).

The Shepard diagram relates these two dimensions: distance and proximity. The validity of the configuration is linked to the shape of the scatter of points (linear pattern) and the stress value (tends towards 0).

#### 7.3.2.2. *From access-time locations to space-time map: bidimensional regression*

To create the anamorphic map, the two steps of the bidimensional regression are processed. The fitting, detailed in Figure 7.17, brings the MDS aspatial configuration (multipolar space-time) back into a reference space (geographic space in general): it combines translation, rotation and scaling. Statistical indices qualify the quality of the fit (R2) and reveal the directions and average displacement values expressed in the map unit (RMSE). The map of displacement vectors highlights the pattern discrepancies between the planimetric locations and the access time positions derived from the fit process. In 1980, there was no high-speed railway in place, apart from the 200 km/h link between Strasbourg and Mulhouse: only the railway network structure centered on Paris explains the spatial variations.

The space-time anamorphic map resulting from the interpolation – step 2 of the bidimensional regression already mentioned in the section on unipolar accessibility – reveals the underlying structure of the transport network. The advantage is that we can compare different networks (by mode or over time) using statistical indices, anamorphic maps and the other visualizations derived from processing (vector field, map of deformation forces, etc.). The successive development of the French railway network can be seen in Figure 7.18: the reinforcement of spatial discrimination via the high-speed links, which coexist with conventional and regional railways, the neglect of cross-country links, which were already underdeveloped, and how intercity trains are disrupting the territory in terms of accessibility. The grid reveals spatial inversions, with central locations in the geographical space increasingly marginalized on the space-time map.

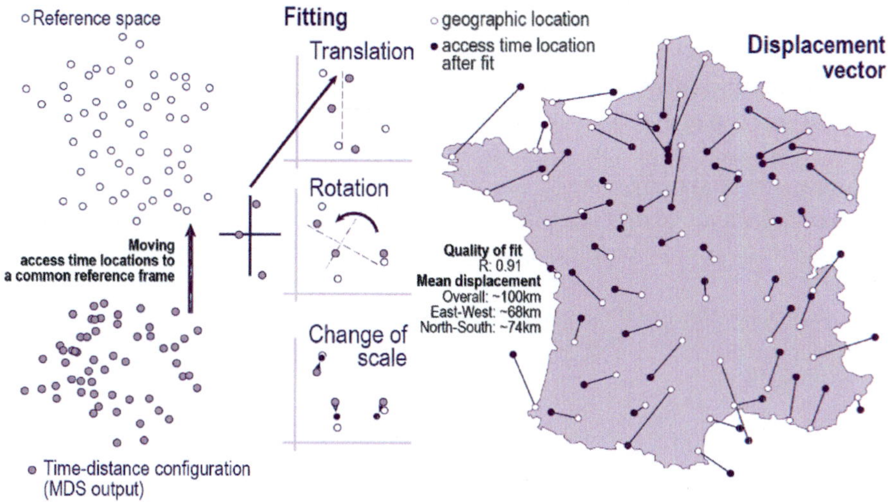


**Figure 7.17.** *Multidimensional regression: fitting. Time-distances by train, 1980. Multipolar accessibility (sources: IGN ADMIN EXPRESS, SNCF official timetable 1980)*

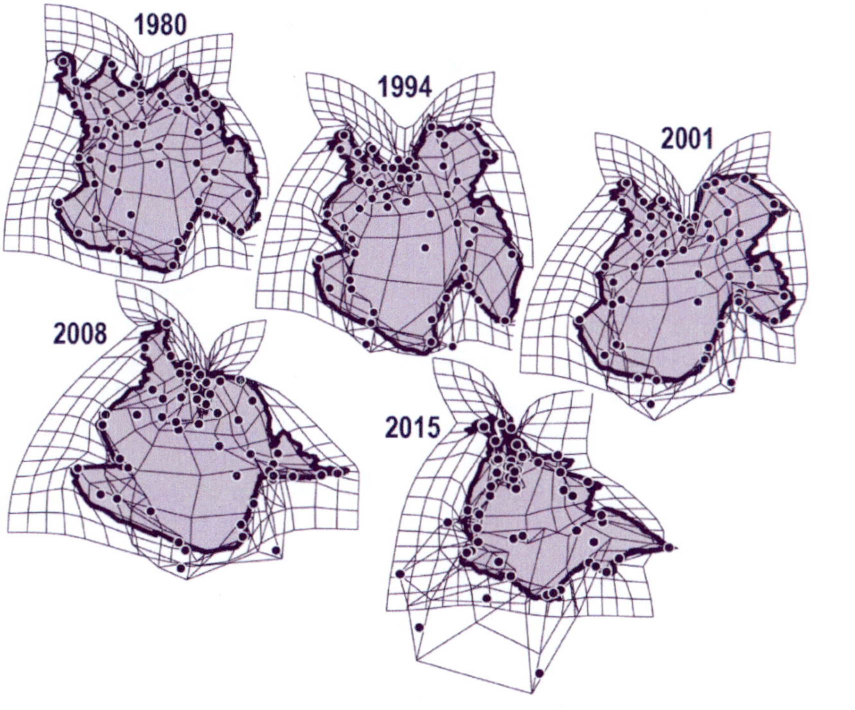


**Figure 7.18.** *Multipolar anamorphic map. Railway functional space (sources: IGN ADMIN EXPRESS, SNCF official timetable 1980, 1994, 2001, cfl.lu 2008, 2015)*

The rendering of anamorphic maps based on distances thus generates a large number of cartographic outputs, statistical indices and quality measures, facilitating the validation of results and the management of the propagation of error calculations in the methodological combination. Space-time maps can be used to reveal the spatial organization of different modes of transport or the evolution of a network over time, to highlight territorial inequalities, to compare different spaces (geographical, time, cost, cognitive, etc.) and to create an alternative background map if required, with the positioning of places based on their accessibility and links.

To complete this overview of methods, and in connection with the idea of distance, we will say a few words about azimuthal projections, also known as fisheye projections.

### 7.3.3. *Azimuthal transformation*

Azimuthal transformation or fisheye projection is best illustrated by references to M.C. Escher's *Balcony* in the field of art, *Le plan Morant* of 1548 for old maps or Hägerstrand's *Asby map* (Hannerberg and Hägerstrand 1957) in cartography. A mathematical function transforms the geographical shape from a single place: Fischauge-Projektionen (Rase 1997), azimuthal projection or transformation (Tobler 1959; Cerny 1971; Boutoura 2006) and variable-scale map projections (Fairbairn and Taylor 1995). In our view, these methods have several advantages.

When the function used adjusts a set of time-distances, the map can be likened to a spatial smoothing, a generalization of space-time (Cauvin 1997; Cauvin et al. 2010). Note that only this type of application classifies these methods in the anamorphosis cartographic system of correspondence and transformations between XY and Z, where the thematic variable is involved in calculating new territorial shapes. In other cases, the function serves to focus on an area for better visualization and understanding of a spatial phenomenon, a purpose close to the idea of the continuous zoom lens developed to manage the concentration of geographical objects, increasing readability by dilating areas of high graphic density (Rase 1981, 1997; Boutoura 2006). This makes it possible to map at several scales while maintaining spatial continuity (an alternative to inset maps). Transformation also serves to illustrate an egocentric point of view, or a metaphorical view of our accelerated connection to distant places, to the detriment of neighborhood relationships.

In Figure 7.19, an illustration of the method that returns to the example of the second round of the 2017 presidential election in the Grand Est region, the continuous zoom on the Haute-Marne area highlights the electoral behavior of a small rural municipality with de facto higher rates, since the number of inhabitants, and therefore votes cast, is smaller.

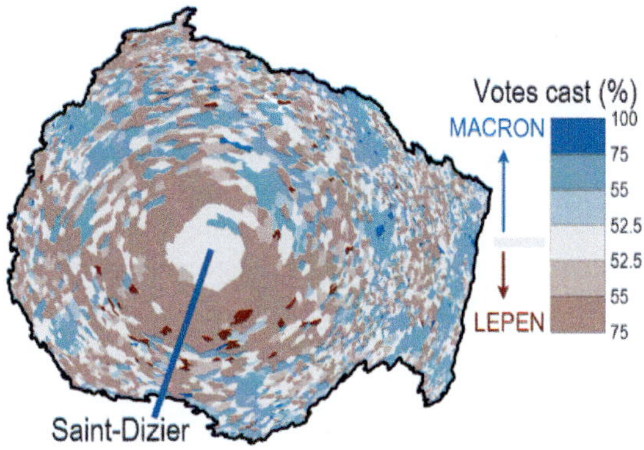


**Figure 7.19.** *Fisheye projection or azimuthal transformation. Second round of 2017 French presidential election. Grand Est region, municipalities (sources: IGN ADMIN EXPRESS, data.gouv.fr)*

Bringing the methods behind the most usual anamorphic maps out of the shadows will, we hope, enable us to better articulate them with each other and with other treatments of spatial data, in particular to multiply cross-views and improve research. In the final section, we propose a number of additions to take a closer look at various aspects of the construction or difficulty of reading and interpreting these maps.

## 7.4. Anamorphic maps, cartograms: cross-cutting reflections on common principles and reading difficulties

Many of these methods are based on different principles, rules or assumptions.

### 7.4.1. *Some principles underlying all methods*

A few common features run through the majority of anamorphic maps.

#### 7.4.1.1. *Continuity of the space*

On a continuous cartogram, an anamorphic map derived from a matrix of distances, or a piezopleth map, the territory appears transformed at each point. By mobilizing this type of method, we agree on a continuous vision of the phenomenon.

While spatial continuity is unquestionable for various physical phenomena such as altitude, atmospheric phenomena (temperature, precipitation, air pollution, etc.), it will be discussed with the thematic cartographer for the representation of accessibility or population distribution.

Spatial continuity requires a sufficient number of representative measures distributed all over the studied area (with no missing data). The dataset used to create an anamorphic map based on distances must cover the whole territory, including remote or underserved areas; any selection operations, such as choosing the best performing links, are problematic. Similarly, a continuous cartogram will have low accuracy if some of the spatial units are not affected by the phenomenon: the iterative process of the algorithm that stretches areas cannot reduce a geographical entity to zero.

#### 7.4.1.2. *Link between form and structure*

The use of cartographic transformations of position requires agreement with a postulate derived from D'Arcy Thompson's *On Growth and Form* (1917), and quoted by Cauvin et al. (2010) as follows: "Form is an expression of a phenomenon's structure. If we are able to characterize a form or compare two forms, then by means of a form we can reveal an invisible structure and compare different structures to each other".

Territories transformed by the thematic variables (a weight, a distance, a quantity, etc.) reveal the underlying structures and processes. By comparing shapes, we can illustrate and analyze the evolution of the railway network over time, territorial organization for different transport modes, and the processes leading to unequal distribution of populations, wealth, health, etc.

#### 7.4.1.3. *Reproducibility*

The majority of anamorphic maps are the result of mathematical processes, the result of a more or less complex algorithm that simultaneously takes into account location and the associated thematic value. These cartographic transformations calculate new locations and shapes, and generally produce a measure of quality (allowing the validation of the map and to manage propagation of error calculations).

Other common features can be found in anamorphic maps.

### 7.4.2. Anamorphic maps layout

Anamorphic maps are characterized by the vanishing of the planimetric background, although it is possible to maintain an inset map or watermark of the original space. The grid is an important graphic element in reading the transformations. Although place names hinder the map's readability and attractiveness, they are necessary for interpreting the relative positioned locations. The map surround includes data sources, the names of specific methods, and, if needed, the names of the geotools used to ensure the reproducibility. If space is available, a methodological inset can be added and, as a map never navigates on its own, interpretation keys appear in the notice or commentary.

#### 7.4.2.1. *Legend, scale and the north*

In the anamorphosis system, the scale often becomes a specific element of the map legend, as the theme is embodied in the spatial dimension. The (plani)metric scale is replaced by:

– a time scale on the unipolar anamorphic map, often only drawn by isochrones;

– reference surface areas for the cartogram, which are often forgotten, especially when the cartogram is used as a base map.

The northern direction is not very relevant and can be neglected.

#### 7.4.2.2. *Anamorphic map as map projection*

The continuous cartogram is often used as a map projection to reshape the base map, whether for mapping rates (see Figure 7.2) or overlaying thematic layers, offering the possibility of improving data exploration and analysis by combining information, such as flood zones and population (Andrieu 2014). This set of graphical outputs also enables aesthetically interesting work using the graphical plot density or the grid to display the distribution of the studied phenomenon (see maps by Dutenkefer (2018)).

For techniques that do not incorporate a change of spatial base (circle, square, hexagon, tessellation), a limited generalization of the initial base map outlines, which are rarely processed in general, and for cartograms in particular, can reduce computational time, file size and graphic density, simplify shapes and improve readability. The input spatial dataset will respect the surface areas of the territories (equivalent projection system).

Despite all the attention paid to the map layout, there are some reading biases specific to anamorphic maps.

### 7.4.3. *Reading and understanding anamorphic maps: some difficulties*

While complex to understand, anamorphic maps are particularly attractive in terms of communication, as they combine two perceptually effective visual variables: size and shape variations, which are said to be more operative than color, value, texture, grain or orientation variations. Indeed, the mapped data will cause dilations, contractions, stretches, constrictions, swellings, shrinkages, etc., of space, which reshape the territory and appeal to readers. However, as with any geodata visualization, the message can escape the addressees. We identify three difficulties in particular[16]:

– Losing one's spatial marks when faced with an anamorphic map of a familiar territory is a classic. Part of the reading time is spent looking for links between familiar geographical features and transformed ones, instead of evaluating surface areas or proximities between places. A display of the original outline or nearby, place names, the watermark of the grid revealing the transition between reference and thematic space, or even the animation between original and transformed space can partially resolve this obstacle to map comprehension. One important point that could not be addressed was that of cartographic animations. They are particularly appropriate to visualizing the transition between two shapes, working with longitudinal data, understanding the transformations between geographic and thematic space or between two thematic spaces, showing the evolution of the world's population or the acceleration of travel times, and so on.

– A second difficulty in map interpretation arises from the widespread use of cartograms as map projection onto which a second variable is plotted – a practice we have deliberately reproduced in this chapter by mapping election results onto cartograms based on the number of registered voters. The variation in shape/surface area is combined with a color scheme, forcing a double reading. Any representation that gathers several variables, such as the controversial use of proportional circles for count data, with a second thematic variable of the ratio or category type inside the circles, creates the same problem of twofold interpretation.

– Another reading error is linked to visual perception and cartographic conventions. By convention, dark areas represent the highest values, light areas the lowest. This is why some people spontaneously attribute to the areas perceived as dark on the map the largest shares of the phenomenon represented and vice versa. However, on some cartograms, the dark areas created by graphic concentration are the least affected by the phenomenon represented.

16 Reading difficulties identified by the author when presenting her work at scientific events open to the public or during cartography training courses.

– Finally, all other things being equal, understanding of the anamorphic map varies according to the extent of the map. Indeed, on a global scale, the continuous cartogram represents a diagram of the distribution of the large masses of the studied phenomenon, and is fairly well read. This is also the case at continental, sub-continental, national and even regional scales. At these scales, depending on the level of detail, the continuous cartogram will gradually reveal the location and weight of metropolitan areas, underpinning the opportunities of visualizing the dynamics of metropolization if longitudinal datasets are available. Spatial units little affected by the phenomenon will be compressed and appear as axons squeezed between two masses or at the periphery of the map. On a local scale, even when the territory is known, the heterogeneity of the spatial entities creates a map that is often less legible. Indeed, while the spatial pattern of a phenomenon in the form of poles and large, continuous areas with few inhabitants is present at the regional level, on a metropolitan scale, the map will show a more dilated central area, with sparsely populated peripheral municipalities reduced to filaments. At the city level, the cartogram reveals the high variability of urban space.

More than a collection of cartograms or anamorphic maps, we have attempted a non-exhaustive panorama of methods, hoping to have provided the keys not only to reading and understanding them, but also the spirit and general framework for producing them. Like all maps, cartograms or anamorphic maps are transformations of reality and form part of a scientific problem, whether in the descriptive, exploratory, predictive or confirmatory phase of the research being carried out.

## 7.5. Conclusion

The aim of this chapter is to anchor the specificity of the anamorphosis cartographic system, a transformation of territories drawn by a thematic variable, which intervenes “directly” in the processing, the map “embodying” the method; it is the product of an equation that links spatial dimension XY and variable Z: $X = f(Z)$ and $Y = g(Z)$.

Work with anamorphic maps feeds discussions around the question of the base map: a planimetric space, an absolute positioning of places, necessary to survey, measure, describe and understand the visible, versus a thematic-geographic space, built on a relative positioning of geographical entities. On a global scale, it is hard to get rid of the immutable reference of Mercator’s planisphere (see Chapter 1), even if the variety of projection systems underlines the fact that the world space has multiple shapes. This shows how difficult it is to transform our planimetric base map, the geographic locations, the shapes of territories, to establish new relationships between mapped entities, which is nevertheless central when working with sociodemographic data (Ballas et al. 2014) or on practiced or "lived" territories. The difficulty lies in the fact that interpreting the results requires a minimum understanding, not only of the data analyzed, but also of the methods behind the maps, which we have attempted to clarify in this chapter.

Although we have used the term metaphor several times in this chapter, we hope to have taken anamorphic map out of its trivial assignation as a mere communication product, used only for its visual impact. The primary use of these methods is to reveal underlying structures, processes and dynamics via the shape of territories.

## 7.6. References

Andrieu, D. (2014). Les cartogrammes : une séquence d'enseignement de la sémiologie à la géographie. *Carte & géomatique*, 222, 123–138 [Online]. Available at: www.lecfc.fr.

Andrieu, D., Kaiser, C., Ourednik, A. (2008). ScapeToad [Online]. Available at: scapetoad.choros.place/.

Ballas, D., Dorling, D., Hennig, B. (2014). *The Social Atlas of Europe*. Policy Press, Bristol.

Bertin, J. (2011). *Semiology of Graphics: Diagrams, Networks, Maps*. Translated by W.J. Berg (University of Wisconsin, 1983). ESRI Press, Redlands, CA.

Bortins, I., Demers, S., Clarke, K.D. (2002). Cartogram types [Online]. Available at: ncgia.ucsb.edu/projects/Cartogram_Central/types.html.

Boutoura, C. (1994). Logarithmic urban thematic mapping in MIS environment. *Cartographica: The International Journal for Geographic Information and Geovisualization*, 31(1), 41–53. doi: 10.3138/C0XG-1260-5808-34HX.

Bunge, W. (1966). *Theoretical Geography*. Gleerup, Lund.

Cano, R.G., Buchin, K., Castermans, T., Pieterse, A., Sonke, W., Speckmann, B. (2015). Mosaic drawings and cartograms. *Computer Graphics Forum*, 34(3), 361–370. doi: 10.1111/cgf.12648.

Cauvin, C. (1997). Au sujet des transformations cartographiques de position. *Cybergeo: European Journal of Geography*. doi: 10.4000/cybergeo.5385.

Cauvin, C. (1998). Usagers et représentations cognitives de la ville : exemples à Stras-bourg. In *L'espace géographique des villes : pour une synergie multistrates*, Cauvin, C. and Kleinschmager, R. (eds). Anthropos-Economica, Paris.

Cauvin, C. (2002). Cognitive and cartographic representations: Towards a comprehensive approach. *Cybergeo: European Journal of Geography*. doi: 10.4000/cybergeo.194.

Cauvin, C. (2005). A systemic approach to transport accessibility. A methodology developed in Strasbourg: 1982-2002. *Cybergeo: European Journal of Geography*. doi: 10.4000/cybergeo.3425.

Cauvin, C. and Schneider, C. (1989). Cartographic transformations and the piezopleth maps method. *The Cartographic Journal*, 26(2), 96–104. doi: 10.1179/caj.1989.26.2.96.

Cauvin, C., Lepetit, B., Reymond, H. (1987). Cartes postales : un espace de relation dans la France pré-industrielle. *Histoire de la mesure*, II(3/4), 89–114.

Cauvin, C., Escobar, F., Serradj, A., Antoni, J.-P. (2010). *Cartography and the Impact of the Quantitative Revolution. Thematic Cartography - Volume 2*. ISTE Ltd, London, and John Wiley & Sons, New York.

Cerny, J.W. (1971). A computer program for azimuthal map transformations. *The Professional Geographer*, 23(2), 144–145. doi: 10.1111/j.0033-0124.1971.00144.x.

Cornillon, P.-A. and Demoraes, F. (2021). Package cartogramR: User guide [Online]. Available at: https://hal.archives-ouvertes.fr/hal-03280009.

D'Arcy Thompson, W. (1917). *On Growth and Form*. Cambridge University Press, Cambridge [Online]. Available at: www.gutenberg.org/files/55264/55264-h/55264-h.htm.

Dorling, D. (1991). The visualization of spatial social structure. PhD Thesis, University of Newcastle upon Tyne, Newcastle upon Tyne [Online]. Available at: www.dannydorling.org.

Dorling, D. (1996). Area cartograms: Their use and creation. CATMOG: Concepts and Techniques in Modern Geography 59, Department of Geography, University of Durham [Online]. Available at: www.dannydorling.org.

Dorling, D. (2007). Worldmapper: The human anatomy of a small planet. *PLoS Medicine*, 4(1), e1. doi: 10.1371/journal.pmed.0040001.

Dougenik, J.A., Chrisman, N.R., Niemeyer, D.R. (1985). An algorithm to construct continuous area cartograms. *The Professional Geographer*, 37(1), 75–81. doi: 10.1111/j.0033-0124.1985.00075.x

Dutenkefer, E. (2018). A cidade e o mapa: representações cartográficas da urbanidade de São Paulo. PhD Thesis, University of São Paulo, São Paulo [Online]. Available at: teses.usp.br/teses/disponiveis/8/8136/tde-04072018-123123/publico/2018_EduardoDutenkefer_VCorr.pdf.

Fairbairn, D. and Taylor, G. (1995). Developing a variable-scale map projection for urban areas. *Computers & Geosciences*, 21(9), 1053–1064.

Gastner, M.T. and Newman, M.E.J. (2004). Diffusion-based method for producing density-equalizing maps. *Proceedings of the National Academy of Sciences*, 101(20), 7499–7504. doi: 10.1073/pnas.0400280101.

Gastner, M.T., Seguy, V., More, P. (2018). Fast flow-based algorithm for creating density-equalizing map projections. *Proceedings of the National Academy of Sciences*, 115(10), E2156–E2164. doi: 10.1073/pnas.1712674115.

Giraud, T., Hernangómez, D., Pecout, H. (2021). *mapsf: Thematic Cartography* [Online]. Available at: CRAN.R-project.org/package=mapsf.

Groenen, P.J.F. and Borg, I. (2015). Multidimensional scaling II. In *International Encyclopedia of the Social & Behavioral Sciences*, 2nd edition, Wright, J.D. (ed.). Elsevier, Oxford. doi: 10.1016/B978-0-08-097086-8.42167-7.

Hannerberg, D. and Hägerstrand, T. (eds) (1957). *Migration in Sweden, a Symposium*. CWK Gleerup, Lund.

Heilmann, R., Keim, D.A., Panse, C., Sips, M. (2004). RecMap: Rectangular map approximations. In *IEEE Symposium on Information Visualization*, Austin. doi: 10.1109/INFVIS.2004.57.

Henriques, R., Bação, F., Lobo, V. (2009). Carto-SOM: Cartogram creation using self-organizing maps. *International Journal of Geographical Information Science*, 23(4), 483–511. doi: 10.1080/13658810801958885.

House, D.H. and Kocmoud, C.J. (1998). Continuous cartogram construction. In *Proceedings Visualization '98*. IEEE, Research Triangle Park. doi: 10.1109/VISUAL.1998.745303.

Jeworutzki, S., Giraud, T., Lambert, N., Bivand, R., Pebesma, E., Nowosad, J. (2020). Cartogram: Create cartograms with R [Online]. Available at: CRAN.R-project.org/package=cartogram.

K-SOL S.r.l. (n.d.). ISO4APP API – Isochrones map (travel time map) and isodistances map [Online]. Available at: www.iso4app.net.

Kadmon, N. and Shlomi, E. (1978). A polyfocal projection for statistical surfaces. *The Cartographic Journal*, 15(1), 36–41. doi: 10.1179/caj.1978.15.1.36.

Keim, D.A., North, S.C., Panse, C. (2004). Cartodraw: A fast algorithm for generating contiguous cartograms. *IEEE Transactions on Visualization and Computer Graphics*, 10(1), 95–110. doi: 10.1109/TVCG.2004.1260761.

van Kreveld, M. and Speckmann, B. (2007). On rectangular cartograms. *Computational Geometry*, 37(3), 175–187. doi: 10.1016/j.comgeo.2006.06.

Kruskal, J.B. (1964). Multidimensional scaling by optimizing goodness of fit to a nonmetric hypothesis. *Psychometrika*, 29, 1–27.

Lambert, N. (2020). COVID-19 : cartographie(s) d'un débordement. *Carnet (néo)cartographique* [Online]. Available at: neocarto.hypotheses.org/10229.

Langlois, P. (2003). Anamorphose par transformations pseudo-équivalentes. In *6^e^ Rencontres de Théo Quant*, Besançon [Online]. Available at: thema.univ-fcomte.fr/theoq/pdf/2003/TQ2003%20ARTICLE%2022.pdf.

Mair, P., Groenen, P.J.F., de Leeuw, J. (2019). More on multidimensional scaling and unfolding in R: smacof version 2 [Online]. Available at: cran.r-project.org/web/packages/smacof/vignettes/smacof.pdf.

Markowska, A. (2019). Cartograms – Classification and terminology. *Polish Cartographical Review*, 51(2), 51–65. doi: 10.2478/pcr-2019-0005.

Nickel, S., Sondag, M., Meulemans, W., Chimani, M., Kobourov, S., Peltonen, J., Nöllenburg, M. (2019). Computing stable Demers cartograms. In *Proceedings of the 27th International Symposium on Graph Drawing and Network Visualization (GD 2019)*, Průhonice/Prague. doi: 10.48550/arXiv.1908.07291.

Nusrat, S. and Kobourov, S. (2016). The state of the art in cartograms. *Computer Graphics Forum*, 35(3), 619–642. doi: 10.48550/arXiv.1605.08485.

Nusrat, S., Alam, M.J., Kobourov, S. (2018). Evaluating cartogram effectiveness. *IEEE Transactions on Visualization and Computer Graphics*, 24(2), 1077–1090. doi: 10.48550/arXiv.1504.02218.

Olson, J.M. (1976). Noncontiguous area cartograms. *The Professional Geographer*, 28(4), 371–380. doi: 10.1111/j.0033-0124.1976.00371.x.

Pitch Interactive (n.d.). Tilegrams [Online]. Available at: pitchinteractiveinc.github.io/tilegrams/.

Raisz, E. (1934). The rectangular statistical cartogram. *Geographical Review*, 24(2), 292–296. doi: 10.1111/cgf.12932.

Rase, W.-D. (1981). Computer graphics for federal planning in West-Germany. In *EUROGRAPHICS' 81*, Darmstadt [Online]. Available at: www.wdrase.de/EUROGRAPHICS_81pp305-319.pdf.

Rase, W.-D. (1997). Fischauge-Projektionen als kartographische Lupen. *Salzburger Geographische Materialien*, 115–122 [Online]. Available at: www.wdrase.de/Agit97.pdf.

Ribecca, S. (2018). Chart combinations: Tile grid maps. *The Data Visualisation Catalogue Blog* [Online]. Available at: datavizcatalogue.com/blog/chart-combinations-tile-grid-maps/.

Roser, M. (2018). The map we need if we want to think about how global living conditions are changing. *Our World in Data* [Online]. Available at: ourworl-dindata.org/world-population-cartogram.

Sagar, B.S.D. (2014). Cartograms via mathematical morphology. *Information Visualization*, 13(1), 42–58. doi: 10.1177/1473871613480061.

Selvin, S., Merrill, D., Sacks, S., Wong, L., Bedell, L., Schulman, J. (1984). Transformations of maps to investigate clusters of disease. In *Annual Meeting of the American Public Health Association*, Anaheim [Online]. Available at: escholarship.org/uc/item/7hc3n11x.

Shepard, R.N. (1957). Stimulus and response generalization: A stochastic model relating generalization to distance in psychological space. *Psychometrika*, 22, 325–345.

Spiekermann, K. and Wegener, M. (1993). Zeitkarten für die Raumplanung. *Informationen zur Raumentwicklung*, 7, 459–487 [Online]. Available at: www.spie-kermann-wegener.com/pub/pdf/IRPUD_AP117.pdf.

Sun, S. (2013). A fast, free-form rubber-sheet algorithm for contiguous area cartograms. *International Journal of Geographical Information Science*, 27(3), 567–593. doi: 10.1080/13658816.2012.709247.

Sun, S. (2020). Applying forces to generate cartograms: A fast and flexible transformation framework. *Cartography and Geographic Information Science*, 47(5), 381–399. doi: 10.1080/15230406.2020.1745092.

Tobler, W.R. (1959). Some notes on azimuthal projections. University of Washington, Seattle.

Tobler, W.R. (1961). Map transformations of geographic space. PhD Thesis, University of Washington, Seattle [Online]. Available at: hdl.handle.net/1773/5629.

Tobler, W.R. (1973). A continuous transformation useful for districting. *Annals of the New York Academy of Sciences*, 219, 215–220. doi: 10.1111/j.1749-6632.1973.tb41401.x.

Tobler, W.R. (1977). Bidimensional regression: A computer program. *Wiley Online Library*, 71.

Tobler, W.R. (1979). A transformational view of cartography. *The American Cartographer*, 6(2), 101–106. doi: 10.1559/152304079784023104.

Tobler, W.R. (1986). Pseudo-cartograms. *The American Cartographer*, 13(1), 43–50. doi: 10.1559/152304086783900194.

Tobler, W.R. (2004). Thirty five years of computer cartograms. *Annals of the Association of American Geographers*, 94(1), 58–73. doi: 10.1111/j.1467-8306.2004.09401004.x.

Vuidel, G. (2011). Anaplaste [Online]. Available at: sourcesup.renater.fr/www/transcarto/anaplaste.

Vuidel, G. (2013). IsoDist/Angle [Online]. Available at: sourcesup.renater.fr/www/transcarto/isodist.

Vuidel, G. (2021). Darcy. Logiciel de transformation cartographique vectorielle [Online]. Available at: sourcesup.renater.fr/www/transcarto/darcy.

# 8

# Exploration, Aggregation and Spatiotemporal Visualization of Big Data

**Claude GRASLAND**[1,2]
[1] *Géographie-cités, Université Paris Cité, France*
[2] *CIST, CNRS, Université Paris Cité, Université Paris 1 Panthéon-Sorbonne, France*

## 8.1. Introduction

Big data poses particularly interesting visualization problems. They require simplifications, in particular aggregations, which can be made according to multiple dimensions (geographic, temporal, thematic, etc.) and create higher level, less big data. In order to be able to aggregate atomic information, these dimensions must therefore be categories. For most big data, unlike constructed data such as census data, these categories need to be defined. More specifically, in the case of data from social networks, we usually have short texts (tweets, messages, really simple syndication [RSS] feeds from newspapers, etc.) for which we know precisely the date of circulation and, in some cases, the author; the thematic dimensions are often derived from analysis of "free texts" using natural language processing (NLP) methods. The geographical dimension can be either a location (X,Y) managed by the system or a toponym cited. Both must be matched to a predefined nomenclature of geographic entities at different scales.

Many data contain references to more than one place, which can be aggregated to create matrices of flows (Bahoken 2022; Chapter 4) or networks linking places. The same applies to individual census data, where several places are generally

*Handling and Mapping Geographic Information*,
coordinated by Claire CUNTY and Hélène MATHIAN. 

attached to each individual (place of residence, place of work, previous place of residence, etc.). By aggregating two locations, we can reconstruct residential mobility (place of residence at previous census × current place of residence) or commuting patterns (place of residence × place of employment). We obtain data at different territorial levels, depending on the spatial granularity chosen to reference these locations (municipalities, départements, regions, etc.). These flows can then be broken down according to other dimensions, such as demographic categories (age, gender, etc.) or social categories (profession, nationality, etc.).

Here, we present an example of big data exploration using visualizations of data aggregated over multiple dimensions. The case of international press news from RSS feeds of daily newspapers is a particularly interesting example of big data with multiple geo-aggregation keys.

In this way, we can reconstruct both the trace of localized events (the foreign countries in question) and the origin-destination flows of information linking the country where the newspaper is located and the countries it reports on. This example shows just how many exploratory processes can be performed on complex data, and how many spatial and temporal visualizations can be created to provide rich, complementary insights. It is by combining these different graphical and cartographic representations that organizational structures can be detected.

The specific example chosen will be that of international news published by major daily newspapers in four Western European countries (France, Germany, Spain, the UK) between 2014 and 2019 in the form of RSS feeds. We shall be looking at the extent to which the press in these countries – which are historically and spatially close – offer similar or different coverage of international events, whatever the nature of these events (political, sporting, cultural, etc.).

We begin by presenting the data (press news) and the various spatial and non-spatial dimensions present, which will serve as a query filter. The following paragraphs will propose various types of explorations and (carto)graphic visualizations: first, we will evaluate foreign news publication profiles by country and their variation over time.

We will then analyze the geography of international news press flows, examining whether there is a relationship between the location of newspapers and the frequency with which foreign countries are cited. Furthermore, we shall also demonstrate that press news also reveals indirect links of co-citations between countries present in the same news. We conclude with a discussion of the general challenges of multidimensional visualization of big data in space and time using appropriate interfaces.

## 8.2. Defining the object of study and selecting the corpus

The *international news* item, which constitutes the basic object, is much less simple to define than it might seem, if we choose not to confine ourselves to the section headings ("World" or "International" pages) proposed by the newspapers, but to put forward an operative definition that can be applied to all the news items published in different daily newspapers in different countries.

### 8.2.1. *Press news: a multidimensional object*

To fully understand the multiple analytical possibilities available to the observer, the simplest way is to start with the atomic unit of a "news item" captured from a newspaper's website via its RSS feed. RSS is a tool for disseminating news published by a newspaper on the Web. Its main aim is to facilitate the work of search engines, encouraging Internet users to click on a URL (uniform resource locator) pointing to the newspaper's site, where the full version of the article can be found. Each RSS item therefore necessarily includes the identity of the publishing newspaper and, where applicable, the newspaper's corresponding rubric (e.g. the international RSS feed for *Le Figaro* newspaper). It also includes a publication date, but as this is not always provided, it can be replaced by a retrieval date when using an automatic harvest tool. Last but not least, an RSS item systematically includes a title followed by a description of variable length, which constitutes a fragment of information designed to attract the reader to the complete news item. The example of a news item published by *Le Figaro* via its international RSS feed will briefly demonstrate the full range of analysis potential offered by a collection of RSS items (see Table 8.1).

| Source | *Le Figaro* (France) – International RSS feed |
|---|---|
| Date/hour | Published on 25/08/2015 at 10:26, updated on 25/08/2015 at 23:37. |
| URL | https://www.lefigaro.fr/international/2015/08/25/01003-20150825ARTFIG00072-l-impressionnant-exode-de-milliers-de-migrantspar-la-route-des-balkans-de-l-ouest.php. |
| Title | The impressive exodus of thousands of migrants along the "La route des Balkans de l'Ouest" (Western Balkan route). |
| Description | PICTURES – Since January, more than 125,000 migrants have entered the European Union via Hungary, Serbia, Macedonia and Greece, as Europe faces one of the most serious refugee crises since the Second World War. |

**Table 8.1.** *Example content of an RSS item*

The information contained in an RSS item can be broken down into six dimensions, summarized by the acronym *5W + H* (Who? What? When? Where? Why? How?). This breakdown into dimensions is found in different forms in various scientific fields. An analogy can be drawn with the *What-When-Where* framework proposed by Peuquet and Duan (1995) in geomatics and discussed at greater length in Chapter 6 of this book. These six dimensions are taught in journalism schools as a reminder of the set of questions that a journalist reporting on an event must answer. It is advisable to introduce them in the title or first lines of the article, so that the reader is immediately aware of the article's general objectives (inverted pyramid technique).

The combination of all or some of these six dimensions makes it possible to define a scientific analysis approach to news published on the Web, whether it emanates from newspaper RSS feeds (Grasland et al. 2016) or from messages exchanged on social networks such as X (formerly Twitter) (Bellmore et al. 2015).

– *Who is talking* (Who?): This dimension characterizes the media publishing the news, as well as the channel used to disseminate it and the language used. It makes it possible to establish a certain number of inferences about the recipients of the news, even if we do not know the actual audience who may have had access to it. It can therefore be used to add auxiliary information about the newspaper, its geographical location, its political orientation and so on. At a later stage of analysis, it can be used as a basis for aggregation by language or country.

– *When?*: This second dimension enables us to identify the time at which a news item appears, provided that the publication date is accurate or that the collection tool visits the site frequently enough to provide a retrieval date with a limited margin of error. This dimension, which is mathematically quasi-continuous, allows multiple forms of aggregation by hour, day, month, week or year.

– *What?*: Unlike the first two dimensions, which are unambiguously defined, the definition of a story's topic always depends on the observer's interpretation of the textual content. Whether using manual or automatic methods, the attribution of themes must always be justified. Here, we have chosen to analyze "foreign" or "international" news as opposed to "domestic" or "national" news.

– *Where?*: As an international short story is by definition about faraway places, the presence of a foreign location will be almost systematic, if not in the title, at least in the first few sentences of the description. However, places are not always cited as elements of localization (e.g. "Earthquake in Turkya"); they can also play a role as geopolitical actors, particularly when states or regional organizations are involved (e.g. "Turkey blames EU").

– *How?*: This dimension, which is generally measured using lexical sentiment analysis tools, would enable us, for example, to assess the relative shares of "positive", "neutral" or "negative" news and the variation in these shares according to the countries concerned (Segev 2016). However, it will not be used here.

– *Why?* is no longer a question of exploration – the subject of this chapter – but instead of explanation and modeling. It consists, for example, of demonstrating that journalists talk more about the most powerful countries and those that are close in terms of geographical, historical or linguistic distance (Grasland 2020).

In this chapter, we shall be focusing more specifically on the geographical dimension of Where (what places are we talking about?), which can be combined with Who (where are the newspapers located?), What (which news items are international in the sense that they mention foreign countries?) or When (when are foreign countries most talked about?).

### 8.2.2. *Defining international news*

For our purposes, the essence of international news is that it is defined in relation to two countries. Denis H. Wu (1998, 2000, 2003), a specialist in international news flows, has proposed the terms host country and foreign country to describe the fundamentally asymmetrical relationship between a foreign country in which the event described in the news item has occurred, and a host country that has selected this news item from among many others to bring it to the attention of its readers (see Box 8.1). An international news item can therefore be conceptualized as an information flow between two countries, linking them in both directions. On the one hand, an event in the foreign country is known to the host country; on the other hand, the foreign country is recognized as newsworthy by the host country.

The challenge of such a definition is to find a method of identifying international news that is both objective and reproducible, that is, that can be applied in both time and space. There are two possible approaches, both of which present varying degrees of difficulty.

The *extensive approach* considers international relations in all their diversity. The identification of international news should therefore take into account the presence of foreign countries and supra-state organizations (NATO, EU, United Nations, etc.), non-governmental organizations (MSF, Greenpeace, etc.), transnational corporations (Google, Facebook, Huawei, etc.) and even individuals who play a role in international relations (Edward Snowden, Bill Gates, George Soros, etc.).

The *restrictive approach* in line with Wu's definition is limited to interstate relations and defines a fixed grid of spatial entities of comparable level that can be mentioned in the news. Typically, this would be the member countries of the United Nations, although exceptions can be made for territories with contested international legitimacy (Taiwan, Kosovo, Western Sahara, etc.).

We can illustrate the difference between the two methods by taking the example of the *Figaro* news item presented in the introduction (see Table 8.2). It is clear that the restrictive approach misses an important piece of information, since it does not mention the fact that the news is about the EU (institution) and Europe (territory).

| | Extensive approach | Restrictive approach |
|---|---|---|
| **Title** | Western Balkans | – |
| **Description** | Hungary, Serbia, Macedonia, Greece, European Union, Europe | Hungary, Serbia, Macedonia, Greece |

**Table 8.2.** *Recognition of international named entities*

Without debating the superiority of either approach, we have opted here for the restrictive approach, which has the advantage of lending itself better to a cartographic analysis of flows and can rely on simple but robust country recognition tools that mobilize gazetteers or dictionaries referencing geographic places based on toponyms (Middleton et al. 2018; Moncla and Gaio 2018). In this way, we can determine in which cases the word "Brussels" refers to the capital of Belgium or the headquarters of the institutions of the European Union.

The daily press is a favored field of investigation for the analysis of international press flows, as it is subject to particularly strong time constraints in the selection of news, which is, by definition, perishable material. The process by which the press selects international news can be imagined as a gigantic exchange of perishable products (corresponding to the news feeds of the major news agencies), from which each newspaper makes its purchases, either upwards (choice of new information which it hopes will gain in importance) or downwards (choice of information already covered by other media, which it is necessary to cover even with a delay). Competition takes place between newspapers sharing the same readership, that is, newspapers in the same country or category (quality press or tabloid), or with the same reach in terms of target audience (regional press, national press, international press, etc.).

**Box 8.1.** *The competition for news*

These dictionaries limit the recognition of countries to three or four main characteristics: the official name and customary names of the country, the associated adjective and, if different from the adjective, the name of the inhabitants, and finally the country's political capital. The dictionary can of course be expanded to include the names of heads of state, major cities, companies and landmarks. But this would introduce bias and, above all, create a temporal variability in the dictionary that would be difficult to manage. It is therefore preferable to stick to a limited list of criteria, even if it means then using learning tools to add specific terms for certain periods, as proposed by the author of the R *newsmap* package, which is one of the leading international country recognition dictionaries (Watanabe 2018).

### 8.2.3. *Corpus definition based on RSS news feeds*

Unlike the large private press databases that make complete press articles available against payment (*Dow Jones Factiva, Europresse, Lexis Nexis*, etc.), newspaper RSS feeds, which can be obtained free of charge by entering them yourself (*Geomedia*) or by using Open Source databases (*MediaCloud*), generally only provide the title of the news item, possibly supplemented by a few sentences of description. The question thus arises as to which part of the text will be used to search for foreign countries in order to define a news item as international. Using the news title alone is a tempting solution, since this information is by definition always available in an RSS feed, and the title is supposed to offer the most relevant summary of the news content. This is the solution adopted by Elad Segev, who has produced one of the most important syntheses of international news feeds in recent years (Segev 2014, 2016).

From the *MediaCloud* database, we selected a corpus of 16 newspapers from four Western European countries (Germany, Spain, France, Great Britain) for which both the news title and the first sentences of the description were available (see Table 8.3). These four European countries, spatially close and of similar size, share a common history, and their newspapers should therefore produce similar worldviews. However, they are also characterized by different economic and political histories, which have led each of them to develop specific relationships with certain parts of the world, notably through colonization and the spread of their respective languages. More contemporarily, each of these countries retains specific economic or political relations with one or other part of the world, notably through business strategies, diaspora locations or tourist destinations. It will therefore be particularly interesting to compare the differences in the coverage of the world's countries in the newspapers of each of these countries.

| Code | Nom | URL |
|---|---|---|
| DEU_tagspi | *Tagespiegel* | www.tagesspiegel.de/ |
| DEU_frankf | *FAZ* | www.faz.net/ |
| DEU_suddeu | *Süddeutsche Zeitung* | www.sueddeutsche.de/ |
| DEU_diewel | *Die Welt* | www.welt.de/ |
| ESP_abcxxx | *ABC* | www.abc.es/ |
| ESP_percat | *Periodico de Catalonia* | www.elperiodico.com/es/ |
| ESP_vangua | *La Vanguardia* | www.lavanguardia.com/ |
| ESP_mundo | *El Mundo* | www.elmundo.es/ |
| FRA_figaro | *Le Figaro* | www.lefigaro.fr/ |
| FRA_lacroi | *La Croix* | www.la-croix.com/ |
| FRA_libera | *Libération* | www.liberation.fr/ |
| FRA_lmonde | *Le Monde* | www.lemonde.fr/ |
| GBR_guardi | *The Guardian* | www.theguardian.com/uk |
| GBR_indept | *The Independent* | www.independent.co.uk/ |
| GBR_mirror | *The Mirror* | www.mirror.co.uk/ |
| GBR_dailyt | *The Telegraph* | www.telegraph.co.uk/ |

**Table 8.3.** *List of newspapers surveyed*

## 8.3. Crossing the "who" and "what" dimensions

Is international interest the same from one newspaper to another? Who talks most about foreign countries? While all the newspapers in the corpus are broad-sheet newspapers with a "world" or "international" section, they do not all have the same degree of internationalization of their news. Are there significant differences between the media in the four countries studied?

Thanks to the data provided by *MediaCloud*, we are able to cover a relatively long-time span, from the beginning of the second half of 2013 to the end of the first half of 2020. Although the collection of RSS feeds from certain newspapers may experience periods of interruption lasting a few days or weeks (either because the newspaper ceases publication, or more often because the collection tool experiences a breakdown), the inclusion of four newspapers in each of the countries studied ensures regular coverage of the target time period, with an average of 500 to 2,000 semantic units (i.e. sentences) per day, corresponding to complete news items of the order of 200–1,000.

As the corpus includes news items in four languages (French, German, English and Spanish), we have built four dictionaries of keywords or regular expressions, enabling us to identify the same list of countries in each case, and to register them using their ISO3 code. These dictionaries include the name of the country and its capital, as well as the lexeme or adjective designating its inhabitants, but exclude the list of its political leaders. Without going into the details of the country recognition method, it should be pointed out that it has been manually verified to be highly reliable in terms of false positives (few cases of incorrect identification of a country), but less reliable in terms of false negatives (non-identification of certain countries).

The presence of foreign news has been calculated separately for the title and each of the first two sentences of the description to facilitate comparisons between newspapers with more or less lengthy descriptions and, subsequently, to identify the co-cited countries within the same sentence. This methodological precaution is important, as it enables us to verify whether or not it makes no difference to limit ourselves to analyzing news headlines alone, as many authors do.

Table 8.4 shows that the proportion of foreign news is significantly higher in the French press than in the other three countries, regardless of the portion of text considered. Therefore, 25% of French newspaper headlines mention a foreign country, compared with 16.7% of British newspaper headlines, 12.4% of Spanish newspaper headlines and 10.3% of German newspaper headlines.

This result might suggest that French newspapers are more open to world news. However, it may also be explained by other factors, such as a lower proportion of domestic news or a focus on the names of countries and their capitals, whereas newspapers published in other countries tend to focus on the leaders of countries and adopt a more personalized view of geopolitical players. A dictionary introducing the names of leaders would perhaps lead to different conclusions.

However, the proportion of foreign news does not just depend on the country and the language of the newspaper. When we break down the proportion of foreign news according to the order of semantic units, we find that, in all the countries considered, it is the first sentence of the description and not the headline that has the highest proportion of foreign countries cited. The proportion of foreign news then falls rapidly in the second sentence of the description and is even lower in subsequent sentences. This important result confirms that journalists do indeed use the inverted pyramid technique, locating events in space at the very beginning of the article, even more so than in the headline.

| Newspaper localization | Type of observation unit | Number of observation units | Percentage of international news |
|---|---|---|---|
| Germany | Title | 1,615,112 | 10.3 |
| | Phrase 1 | 1,196,833 | 18.4 |
| | Phrase 2 | 1,533,361 | 7 |
| | Total | 3,818,315 | 12.2 |
| Spain | Title | 1,803,768 | 12.4 |
| | Phrase 1 | 1,459,774 | 19.7 |
| | Phrase 2 | 2,745,153 | 11.7 |
| | Total | 4,379,705 | 15.1 |
| France | Title | 1,192,461 | 25.1 |
| | Phrase 1 | 945,886 | 30.2 |
| | Phrase 2 | 978,240 | 18 |
| | Total | 2,695,845 | 25.7 |
| United Kingdom | Title | 2,202,787 | 16.7 |
| | Phrase 1 | 1,082,827 | 17.3 |
| | Phrase 2 | 748,045 | 11.7 |
| | Total | 3,623,089 | 16.8 |

**Table 8.4.** *Share of international news by country and observation unit*

## 8.4. Crossing the "who", "what" and "when" dimensions

In the same way as we combined the Who and What questions in the previous paragraph, we shall now add the When dimension, looking at the share of international news in the flow of information produced by newspapers in different West European countries over time: When are foreign countries most talked about?

Here, we will carry out an exploratory data analysis (Tukey 1977) to compare variations in the share of international news at different levels of time aggregation, which may reflect long-term trends or cycles. The availability of a large, multi-year corpus of newspapers from different countries and using different languages will enable us to verify the extent to which the observations made are generalizable or not.

### 8.4.1. *Detecting trends*

The first question to ask is whether the share of international news is stable throughout the observation period and whether, regardless of exceptional events, its distribution appears homogeneous over time.

The monthly distribution of the share of foreign news over the entire period shows significant variations in each country (see Figure 8.1). It can be assumed that peaks are due to exceptional events outside the country's borders (war, crisis, earthquake), while troughs correspond to a heightened focus on national news, for example, during elections or political crises. Nevertheless, irrespective of these ad hoc explanations, there is a general and statistically significant trend in each country toward a reduction in the share of foreign news between 2013 and 2020.

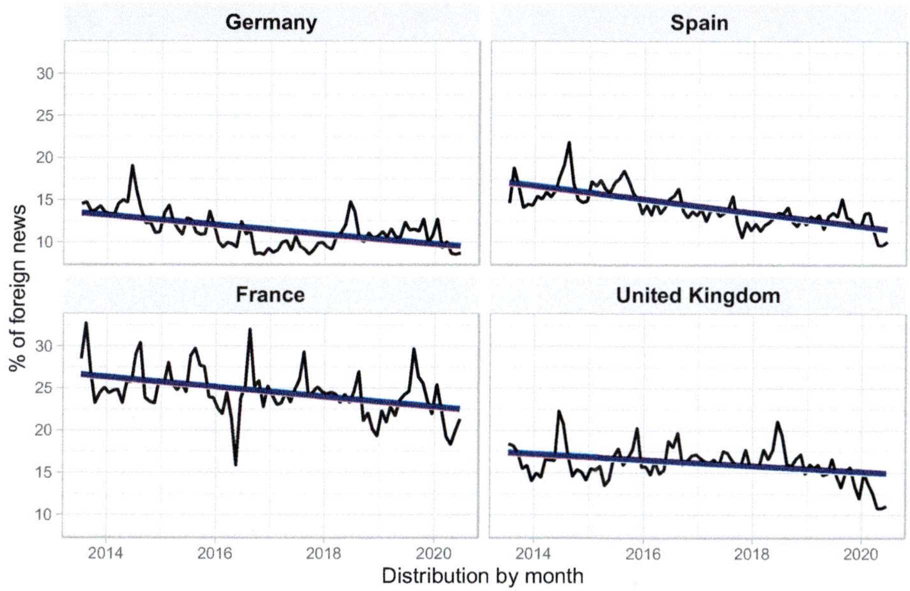


**Figure 8.1.** *Change in the proportion of foreign news*

The interpretation of the results (Why?) goes beyond the deliberately exploratory scope of this chapter. However, it does show how the exploratory analysis of big data can lead to new avenues of research that are more focused on a particular issue and for which the analysis of headlines is combined with that of full articles. This raises the question of whether the decline in international news is specific to Western Europe. Has it been accentuated by the coronavirus crisis? Does it reflect a form of geopolitical withdrawal that could have long-term international consequences?

### 8.4.2. *Analysis of seasonal variations*

As predicted by Figure 8.1, there are significant seasonal variations in the propensity of newspapers to talk about foreign countries. Figure 8.2 confirms that the summer period is apparently the most favorable for the production of international news, since all newspapers in each of the four countries experience a significant peak during the months of June, July or August, with slight shifts that seem to be linked to differences in the vacation calendar in each country.

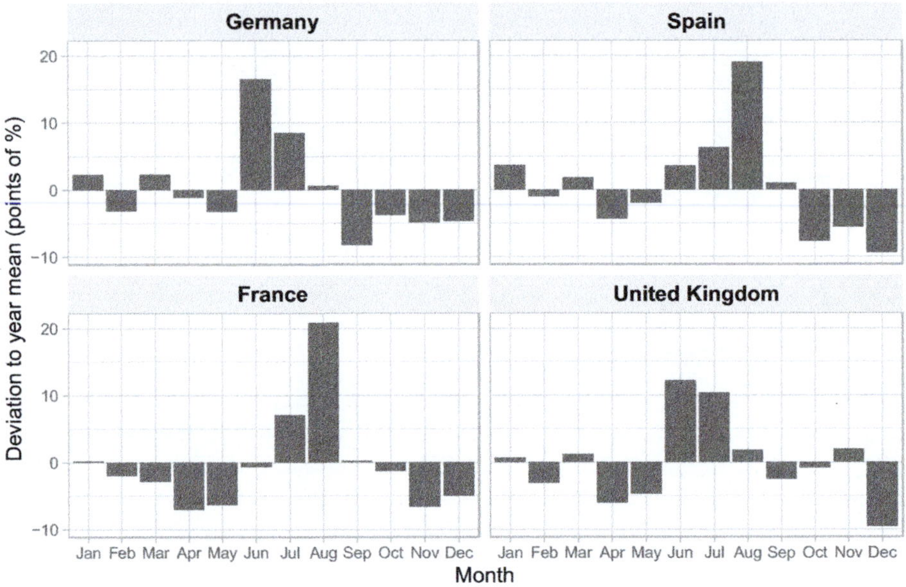


**Figure 8.2.** *Monthly variations in the proportion of foreign news*

As in the previous case, the exploratory analysis highlights a very clear statistical anomaly, all the more significant as it was established over a period of six consecutive years, which rules out the effect of an isolated event. This fascinating result leads to two sets of hypotheses:

– "Realistic" hypothesis: summer in the northern hemisphere is the period when most international events take place outside the countries of the media studied, and Western European newspapers are more likely to report on foreign countries in summer because they are a faithful reflection of these events. This is not an absurd hypothesis, given the timing of wars, migrant arrivals and droughts in poor countries, for example.

– "Constructivist" hypothesis: the summer period corresponds to a drop in the publication of national news due to the summer break, which affects both politics and sport. Journalists would then have more room to publish international news, regardless of the temporality of events. A natural disaster affecting a poor country would therefore be more likely to make the news in Europe if it occurred in summer rather than autumn.

No doubt complementary, these two hypotheses cannot be validated using a simple quantitative exploratory analysis and deserve a qualitative content analysis. However, the latter at least has the merit of posing the problem, visualizing the data and suggesting hypotheses for further analysis.

### 8.4.3. *Analysis of weekly variations*

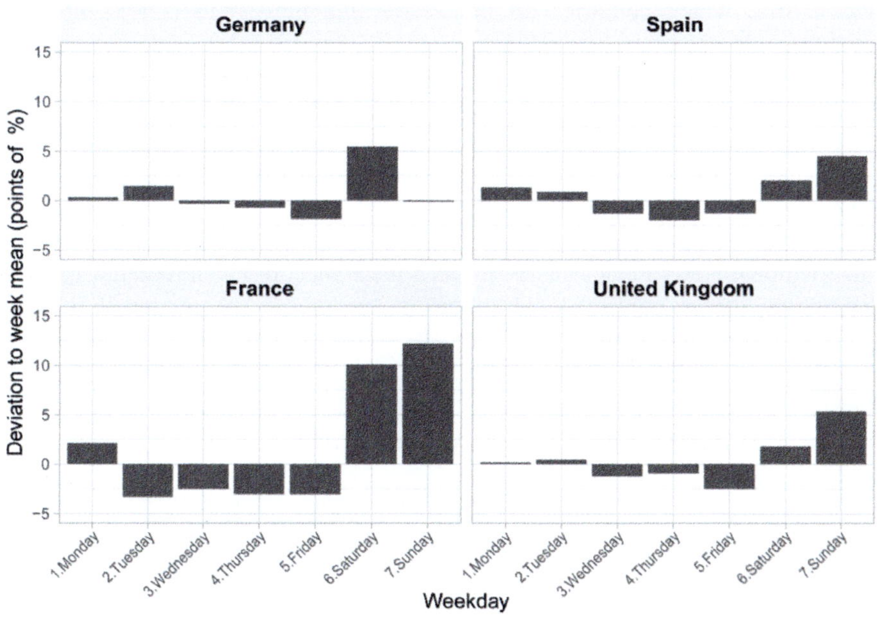


**Figure 8.3.** *Weekly variations in the proportion of foreign news*

Less accentuated but just as statistically significant are the weekly variations in interest in foreign news, with a marked peak over the weekend (see Figure 8.3). The case of French newspapers is particularly spectacular, with an average increase in foreign news of over 10% on Saturday and Sunday. Variations are less marked in other countries, with positive anomalies on Saturdays but not on Sundays in Germany.

Interpreting this result would require an in-depth reading of the news in order to better grasp whether the weekend peak is linked to an overall different agenda or whether it simply reflects a greater weighting for certain sections such as sports on weekends. This singularity has led some authors to exclude news published on Sundays or during the vacations from their analyses of the media agenda (Geiß 2018).

## 8.5. Crossing the "who", "what" and "where" dimensions

Newspapers in the four countries studied therefore follow the same temporal rhythms in their coverage of events relating to foreign countries, albeit with a few nuances. Still, which countries do they cover the most, and is this the same depending on the location of the newspaper? Do they offer the same geographical coverage of international news? In terms of multi-dimensional analysis, we are talking here about the relationship between the media's places of origin (Who–Where) and the places associated with the events reported by the media (What–Where). We can thus truly speak of international information flows linking "host countries" and "guest countries".

International News Flow Theory has been the subject of much research since the two founding articles published in 1969 in the *Journal of Peace Research* (Galtung and Ruge 1965; Östgaard 1965). The 12 laws governing the circulation of international news flows proposed by Galtung have since been the subject of numerous verifications (Segev 2014, 2016), although they have also been the subject of criticism and proposals for improvement (Harcup and O'Neill 2001).

These laws are of particular interest to geography because, in many respects, they re-cut the assumptions of spatial interaction models used to model trade or migratory movements, a point already made by Zipf (1946) when studying the distribution of places cited in the obituaries of the *New York Times*. In this respect, it is particularly striking to note that the development of the Internet has not fundamentally altered the role of different forms of distance which, combined with the size and power of countries, continue to explain to a large extent the degree of prominence of countries in international news (Beaugitte et al. 2016; Grasland 2020). At least three main hypotheses can be deduced from this work.

– *H1*: we talk more about large countries than small ones. More precisely, rich and powerful countries are more talked about than poor and weak ones. When the latter are mentioned in the newspapers, it is usually in connection with catastrophic events, wars or crises.

– *H2*: all things being equal in terms of size criteria (H1), we tend to talk more about near countries than distant ones. Proximity is understood here in a geographical sense (distance as the crow flies or common border), a historical sense (colonial relationship, historical belonging to the same state) and a sociological and cultural sense (common language, diaspora).

– *H3*: all things being equal in terms of size (H1) and proximity (H2), we tend to talk more about countries belonging to the same region of the world where the newspaper is located. The notion of world region can be understood here in either a political sense (member countries of the European Union) or a psychological sense (sense of belonging to Europe or Western Europe).

As the four West European countries selected for this analysis are spatially, demographically, economically and historically close, we can expect their respective coverage of international news to be fairly similar; in any case, closer to each other than to the world coverage of a media outlet located in Asia, sub-Saharan Africa or Latin America.

### 8.5.1. *Mapping the distribution of international news by country*

The addition of the "Where" question brings data mining into the realm of spatial analysis. Additional information on the location of the news items (world map backgrounds) and on the distances separating the host and foreign countries will be added to the data. Rendering the results in cartographic form thus becomes fundamental and involves making reasoned choices in terms of projection and graphic semiology.

In view of the three hypotheses formulated above, we can expect foreign news from French, German, English and Italian newspapers to be particularly spatially concentrated within the European Union (H2 and H3), and more specifically in its western part, where the largest, richest and longest standing members of this political space are located (H1). We have therefore opted for an azimuthal map projection centered on Brussels, which highlights the magnifying glass effect and the decrease in interactions with distance (see Figure 8.4).

We have also expressed the importance of flows as a proportion of total foreign news, so as to be able to compare the maps visually, independently of each country's total foreign news production. By taking into account the entire study period, we can limit the impact of exceptional events that temporarily highlight a country (e.g. Nepal in 2015), but maintain the importance of countries that suffer longer term events (Syrian war, Ukrainian crisis, etc.).

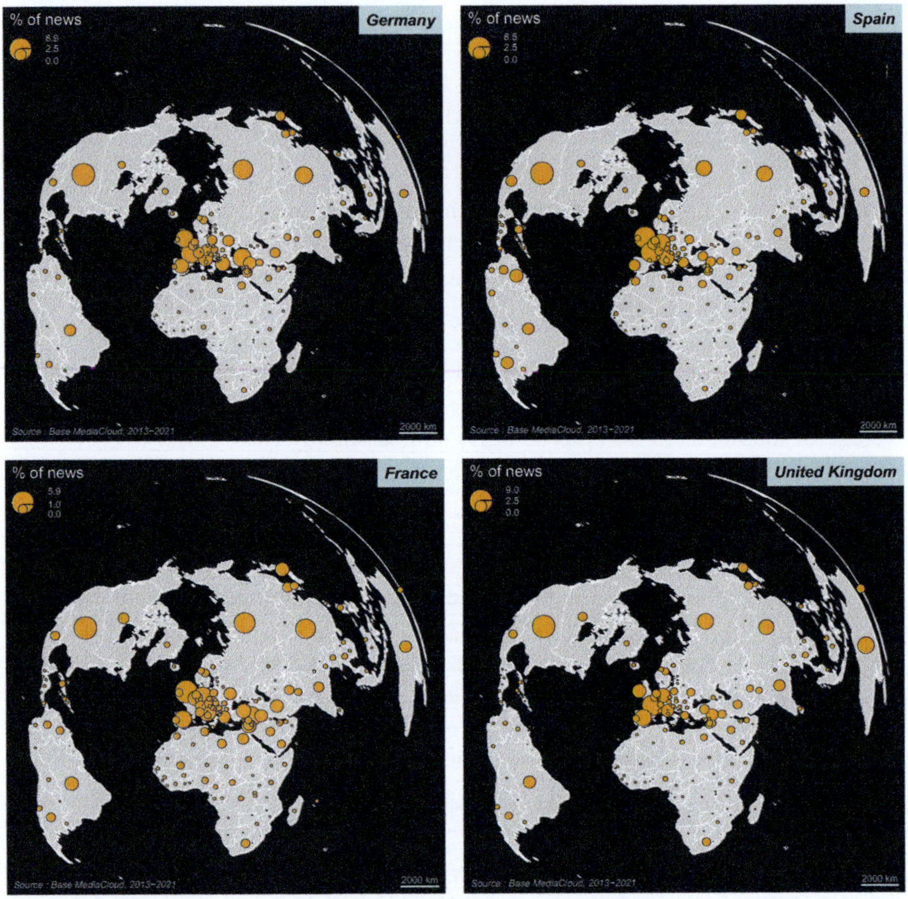


**Figure 8.4.** *Spatial distribution of foreign news published by daily newspapers in Germany, Spain, France and the UK (July 2013–June 2020) (source: MediaCloud database, 2013–2021)*

All these choices allow us to read the four maps together. The four countries appear at first glance to have relatively similar distributions of foreign news, in line with the three hypotheses of international news circulation:

– the major powers, in particular the five permanent members of the UN Security Council (the United States, Russia, China, France and the UK), are present in a far greater number of foreign news stories than the smaller countries. The US enjoys a specific premium, which can be explained both by its major hard and soft power and by its greater linguistic, cultural and historical proximity to the four countries studied;

– Countries that are members of the European Union or located in its immediate vicinity (North Africa, Middle East) are much better covered than countries of a similar size but further away.

– Wealthy countries and European colonial countries that have inherited the language are also more frequently mentioned, as are Australia, Canada and Latin American countries.

### 8.5.2. *Identifying and mapping specific national features*

However, we should not immediately conclude that the international agenda of Western European countries is homogeneous. A more detailed analysis may reveal preferential relations or, on the contrary, deficits between foreign countries and certain host countries.

Such an analysis is based on a table cross-referencing the number of news items per pair of "host country" and "foreign country" (contingency table). An appropriate statistical procedure (chi-square calculation and test) is used to identify the most significant anomalies, both in absolute and relative terms.

The map corresponding to these anomalies[1] (see Figure 8.5) offers a striking vision of a veritable "division of the world" between the four European countries studied, which cannot fail to hark back to geohistory and the colonial division of the world (Grataloup 2015). Like trade, press flows remain strongly influenced by linguistic proximities and colonial legacies (Bröcker and Rohweder 1990; Lohman 2011).

1 From a statistical point of view, this "anomaly" corresponds to "a deviation from the model". The model here is that of an equal distribution of news by "foreign country" for each of the "host countries". The differences observed are therefore internal to the media space of the four Western European countries. They would be different if the sample had been extended to include newspapers from other countries around the world.

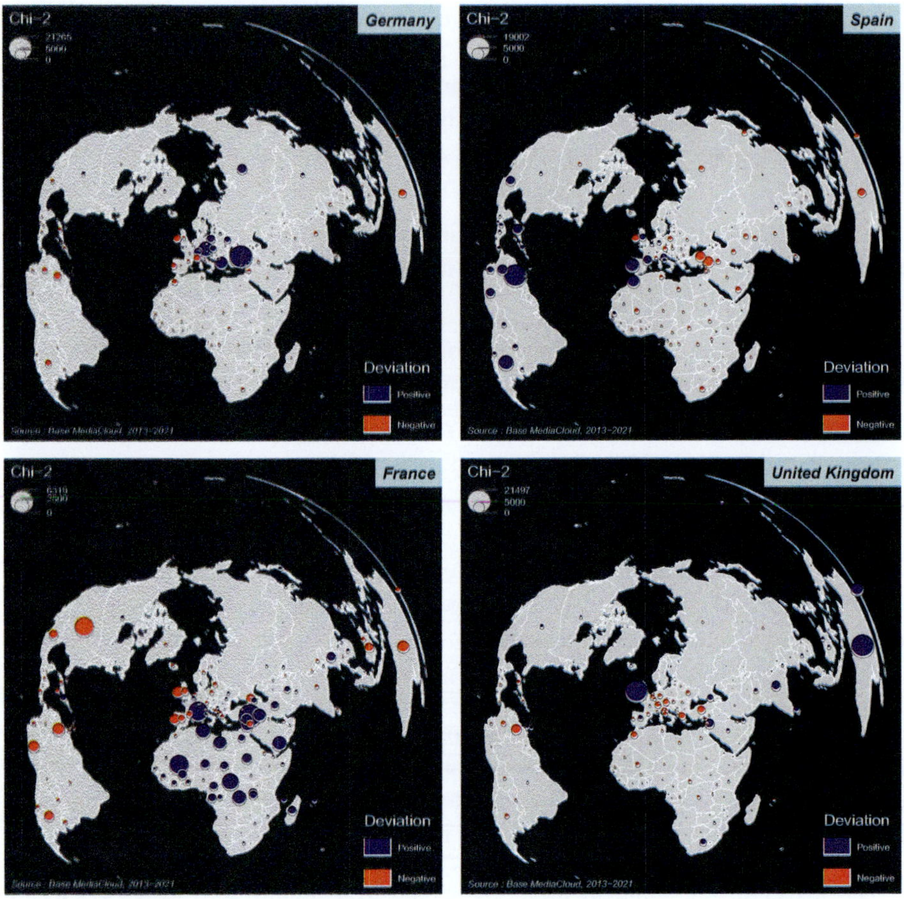


**Figure 8.5.** *Specificity of foreign news published by daily newspapers in Germany, Spain, France and the UK (July 2013–June 2020) (source: MediaCloud Base, 2013–2021)*

– In general, German newspapers stand out for the greater attention they pay to news from Central and Eastern Europe or Asia. Events in Turkey, Austria and Greece are mentioned much more prominently than in other newspapers. Due to their limited colonial heritage and their language, which is not widely spoken around the world, they also more frequently mention the world's major powers (the United States, Russia and China).

– French newspapers show a remarkable tropism toward African countries, with the exception of the southern part and Morocco, which is covered more by Spain.

They are also strongly oriented toward the Middle East, with extensive coverage of Syria, Yemen, Iraq, Iran and the Israeli-Palestinian conflict. Colonial heritage and the spread of the French language around the world are obviously important factors in explaining these specificities.

– Symmetrically, Spanish newspapers offer a remarkable peculiarity when it comes to Spanish-speaking Latin American countries. In this respect, it is interesting to note that Portuguese-speaking Brazil is not specifically mentioned more by Spanish newspapers than by their English, French or German counterparts. Spain is also characterized by extensive coverage of neighboring countries (Portugal, Morocco) and, more generally, of Western European countries, notably Italy and the UK.

– Finally, British newspapers are a remarkable illustration of the image of a country that asserts its insularity in relation to the European "continent" and wants to be open to the "world at large". In fact, even before Brexit, British newspapers quoted the countries of the European Union far less than other newspapers, with the exception of Ireland and Spain. For both historical and linguistic reasons, there is an exceptional over-representation of Australia and New Zealand (and Ireland) in the news, as well as other Commonwealth countries (India, Pakistan, South Africa, etc.), not forgetting Canada and the United States, which are only slightly over-represented.

All in all, the hypothesis of homogeneous geographical coverage of the world by Western European newspapers appears highly questionable and confirms the strong resistance of national arenas of information and debate. Some authors even conclude that the hypothesis of a global media space is a "myth" (Hafez 2007).

## 8.6. Graphs to represent co-location relationships

The analysis of news flows linking host countries to foreign countries can be usefully complemented by an analysis of the links defined by the co-occurency of foreign countries within the same news item. In the example given at the start of this chapter, the host country (France) was describing an event involving four foreign countries (Hungary, Serbia, Macedonia and Greece). Initially, we analyzed this news item in terms of information flows (France-Hungary, France-Serbia, etc.); however, we can also carry out a secondary analysis in terms of information links (Hungary–Serbia) to understand which countries are most frequently linked in a media outlet's news and what these links tell us about that outlet's representations of the world.

### 8.6.1. *Measuring and visualizing association links*

As shown by the work of Segev (2008, 2010), who was a precursor of this approach, networks of co-presence of countries in the news often highlight the host country itself, insofar as many foreign news items actually concern relations between the country where the media is located and the rest of the world. However, they can also reveal polycentric networks, as in the case of the Russian media, which in 2009 gave a major role to the great American rival (Figure 8.6).

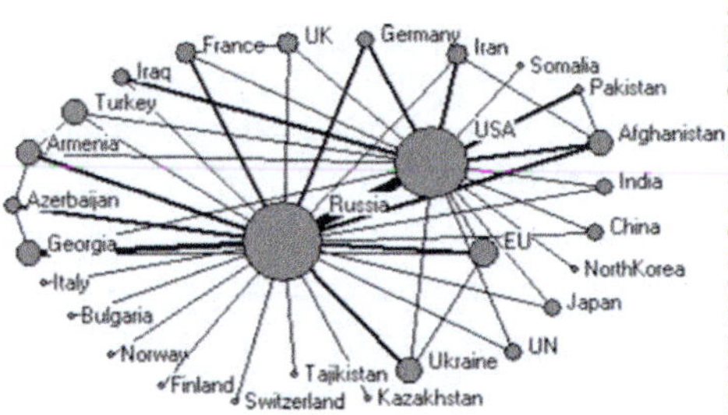


| Countries mentioned | Items | Countries mentioned | Items |
|---|---|---|---|
| Russia – U.S. | 266 | Armenia – Turkey | 31 |
| Georgia – Russia | 123 | China – U.S. | 31 |
| Russia – Ukraine | 89 | EU – Georgia | 29 |
| EU – Russia | 87 | Finland – Russia | 29 |
| Afghanistan – U.S. | 86 | Turkey – U.S. | 28 |
| Iran – U.S. | 73 | Somalia – U.S. | 28 |
| France – Russia | 65 | UN – U.S. | 26 |
| Germany – Russia | 57 | EU – U.S. | 25 |
| Azerbaijan – Russia | 54 | Armenia – Azerbaijan | 25 |
| Pakistan – U.S. | 52 | Azerbaijan – Georgia | 25 |
| Afghanistan – Russia | 50 | India – U.S. | 23 |
| Armenia – Russia | 49 | Georgia – U.S. | 23 |
| Germany – U.S. | 48 | U.S. – Ukraine | 22 |
| Iraq – U.S. | 43 | Japan – Russia | 22 |
| India – Russia | 40 | Bulgaria – Russia | 21 |
| UK – U.S. | 39 | Armenia – U.S. | 21 |
| Israel – Palestine | 39 | Kazakhstan – Russia | 20 |
| France – U.S. | 37 | Afghanistan – Iran | 20 |
| Russia – Turkey | 36 | France – UK | 20 |
| Japan – U.S. | 36 | EU – Ukraine | 19 |
| Afghanistan – Pakistan | 35 | Norway – Russia | 18 |
| Russia – UK | 35 | Italy – Russia | 18 |
| Russia – UN | 34 | N. Korea – U.S. | 18 |
| China – Russia | 33 | Iraq – Russia | 17 |
| Iran – Russia | 31 | Russia – Switzerland | 17 |

**Figure 8.6.** *Example of a relational analysis of international flows: Russian media in 2009 (source: Segev 2010, reproduced with the author's permission)*

Here, we propose to adopt this approach by merging all media from the four countries in our corpus, but excluding links between host and foreign countries. This means that a France–China link will not be counted if it emanates from a French newspaper, but will be counted if it is produced by a Spanish, German or British newspaper. We also introduce an inverse weighting according to the number of countries mentioned to reduce the influence of news items mentioning an exceptional number of countries. In the case of a news item featuring four countries, each pair of countries will thus receive a weight of 1/16 of the total, whereas in the case of a news item featuring only two countries, the weight would be 1/2. We can thus draw up a matrix of co-citation links between the countries of the world and the Western European press and deduce the most frequent cases of association over the period from June 2013 to June 2020 (Table 8.5). These links and their weights can be represented on a map as a spatialized graph (Figure 8.7).

| Rank | Foreign country No. 1 | Foreign country No. 2 | Link frequency | Percentage of total |
|---|---|---|---|---|
| 1 | Israel | Palestine | 3,989 | 3.50 |
| 2 | Russia | Ukraine | 3,268 | 2.86 |
| 3 | Russia | Syria | 2,169 | 1.90 |
| 4 | China | USA | 2,052 | 1.80 |
| 5 | Russia | USA | 1,923 | 1.69 |
| 6 | Syria | Turkey | 1,750 | 1.53 |
| 7 | Iran | USA | 1,178 | 1.03 |
| 8 | Iraq | Syria | 1,160 | 1.02 |
| 9 | Syria | USA | 1,074 | 0.94 |
| 10 | South Korea | North Korea | 968 | 0.85 |
| 11 | North Korea | USA | 949 | 0.83 |
| 12 | Germany | France | 874 | 0.77 |
| 13 | United Kingdom | USA | 856 | 0.75 |
| 14 | Mexico | USA | 813 | 0.71 |
| 15 | Cuba | USA | 687 | 0.60 |
| 16 | Russia | Turkey | 675 | 0.59 |
| 17 | France | Italy | 671 | 0.59 |
| 18 | France | United Kingdom | 653 | 0.57 |
| 19 | United Kingdom | Russia | 635 | 0.56 |
| 20 | Germany | United Kingdom | 615 | 0.54 |

**Table 8.5.** *The 20 most frequent foreign country associations*

The analysis of the strongest links differs from the analysis of the most cited countries in that it is not necessarily limited to associations between the most powerful countries, although such links are obviously present. The strongest link, for example, is to the Israeli–Palestinian conflict, while the second is to the Russian–Ukrainian geopolitical crisis that led to the annexation of Crimea in 2014. As for the Syrian conflict, it is omnipresent and generates a high number of links with both neighboring countries (Turkey and Iraq) and major geopolitical players (the United States and Russia). We might have feared that the graph of co-citations would be dominated by international sporting events (Olympic Games, World or Euro soccer tournaments, etc.), but their weight seems finally limited in the face of major geopolitical crises, which seem to play the major role in the creation of structural links between countries.

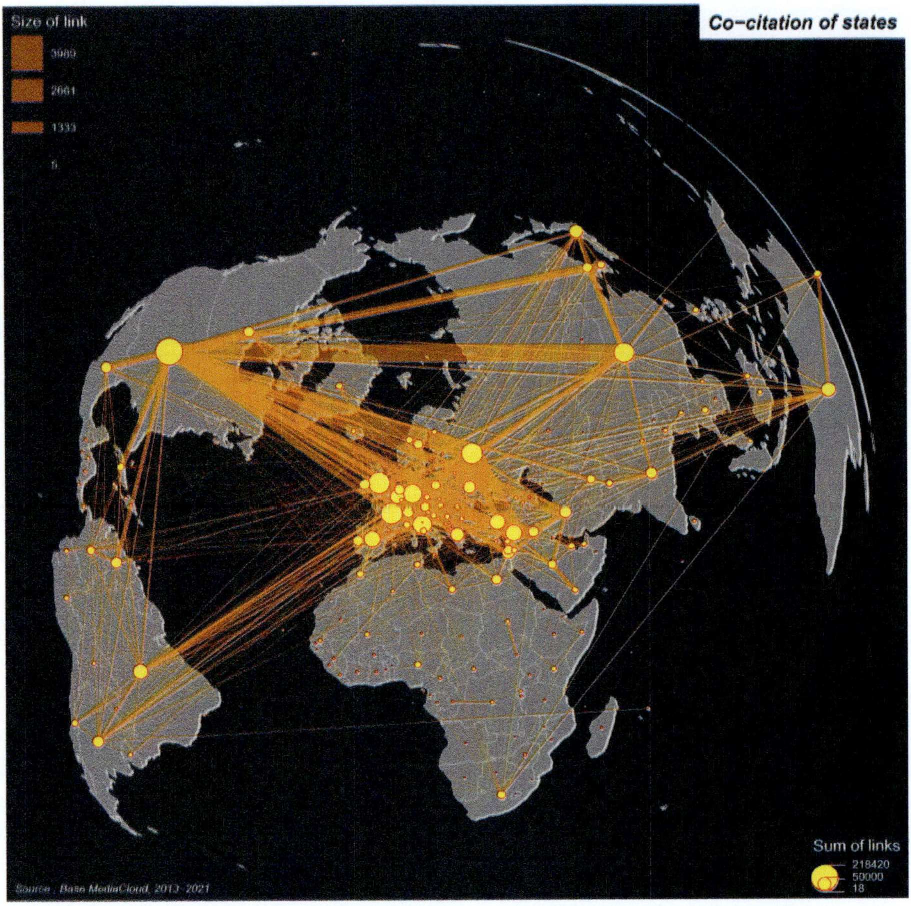


**Figure 8.7.** *Graphs of foreign country co-citation links in daily newspapers from Germany, Spain, France and the UK (July 2013–June 2020)*

### 8.6.2. *From co-citations to global regionalization*

One aim of exploratory analysis is to produce a simplified map that synthesizes the complexity of the complete information. Here, we propose a method adapted to link data between pairs of locations. Although the associations between countries are symmetrical, a media hierarchy of countries and a regionalization of the world can be deduced from the co-presence matrix using the dominant flows method proposed by Nyusten and Dacey (1961). These two authors sought to define an urban hierarchy and nodal regions from a matrix of telephone flows between cities that was precisely as symmetrical as the co-presence graph studied here. For an association matrix between locations $N_{ij}$, these authors propose to say that $i$ is dominated by $j$ if and only if two conditions are verified:

– *Condition 1*: the strongest association of $i$ is with $j$.

– *Condition 2*: the sum of $j$'s associations is greater than that of $i$.

Therefore, the country most associated with Ukraine in our newspaper sample and for the period under consideration is Russia (3,268), and the sum of news items associating Russia with other countries (14,890) is greater than the sum of Ukraine's associations with other countries (5,089). Ukraine is therefore media-dominated by Russia in the sense of Nyusten and Dacey's method. In the case of Australia, the most frequent association is with New Zealand (386), but the sum of New Zealand's associations with other countries (1,276) is less than that of Australia (3,908). In this case, Nyusten and Dacey's method leads to the conclusion that Australia is not dominated and constitutes a central node in the network around which a tree of direct or indirect dominations is organized (e.g. New Zealand is dominated by Australia but dominates the Kiribati Islands). We can thus distinguish three types of country according to their position in the space of media associations:

– Dominant countries: these are countries that form the core of media regions defined by the oriented tree at their center. Their strongest link of association is with a country with a smaller number of associations. Some dominant countries define very small regions, such as Sweden, which dominates only Norway and Denmark.

– Relay countries: these are countries that are both dominant and dominated, serving as relays for the domination of a dominant country and forming sub-regions within the main regions. A good example is China, which is media dominated by the United States but polarizes a large part of East Asia.

– Dominated countries: these are the exclusively dominated countries that form the terminal ends of the graphs.

To fully grasp the geopolitical value of this method, it is worth drawing a parallel with the analysis of international trade. A country whose main flow of goods or services is directed toward a larger country is vulnerable to asymmetric trade retaliation (Serrano et al. 2007). In the case of international news flow, a country that is most often associated by the press with another, more frequently cited country will lose international visibility and will not be the subject of stories that concern it alone. It will be dominated in the sense that it will generally only be talked about in relation to countries higher up in the media hierarchy. If we take Ukraine as an example, we can see that the Western European press almost never mentions this country in isolation, but more often than not in connection with its large Russian neighbor. Conversely, Saudi Arabia is associated with many news items involving its less media-salient neighbors, but is not systematically linked to the United States, China or Russia, making it the focus of a small, autonomous media space.

Beyond bilateral power relations, Nyusten and Dacey's method has the major geographical advantage of regionalizing the world into zones polarized by a central country and its various relays. In this respect, the regions highlighted in Figure 8.8 are a remarkable indicator of the way in which Western European media divide the world according to foreign news:

– The "Russian" region, for example, highlights the ghost of the USSR, whose former republics, including those that are members of the European Union, appear to be dominated by Russia. But this region also extends far into the Middle East, underlining Russia's activism in this area during the observation period, particularly in the context of the war in Syria.

– The "American" region covers North and Central America, but not the whole of South America, where Brazil polarizes an autonomous media space. The growing confrontation with China is resulting in the inclusion of the entire sub-region dominated by China in the "American" region. It is hard not to think of the tendency for American strategic priorities to shift from Atlantic (NATO) toward the Asia-Pacific zone.

– The "European" region, with France and Germany at its heart, appears to be symbolically deprived of Ireland and the UK, which are attached to the American region due to the large number of news items linking Ireland to the UK, and then the UK to the United States. This "European" zone extends south of the Mediterranean, encompassing the Maghreb and part of West Africa, no doubt in connection with the migration crisis and French military interventions in the Sahel.

– The "Australian" region covers most of Oceania, as well as Southern Africa via South Africa. These connections are specific to British newspapers, but are sufficiently intense to influence the overall composition of the graph. However, the Commonwealth effect does not prevent India from forming an autonomous region with neighboring countries on the Indian subcontinent.

Of course, we can debate the relevance of aggregating the newspapers of the four countries studied and suggest that they be analyzed separately in more focused analysis. The aggregate result is nonetheless fascinating, as in many ways it reflects preconceived ideas about the geopolitical organization of the world as seen through the eyes of the Western European media. For sure, applying the same method to another sample of newspapers located in other parts of the world would certainly not have led to the same result. What is shown here is the Western European media's situated perception of international relations.

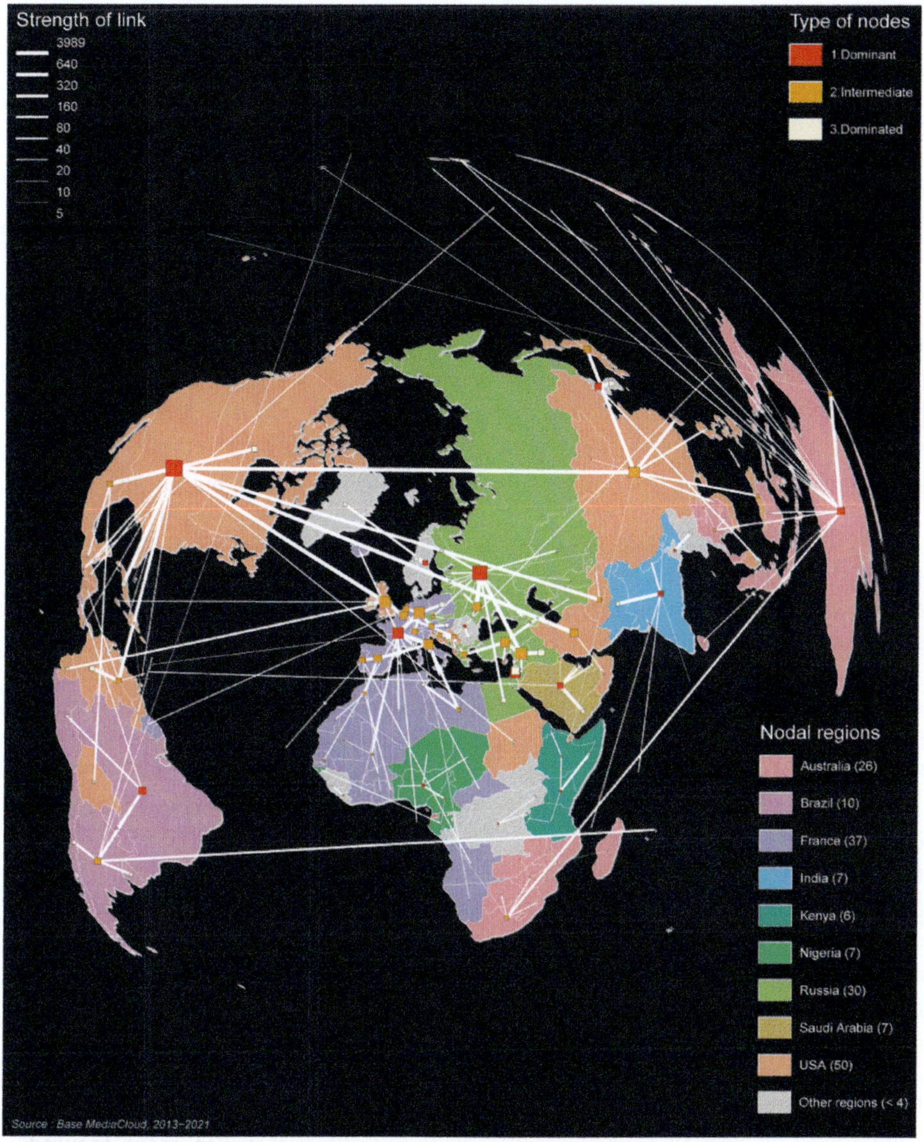


**Figure 8.8.** *Dominant link graphs of foreign country co-citations in daily newspapers from Germany, Spain, France and the UK (July 2013–June 2020) and regionalizations derived from dominant node dependencies*

## 8.7. Conclusion

News produced by newspapers and disseminated on their websites in the form of RSS feeds are called "feeds" because of their close relationship with time. A "news item" is quickly out of date and replaced by others, unless it is clarified, developed and updated. However, news stories are more often than not linked to events that are themselves localized in time and space, and brought to the attention of the newspaper's readership.

They involve therefore a spatial transfer of information between two points on the globe and constitute a flow in the spatial sense of the term. The spatiotemporal nature of press news is therefore not fundamentally different from flows of people or goods, which are usually studied at an aggregate level. At the end of this analysis, it seems possible to draw a number of more general conclusions concerning the (carto-) graphic representation of flows, whatever their nature.

– *A flow is an aggregation of atomic events*: flow matrices used in geography are most often presented as quantities $F_{ijt1t2}$ of relationships between places *i* and places *j* over a period of time $[t_1,t_2]$, giving the user a misleading illusion of temporal continuity and homogeneity of the phenomenon. In reality, most flow matrices are the result of an aggregation of singular events, providing prospective or retrospective information at a given instant *t* on movements that are due to occur or have occurred in space and time. Like press news, measures of trade flows between two countries over the course of a year and for a given product are the result of an aggregation of information provided by the customs bulletins of the exporting and importing countries. Just as two newspapers will not report on media events in the same way, two countries will not describe their trade flows in the same way, even if they are theoretically the same. Users of trade databases are well aware that they usually have to choose between data from the country of departure and data from the country of arrival. And this type of asymmetry is just as obvious for international migration.

– *All the dimensions of a flow must be taken into account*: a flow matrix is a two-dimensional simplification of elementary information that is initially far more complex. In the case of a customs entry, there is in fact a wealth of information on the nature of the goods, their price, tonnage, mode of transport as well as the country or department responsible for issuing the entry. The Who–What–When–Where–Why–How analysis framework we presented at the beginning of this chapter for press news could be perfectly applied here: "French customs (Who) declare that 50 tons of potatoes (What) were exported on February 7, 2021 (When) from Lannion (Where-1) to Southampton (Where-2) by ship (How) to supply British supermarkets (Why)". Of course, atomic information is not always available, but with the development of Big Data this is increasingly the case, as shown by the example of the individual mobility databases produced by INSEE in France. As a result, the researcher is faced with an ever-increasing number of choices, since in practice they can generate an infinite number of different flow matrices.

– *The different dimensions form a complex system*: aggregation procedures leading to a simple one- or two-dimensional object result in the loss of a considerable amount of information. Yet hidden structures may exist that are only visible in higher dimensional spaces, and which will be irretrievably lost if the analyst does not take the trouble to change the points of observation. The famous "Triplet" on the cover of Gödel Escher Bach (Hofstadter 1979) is a striking illustration of this phenomenon of variable projections of the same three-dimensional object in two dimensions. An analysis of world trade that did not take time into account would be just as futile as one that did not take space or the types of products traded into account.

– *Mapping is only one of the most efficient, but not exclusive, visualization tools for exploring the multidimensional space of interactions*: while maps are a fundamental element of flow analysis, they are not the only means of exploration. It is interesting to note that, in the case of trade flow visualization, network representations without consideration of spatial location of nodes have become increasingly popular in recent years. In the eyes of the proponents of this approach, this implies that distance as the crow flies may no longer play the same role as in the past, and that its visualization via geographical space can therefore be dispensed with to represent flows.

– *The intersection of more than two dimensions calls for the development of new modes of representation and dynamic visualization interfaces*: if, for example, we add a thematic dimension to international press news, such as the presence of the terms "migrants", "refugees" or "asylum seekers", we can analyze, no longer the share of news that speaks of a foreign country, but the share of foreign news that addresses issues of international mobility, and follow the timeline of what is erroneously referred to as the "migrant crisis" (Leconte et al. 2019). The complexity of the subject then becomes formidable as the What dimension becomes dual, both geographical and thematic. We can, for example, examine in the form of a What–Where–Where graph the foreign countries that are present in the same news items about international mobility issues (see Figure 8.9). But this is only one of many cross-references, and it is necessary to build suitable interfaces to examine such objects in all their richness (Grasland et al. 2021).

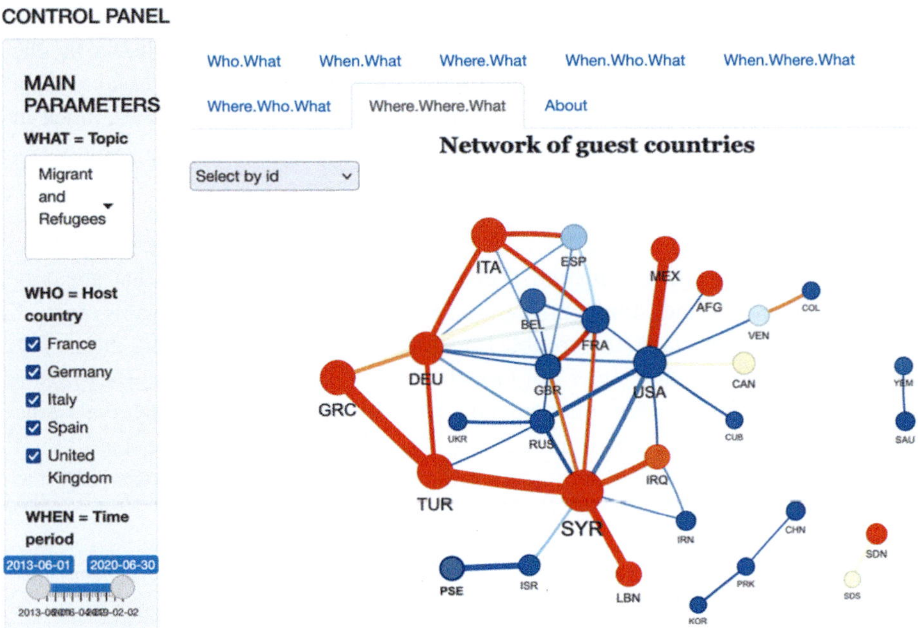


**Figure 8.9.** *TELEMAC: a multidimensional visualization interface (source: Grasland et al. 2021; available at: www.analytics.huma-num.fr/Claude.Grasland/telemac/)*

## 8.8. References

Beaugitte, L., Grasland, C., Sévère, M. (2016). Espaces géographiques et représentations médiatiques. *L'Espace géographique*, 45(1), 1–4.

Bellmore, A., Calvin, A.J., Xu, J.M., Zhu, X. (2015). The five w's of "bullying" on Twitter: Who, what, why, where, and when. *Computers in Human Behavior*, 44, 305–314.

Bertin, J. (1967). *Sémiologie graphique*. Mouton/Gauthier-Villars, Paris.

Bröcker, J. and Rohweder, H.C. (1990). Barriers to international trade. *The Annals of Regional Science*, 24(4), 289–305.

Galtung, J. and Ruge, M.H. (1965). The structure of foreign news: The presentation of the Congo, Cuba and Cyprus crises in four Norwegian newspapers. *Journal of Peace Research*, 2(1), 64–90.

Geiß, S. (2018). The dynamics of media attention to issues. In *From Media Hype to Twitter Storm*, Vasterman, P. (ed.). Amsterdam University Press, Amsterdam.

Grasland, C. (2020). International news flow theory revisited through a spacetime interaction model: Application to a sample of 320,000 international news stories published through RSS flows by 31 daily newspapers in 2015. *International Communication Gazette*, 82(3), 231–259.

Grasland, C., Lamarche-Perrin, R., Loveluck, B., Pecout, H. (2016). L'agenda géomédiatique international : analyse multidimensionnelle des flux d'actualité. *L'Espace géographique*, 45(1), 25–43.

Grasland, C., Leconte, R., Toureille, E., Severo, M. (2021). Mapping international geopolitical agenda. Continuing national conceptions of the emerging European crisis. *Frontiers in Big Data*, 4.

Grataloup, C. (2015). *Géohistoire de la mondialisation* : *le temps long du monde*, 3rd edition. Armand Colin, Paris.

Hafez, K. (2007). *The Myth of Media Globalization*. Polity Press, Cambridge.

Harcup, T. and O'Neill, D. (2001). What is news? Galtung and Ruge revisited. *Journalism Studies*, 2(2), 261–280.

Hofstader, D. (1979). *Gödel, Escher, Bach: An Eternal Golden Braid*. Basic Books, New York.

Leconte, R., Toureille, E., Grasland, C. (2019). La production médiatique d'une "crise migratoire". Dynamiques spatio-temporelles de l'agenda global de la presse en 2015. *Socio-anthropologie*, 40, 181–199.

Lohmann, J. (2011). Do language barriers affect trade? *Economics Letters*, 110(2), 159–162.

Middleton, S.E., Kordopatis-Zilos, G., Papadopoulos, S., Kompatsiaris, Y. (2018). Location extraction from social media: Geoparsing, location disambiguation, and geotagging. *ACM Transactions on Information Systems (TOIS)*, 36(4), 1–27.

Moncla, L. and Gaio, M. (2018). Services Web pour l'annotation sémantique d'information spatiale à partir de corpus textuels. *Revue internationale de géomatique*, 28(4), 439–459.

Nystuen, J.D. and Dacey, M.F. (1961). A graph theory interpretation of nodal regions. *Papers of the Regional Science Association*, 7(1), 29–34.

Östgaard, E. (1965). Factors influencing the flow of news. *Journal of Peace Research*, 2(1), 39–63.

Peuquet, D.J. and Duan, N. (1995). An event-based spatiotemporal data model (ESTDM) for temporal analysis of geographical data. *International Journal of Geographical Information Systems*, 9(1), 7–24.

Segev, E. (2008). The imagined international community: Dominant American priorities and agendas in Google News. *Global Media Journal*, 7.

Segev, E. (2010). Mapping the international: Global and local salience and news-links between countries in popular news sites worldwide. *International Journal of Internet Science*, 5(1), 48–71.

Segev, E. (2014). Visible and invisible countries: News flow theory revised. *Journalism*, 16(3), 412–428.

Segev, E. (2016). *International News Flow Online. Global Views with Local Perspectives.* Peter Lang, New York.

Serrano, M.Á., Boguñá, M., Vespignani, A. (2007). Patterns of dominant flows in the world trade web. *Journal of Economic Interaction and Coordination*, 2(2), 111–124.

Tukey, J.W. (1977). *Exploratory Data Analysis*, Volume 2. Addison-Wesley, Boston.

Watanabe, K. (2018). Newsmap – A semi-supervised approach to geographical news classification. *Digital Journalism*, 6(3), 294–309.

Wu, H.D. (1998). Investigating the determinants of international news flow a meta-analysis. *International Communication Gazette*, 60(6), 493–512.

Wu, H.D. (2000). Systemic determinants of international news coverage: A comparison of 38 countries. *Journal of Communication*, 50(2), 110–130.

Wu, H.D. (2003). Homogeneity around the world? Comparing the systemic determinants of international news flow between developed and developing countries. *Gazette Leiden, Netherlands*, 65(1), 9–24.

Zipf, G.K. (1946). Some determinants of the circulation of information. *The American Journal of Psychology*, 59(3), 401–421.

# Conclusion

**Claire CUNTY[1] and Hélène MATHIAN[2]**
[1] *EVS, Université Lumière Lyon 2, France*
[2] *EVS, CNRS, Lyon, France*

This book covers various aspects of the fundamental step that enables the transition from data to the map for geographic analysis; all the chapters herein demonstrate that this transition requires a number of operations. These operations are often invisible, whether the maps produced are static or interactive. The thoughts and questions that arise during the various stages of mapping data often remain hidden. The endeavor of the authors of these eight chapters has been to make explicit what often remains hidden, because either only the result – the map – takes precedence, or the map is merely an element that supports a demonstration which, in turn, takes precedence. These operations vary according to the question, the context and the data, and the book aims to shed light on this, drawing upon a range of experiences to help anyone dealing with a geographical subject involving the production of maps. The authors have broken down the various phases of analysis, highlighting what they are learning and what additional information they are seeking in order to understand the result obtained (e.g. testing a correlation with another variable, measuring the effect of distance). In this way, readers were able to see different types of analysis unfold, incorporating the many questions and choices they entailed.

Operational issues are intertwined. From the geographic data that enable us to go from geographic objects to cartographic objects, to questions of cartographic communication, it seems difficult, even artificial, to completely isolate operations when they only concern a treatment that is both mathematical and semiotic. Indeed, on the one hand, the whole must be taken into account when dealing with a subject from the point of view of spatial analysis. On the other hand, the analyst does not work in a linear way; they reproduce a cycle several times over, starting with hypotheses to create pertinent cartographic objects and continuing with processing to identify or understand the spatial structure of the phenomenon. This identification or understanding often leads to the formulation of further hypotheses and the repetition of all or part of the cycle.

This book looks at the heuristic dimension of the map, which has grown in importance with the development of data visualization in exploratory approaches. In these approaches, the map acts as an interface and becomes the temporary result of transferring thematic information onto a base map in order to determine its spatial distribution. There is no denying the knowledge-generating and sometimes revealing power of these devices. Nevertheless, the systematization of this type of interface, developed by non-mapmakers, can trivialize, or even impoverish, cartographic production by introducing "generic" operations. It was therefore particularly important to present and illustrate how the transition to the map forces the analyst to question the nature and relevant scale of the objects that are represented and studied, and to imagine original modes of cartographic representation. All the chapters present these questions, which lead them to use methods that have one thing in common: they raise questions over the role of space, and integrate a variety of maps into their approach to shed light on their subject.

It is rare to see cartographic operations in such detail alongside a breakdown of the analyst's choices and accompanied by discussion. With the widespread application of GIS (geographic information system) tools since the late 1990s, and today's web-based environments, the geographic dimension of data is more systematically exploited than ever before. This democratization has sometimes been to the detriment of reflection on the significance of cartographic objects in relation to the terrain, and to the hypotheses underlying the automatic, or even systematic, processing proposed. The current trend toward reproducible science is certainly forcing us to formalize the choices made in the various processes, by replacing series of interactions with command lines; the latter make it possible to reproduce identically and, in the best of cases, to identify a set of parameters corresponding to the choices made and which can be varied. However, the traceability of choices is no substitute for the necessary discussion of them, as has been done here over the course of the chapters and in light of geographical questions. The chapters that propose developments here are part of this movement toward reproducible science. And those analyzing production and manufacturing methods propose formalizing treatments and operations to extract their genericity.

The chapters have also shown the inherent porosity between different steps. Focusing specifically on the processing of geographic information always leads to a discussion about data, on the one hand, and cartographic visualization, on the other hand. In a context of data massification and systematization of the geographic dimension of data, the sole aim of processing may be to make data interoperable or "visible" (particularly on the Web). Processing may be identical, but have different objectives: construction of information or representation. Whatever the tools used – and only a few of them have been presented here – this volume demonstrates the essential and indispensable nature of making operations explicit. It also underlines the fact that the proliferation of computer libraries specializing in geographic information processing is increasing the number of cartographic experiments. While it is of course necessary to test the comprehensibility of the maps produced, these experiments open up new avenues for reflection.

As the volume focuses on the processes involved in transforming data into maps, the discussion of visual rendering and the message that maps can convey remains limited in scope. On the other hand, the question of "objectifying space" by producing "legitimate academic representations" may be raised by some researchers. Without being naive enough to think that transparency of the choices taken translates as objectivity in cartographic production, making them explicit is nonetheless essential to reducing subjectivity in the scientific process. There are, of course, many ways of portraying the reality of a phenomenon. Some depend on processing, others on data selection and critical analysis. Nevertheless, it is undeniable that re-examining each of these stages in relation to the question posed would provide, if not a critical posture, at least a reflexive view of cartographic productions and their uses.

# List of Authors


Myriam BARON
Lab'Urba
Université Paris-Est Créteil
Créteil
and
IDEES, CNRS
Université de Rouen Normandie
Rouen
France

Jean-Benoît BOURON
Géoconfluences
DGESCO
ENS Lyon
France

Anne-Christine BRONNER
SAGE, CNRS
Université de Strasbourg
France

Colette CAUVIN-REYMOND
LIVE, CNRS
Université de Strasbourg
France

Étienne COSSART
EVS
Université Jean-Moulin Lyon 3
France

Claire CUNTY
EVS
Université Lumière Lyon 2
France

Sylvestre DUROUDIER
Géographie-cités
Université Paris 1 Panthéon-Sorbonne
France

Claude GRASLAND
Géographie-cités
Université Paris Cité
and
CIST, CNRS
Université Paris Cité
Université Paris 1 Panthéon-Sorbonne
France

Marianne GUÉROIS
Géographie-cités
RIATE
Université Paris Cité
France

Mohamed HILAL
CESAER
INRAE
L'Institut Agro
Université Bourgogne Franche-Comté
Dijon
France

Julie LE GALLO
CESAER
INRAE
L'Institut Agro
Université Bourgogne Franche-Comté
Dijon
France

Malika MADELIN
PRODIG
Université Paris Cité
France

Hélène MATHIAN
EVS, CNRS
Lyon
France

# Index


3D, 195, 202, 207, 208

## A, B

accessibility, 48, 95, 144, 152, 154–156, 162, 204, 206, 246, 248
aggregation, 7, 30, 41, 44, 49, 52–54, 73, 93, 95, 99, 104, 114, 132, 151, 181, 195, 196, 202, 257, 258, 260, 266, 280, 282, 283
anamorphosis, *see also* cartogram, 101, 103, 174, 217–219, 222, 246–252
animation, 182–186, 188, 190, 195, 197, 198, 202, 206–208, 222, 250
biophysical, 55, 66, 74, 82
box-plot, 144, 145

## C

cartogram, *see also* anamorphosis, 101, 218, 219, 222, 223, 247–251
  continuous, 218, 223, 247–249, 251
  Demers, 223, 230, 231
  Dorling, 219, 223, 229, 230
  rectangular, 223, 227, 231, 232
category, 33, 44, 47, 52, 55, 181, 182, 204, 250, 262
centralization, 109, 110
change of state, 72, 86, 151, 173, 177–181, 190, 191, 197, 219, 249
checkerboard problem, 110, 118
complementarity, 71, 72, 109
component
  spatial, 176–178, 180, 181, 184, 191
  temporal, 174–176, 180, 190, 202, 206
  thematic, 177, 180, 190
contiguity, 109, 118, 132, 146, 223

## D

data
  analysis (methods), 19, 38, 91, 132
    classification, 19, 20, 24, 47, 50, 84, 112–114, 125, 190, 193
    factorial, 112
    multivariate, 18, 112
  big, 32, 257, 258, 267
  contributory, 33, 50, 51
  conventional/traditional, 29, 32, 33, 38, 45, 49, 59, 60
  individual, 139, 257
  model, 135
  new, 29, 30, 32, 33, 35, 37, 38, 40, 41, 45, 48–51, 53, 59, 60
  open
    Data, 35
    Source, 132, 133, 263
deformation, 6, 177, 191, 196, 197, 240, 244
delimitation, 34, 177, 178, 191, 196
discontinuity, 120
discretization, 57, 103, 186
discrimination, 121
displacement, 38, 173, 191–193, 202, 250
dissimilarity, 120, 121
distance, 50, 55, 80, 81, 101, 109, 117, 118, 120, 122, 125, 132, 139, 142, 150, 151, 156, 162, 166, 167, 218, 219, 246, 248, 270, 271
  matrix (of distances), 206
duration, 53, 178–181, 183, 197, 198, 202

## E, F

effects
  neighborhood, 116, 117, 121, 144, 146
equal repartition, 108, 110, 111
event, 179, 180, 198, 199, 260, 261, 267, 268, 275
evolution, 16, 37, 82, 126, 177, 179, 181, 184–186, 190, 192, 196, 207, 222, 246, 248, 250
exploration, 3, 12, 14, 15, 25, 29, 30, 65, 169, 186, 187, 193, 199, 208, 249, 258, 261, 278, 283
flows/feeds, 60, 196, 257–264, 266, 270, 271, 273, 275, 276, 278, 279, 282, 283
  dominant, 278
  RSS, 257–260, 263, 264, 282
fragmentation, 80–82, 95, 121, 125

## G, H

geographic information, 30, 32, 35, 45, 48, 66–68, 71, 72, 112, 126, 133, 134
geovisualization, 202
ghetto, 93, 95, 100, 101, 107, 109, 112, 113, 120, 122, 125
grid, *see also* mesh, 4, 7, 80, 83, 93, 95, 98, 99, 146, 181, 184, 195
heterogeneity
  spatial, 2, 45–47, 51, 54, 60, 110, 131, 138, 149, 157–159, 162, 169, 190, 251
  statistical, 2, 45–47, 51, 54, 60, 110, 131, 138, 149, 157–159, 162, 169, 190, 251
homogeneity, 95, 97, 99, 101, 110, 112, 114, 118, 273, 275, 282

## I


index
- dissimilarity, *see also* segregation, 107, 108, 110
- global, *see also* local, 109–111, 117, 118
- local, *see also* global, 110, 147, 148

integration (data), 30, 45, 46, 48–50, 53, 54, 59, 60, 74, 107, 109, 116, 120, 125

interaction
- (man-machine), 14, 110, 117, 120–122, 125, 132, 182, 186, 187, 206, 208, 270
- social, 109, 117
- spatial, 121, 125, 270

interoperability, 46, 48, 49

interpolation, 55–57, 76, 183

intersection (geometric processing), 49, 68, 69, 71–74, 78, 80


## L, M


level of organization, 89, 95

life (of an object), 177

local contexts, 67, 82, 95, 101, 117, 122

map
- choropleth, 100, 111, 114, 121, 221
- piezopleth, 233, 234, 247

matching, 46, 50, 51, 54

matrix
- flow, 196, 278, 282
- neighborhood, 169
- of distances, 206, 247
- of spatial weights, 146

MAUP, 2, 74, 95, 97

mesh, *see also* grid, 52, 75, 80, 95, 195

movement/motion, 174, 177–180, 183, 191–193, 196, 204


## N, P


neighborhood, 7, 45, 75, 90, 93, 117–122, 124, 132, 139, 142, 146, 147, 150, 151, 158, 190, 275, 277, 280

network, 33, 97, 195, 246, 248, 279

partitioning, *see also* world dissection, 70, 72

population potential (method), 122

projection, 2, 4, 6, 48, 50, 121, 224, 249, 271

proximity, 44, 82, 86, 109, 116, 117, 120, 124, 142, 158, 271, 273


## R


regression, 16, 76, 132, 149–152, 169
- bidimensional, 236–238, 240, 244
- coefficient
  - local, 158, 163, 164, 165
  - of determination, 152, 156
- Geographically Weighted, 76, 150
- hedonic, 132, 149–151, 169
- Ordinary Least Squares, 149

relationships
- spatial, 12, 13, 146
- statistical, 1, 3, 12, 14, 18

resolution
- spatial, 35, 37, 38, 47, 51, 52, 56, 59
- temporal/time, 35, 57

## S


scale, 2, 3, 6, 7, 9, 10, 73, 75, 78–80, 93, 95, 99, 101, 125, 132, 136, 139, 177, 179, 184, 198, 222, 249, 251
segregation, *see also* dissimilarity index, 89–91, 93, 95–97, 99–101, 104, 107–114, 116–118, 120–122, 125
semiology, 3, 199, 220, 271
set theory, 67–71, 75, 86
similarity, 50, 117, 120
small multiple maps, *see also* small multiples, 182, 184–186, 197
small multiples, *see also* small multiple maps, 182
smoothing, 45, 122, 125, 188
spatial
  autocorrelation, 117–119, 121, 132, 144, 146–148, 156, 157, 159, 169, 190
    Moran diagram, 147
    Moran index, 117, 118, 146, 147
  concentration, 37, 40, 95, 99–101, 107–112, 116, 120–122, 125, 221, 246
spatiotemporal, 30, 35, 44, 139, 179, 207
subset of points, 41, 75, 77, 101


## T


time
  calculation, 180, 197
  temporality, 269
  weekly variations, 269
timeline, 187, 202, 204
trajectory, 82, 177–180, 184, 191–193
  spatial, 177–179
  thematic, 177, 179, 184
transformation
  cartographic, 218
  geometric, 101
typology, 113, 120, 180, 190


## U, V, W


union, 69, 71, 73
urban ecology, 90, 104, 112, 114
variable
  visual, 84, 104, 181, 183, 190, 196, 198, 208, 220, 250
Venn diagram, 68, 70
world dissection, *see also* partitioning, 3, 16, 19, 24, 95, 97, 99, 184